# Advanced Mathematical Techniques in Engineering Sciences

# Science, Technology, and Management Series

*Series Editor*
J. Paulo Davim

Advanced Mathematical Techniques in Engineering Sciences
*Mangey Ram and J. Paulo Davim*

Optimizing Engineering Problems through Heuristic Techniques
*Kaushik Kumar, Nadeem Faisal, and J. Paulo Davim*

# Advanced Mathematical Techniques in Engineering Sciences

Edited by
Mangey Ram and J. Paulo Davim

**CRC Press**
Taylor & Francis Group
Boca Raton London New York

CRC Press is an imprint of the
Taylor & Francis Group, an **informa** business

CRC Press
Taylor & Francis Group
6000 Broken Sound Parkway NW, Suite 300
Boca Raton, FL 33487-2742

First issued in paperback 2020

© 2018 by Taylor & Francis Group, LLC
CRC Press is an imprint of Taylor & Francis Group, an Informa business

No claim to original U.S. Government works

ISBN-13: 978-1-138–55439-9 (hbk)
ISBN-13: 978-0-367–78128-6 (pbk)

**Visit the Taylor & Francis Web site at**
**http://www.taylorandfrancis.com**

**and the CRC Press Web site at**
**http://www.crcpress.com**

# Contents

# *Preface*

Mathematical techniques are the strength of engineering sciences and form the common foundation of all novel disciplines as engineering sciences. The book *Advanced Mathematical Techniques in Engineering Sciences* involved an ample range of mathematical tools and techniques applied in various fields of engineering sciences. Through this book, engineers have the opportunity to gain a greater knowledge and it may help them in the applications of mathematics in engineering sciences.

Chapter 1 presents the rules and methods for applying the Laplace transform. Three sections of the mathematical investigation of applied questions are distinguished: the rules for performing operations in the Laplace transform; Laplace transform in research tasks of the vibrations of a rod; and application of the Laplace transform in engineering technology. Specific examples of solving differential equations are presented, applied to problems of mechanics and the theory of oscillations. The essence of Kondratenko's method is described, on the basis of which mathematical modeling of some technological operations of mechanical engineering was carried out and features of dynamic phenomena during the work of equipment and interaction of the tool with detail were revealed.

Chapter 2 investigates the history, nature, and importance of the Fourier series. This chapter describes the periodic function, orthogonal function, Fourier series, Fourier approximation, Dirichlet's theorem, Riemann–Lebesgue lemma for Fourier series, differentiation of Fourier series, convergence of the Fourier series of the functions, Fourier transform, Fourier analysis with Fourier transform, and Gibbs phenomenon. Also, some of the summability methods (Cesàro, Nörlund, Riesz, weighted mean, etc.), absolute summability methods, strong summability methods, necessary and sufficient conditions for regularity of matrix of summability methods, uses of summability, norm, modulus of continuity, Lipschitz condition, various Lipschitz classes in trigonometric Fourier approximation, and importance of degree of approximation have been explained. The applications of summability methods in approximation of the signals, the behavior of the Fourier series of a piecewise smooth function (Gibbs phenomenon), Fourier series of signals of bounded bandwidth, filtering by Fourier transforms, and applications of summability technique and Fourier series have been described.

Chapter 3 describes the basics of soft computing and their applications. The key goal of soft computing is to develop intelligent machines to provide solutions to real-world problems, which are difficult to model mathematically.

Chapter 4 describes the study on solving transportation problems under a multi-objective environment. The main focus of this chapter is to introduce a new approach for solving multi-objective transportation problems in addition to the existing approaches such as goal programming, fuzzy programming, and revised multi-choice goal programming. In the proposed approach a procedure to obtain a Pareto-optimal solution of a multi-objective transportation problem using the Vogel approximation method is incorporated.

The merits and demerits of the approaches goal programming, fuzzy programming, and revised multi-choice goal programming compared to our new approach to solving a multi-objective transportation problem are presented.

Chapter 5 provides the study of simultaneous optimization of yield and viscosity of the pulp cooking process using the dual-response surface methodology. The pulp cooking process is an important step in the manufacturing of rayon grade pulp. The pulp is the cellulose component of the wood. The cellulose is separated from other components and impurities of wood by cooking the wood chips in a highly pressurized chamber followed by multiple stages of washing and chemical treatments. The study is undertaken to increase the pulp yield as far as possible without increasing the viscosity beyond the specified upper limit.

Chapter 6 gives the concept of a time-dependent conflicting bifuzzy set (CBFS), and a new procedure to construct the membership and nonmembership functions of the fuzzy reliability function is proposed with the help of time-dependent CBFS. The concept of triangular CBFS has been developed, and triangular CBFS is used to represent the failure rate function of the system.

Chapter 7 focuses on the failure time data analysis based on the nonhomogeneous Poisson process (NHPP) and discusses several statistical estimation methods for a periodic replacement problem with minimal repair as the simplest application of life data analysis with NHPP. Not only the parametric maximum likelihood estimation of the power law process is applied, but also constrained nonparametric maximum likelihood estimation (CNPMLE) and kernel-based estimation methods for estimating the cost-optimal periodic replacement problem with the minimal repair, where single or multiple minimal repair data are assumed.

Chapter 8 extends the available literature and discusses the important attribute "view-count" of content dynamically. With the Internet emerging as a rapidly growing new market, the netizens are also growing at a fast pace. Making use of this ideology, a modeling framework whose utility has been highlighted through three models is proposed that describes the growing Internet market size and repeat viewers. The models have been validated on a YouTube entertainment video data set.

Chapter 9 analyzes dual-market modeling. Dual-market modeling is an increasingly important concept in marketing, and in order to inculcate heterogeneity and the requirement of some specific characteristics for some technologies to be adopted differently in different geographical locations, the authors intend to study this behavior through a mathematical framework that exhibits the dual-market phenomenon.

Chapter 10 presents a uniform methodology for three fundamental problems in data analysis: identification/detection of atypical elements (outliers), clustering, and classification. Such a unification facilitates understanding of the material and adapting it to the individual needs and preferences of particular users. The investigated material is ready to use, is practically parameter free, and does not require laborious exploration from the researcher. This has been illustrated with a number of applications in the fields of engineering, management, medicine and biology, as well as supplemented by a thematic bibliography extending the issues presented.

Chapter 11 analyzes the statistical tolerance limits used today in both production and research. It is often desirable to have statistical tolerance limits available for the distributions used to describe time-to-failure data in reliability problems. For example, one might wish to know if at least a certain proportion of a manufactured product will operate for at least, say, the warranty period. This question cannot usually be answered exactly, but it may be possible to determine a lower tolerance limit, based on a preliminary random sample, such that one can say with a certain confidence that at least a specified proportion or more of the product will operate longer than the lower tolerance limit. Then reliability statements can be made based on the lower tolerance limit, or, decisions can be reached by comparing the tolerance limit to the warranty period.

Chapter 11 analyzes the lower tolerance limit, based on a preliminary random sample, such that one can say with a certain confidence that at least a specified proportion or more of the product will operate longer than the lower tolerance limit, which presents a new technique for constructing exact lower and upper tolerance limits on outcomes (for example, on order statistics) in future samples. The technique used here emphasizes pivotal quantities relevant for obtaining tolerance factors and is applicable whenever the statistical problem is invariant under a group of transformations that acts transitively on the parameter space. The proposed technique is based on a probability transformation and pivotal quantity averaging. It is conceptually simple and easy to use. The discussion is restricted to one-sided tolerance limits.

Chapter 12 deals with the design of torque-based PID controller and tuning its gains with the help of two algorithms, namely, modified chaotic invasive weed optimization (MCIWO) and modified chaotic invasive weed optimization-neural network (MCIWO-NN) algorithms for the biped robot while walking on a staircase. An analytical method has been developed to generate the gaits and design the torque-based PID controller. The dynamics of the biped robot utilized in the said controller have been derived after utilizing the Lagrange–Euler formulation. Alongside, the authors utilized the MCIWO algorithm to optimize the gains of the PID controller. Further, in the MCIWO-NN algorithm, the MCIWO algorithm is used to evolve the architecture of the NN, which helped in predicting the gains of the PID controller in an adaptive manner. The developed algorithms are tested in computer simulations and on a real biped robot.

In Chapter 13, intelligent predictive models for modeling fertility of Murrah bulls using the various emerging machine learning (ML) algorithms, namely, neural networks (NNs), support vector regression (SVR), decision trees (DTs), and random forests (RFs), and a conventional linear model (LM) for regression have been described. These intelligent ML models would provide decision support to organized dairy farms for selecting good bulls. Hence, the ML models can be employed as a plausible alternative to linear regression models to assess more accurately the conception rate in Murrah breeding bulls at the organized farms.

In Chapter 14, the computational study has been performed on the two-jet vectoring through the Coanda surface, realizing the future concept of the vertical and short takeoff and landing (V/STOL) of an aircraft for civil aviation purposes. The set of computations has been performed from the incompressible flow regime to the incipient of the compressible flow regime. This computational study aims to identify the design parameters through the flow characteristics using the computational fluid dynamics technique.

Chapter 15 deals with the discussion on the collocation method which is a very well-known numerical technique. Along with the description of the methodology adopted, details are given for the used B-spline basis function in the collocation method. The properties of B-spline basis functions are discussed in this chapter with a description of the types and degrees of B-spline basis functions.

In Chapter 16, utilizing Rayleigh's approximation method, an attempt has been made to study the reflection and refraction patterns in a corrugated interface sandwiched between an initially stressed fluid-saturated poroelastic half-space and a highly anisotropic half-space. The highly anisotropic half-space is considered as triclinic. Various two-dimensional plots have been drawn to show the effects of some affecting parameters such as initial stress parameter, corrugation amplitude, wavelength, and frequency factor.

This book can be used as a support book for the final undergraduate engineering course (for example, mechanical, mechatronics, industrial, computer science, information technology, mathematics, etc.). Also, this book can serve as a valuable reference for academics, mechanical, mechatronics, computer science, information technology and industrial engineers, environmental sciences as well as researchers in related subjects.

Mangey Ram
*Dehradun, India*

J. Paulo Davim
*Aveiro, Portugal*

# Acknowledgments

The editors acknowledge CRC Press for this opportunity and professional support. Also, we would like to thank all the chapter authors and reviewers for their availability for this work.

# Editors

**Mangey Ram** received the PhD in mathematics and minor in computer science from G. B. Pant University of Agriculture and Technology, Pantnagar, India, in 2008. He has been a faculty member for around 10 years and has taught several core courses in pure and applied mathematics at undergraduate, postgraduate, and doctorate levels. He is currently a professor at Graphic Era (Deemed to be University), Dehradun, India. Before joining Graphic Era (Deemed to be University), he was a deputy manager (probationary officer) with Syndicate Bank for a short period. He is editor-in-chief of *International Journal of Mathematical, Engineering and Management Sciences*; and the guest editor and member of the editorial boards of many journals. He is a regular reviewer for international journals, including those published by the Institute of Electrical and Electronics Engineers, Elsevier, Springer, Emerald, John Wiley, Taylor & Francis Group, and many other publishers. He has published 125 research publications (published by Institute of Electrical and Electronics Engineers, Springer, Emerald, World Scientific, among others) and in many other national and international journals of repute and also presented his works at national and international conferences. His fields of research are reliability theory and applied mathematics. Ram is a senior member of the Institute of Electrical and Electronics Engineers, member of the Operational Research Society of India, the Society for Reliability Engineering, Quality and Operations Management in India, the International Association of Engineers in Hong Kong, and the Emerald Literati Network in the United Kingdom. He has been a member of the organizing committees of a number of international and national conferences, seminars, and workshops. He has been conferred with the Young Scientist Award by the Uttarakhand State Council for Science and Technology, Dehradun, in 2009. He has been awarded the Best Faculty Award in 2011 and recently the Research Excellence Award in 2015 for his significant contributions in academics and research at Graphic Era (Deemed to be University).

**J. Paulo Davim** received his PhD in mechanical engineering in 1997, and MSc in mechanical engineering (materials and manufacturing processes) in 1991, the Dipl-Ing engineer's degree (5 years) in mechanical engineering in 1986, from the University of Porto (FEUP), the Aggregate title (Full Habilitation) from the University of Coimbra in 2005, and the DSc from London Metropolitan University in 2013. He is Eur Ing by FEANI-Brussels and senior chartered engineer by the Portuguese Institution of Engineers with a MBA and Specialist title in engineering and industrial management. Currently, he is a professor at the Department of Mechanical Engineering of the University of Aveiro, Portugal. He has more than 30 years of teaching and research experience in manufacturing, materials and mechanical engineering with special emphasis in machining and tribology. He also has interest in management and industrial engineering and higher education for sustainability and engineering education. He has guided large numbers of postdoc, PhD, and master's degree students. He has received several scientific awards. He has worked as the evaluator of projects for international research agencies as well as examiner of PhD theses for many universities. He is the editor-in-chief of several international journals, guest editor of journals, books editor, book series editor, and scientific advisor for many international journals and conferences. Presently, he is an editorial board member of 25 international journals and acts as a reviewer for more than 80 prestigious Web of Science journals. In addition, he has published as editor (and co-editor) more than 100 books and as author (and co-author) more than 10 books, 70 book chapters, and 400 articles in journals and conferences (more than 200 articles in journals indexed in Web of Science core collection/h-index 41+/5000+ citations and SCOPUS/h-index 51+/7500+ citations).

# Contributors

**N. Aggrawal**
Department of Computer Science &
  Information Technology
Jaypee Institute of Information Technology
Noida, Uttar Pradesh, India

**R. Aggarwal**
Department of Operational Research
University of Delhi
Delhi, India

**A. Anand**
Department of Operational Research
University of Delhi
Delhi, India

**A. Arora**
Department of Computer Science &
  Information Technology
Jaypee Institute of Information Technology
Noida, Uttar Pradesh, India

**Geeta Arora**
Department of Mathematics
Lovely Professional University
Phagwara, Punjab, India

**G. Berzins**
BVEF Research Institute
University of Latvia
Riga, Latvia

**Neelima Bhengra**
Department of Applied Mathematics
Indian Institute of Technology (ISM)
Dhanbad, Jharkhand, India

**Dinesh Bisht**
Department of Mathematics
Jaypee Institute of Information Technology
Noida, Uttar Pradesh, India

**Atish Kumar Chakravarty**
Computer Centre & Dairy Economics,
  Statistics & Management Division
ICAR-National Dairy Research Institute
Karnal, Haryana, India

**Shshank Chaube**
Department of Mathematics
University of Petroleum & Energy Studies
Dehradun, Uttarakhand, India

**K.K. Chowdhury**
SQC & OR Unit
Indian Statistical Institute
Bangalore, Karnataka, India

**Tadashi Dohi**
Department of Information Engineering
Graduate School of Engineering
Hiroshima University
Higashihiroshima, Japan

**M.S. Irshad**
Department of Operational Research
University of Delhi
Delhi, India

**Boby John**
SQC & OR Unit
Indian Statistical Institute
Bangalore, Karnataka, India

**Leonid Kondratenko**
Department of Machine Science and
   Machine Components
Moscow Aviation Institute (State National
   Research University)
Moscow, Russia

**Piotr Kulczycki**
Systems Research Institute
Polish Academy of Sciences
Warsaw, Poland
and
Division for Information Technology and
   Systems Research
AGH University of Science and Technology
Cracow, Poland

**Anuj Kumar**
Department of Mathematics
University of Petroleum & Energy
   Studies
Dehradun, Uttarakhand, India

**Gurupada Maity**
Department of Applied Mathematics
   with Oceanology and Computer
   Programming
Vidyasagar University
Midnapore, West Bengal, India

**Ravinder Malhotra**
Computer Centre & Dairy Economics,
   Statistics & Management Division
ICAR-National Dairy Research
   Institute
Karnal, Haryana, India

**Ravi Kumar Mandava**
School of Mechanical Sciences
IIT Bhubaneswar
Bhubaneswar, Odisha, India

**Lubov Mironova**
Institute of Applied Technology
Russian University of Transport (MIIT)
Moscow, Russia

**A. Munjal**
Department of Mathematics
National Institute of Technology
   Kurukshetra
Kurukshetra, Haryana, India

**N.A. Nechval**
BVEF Research Institute
University of Latvia
Riga, Latvia

**K.N. Nechval**
Aviation Department
Transport and Telecommunication
   Institute
Riga, Latvia

**Sangeeta Pant**
Department of Mathematics
University of Petroleum & Energy
   Studies
Dehradun, Uttarakhand, India

**Mangey Ram**
Department of Mathematics, Computer
   Science & Engineering
Graphic Era (Deemed to be University)
Dehradun, Uttarakhand, India

**Sankar Kumar Roy**
Department of Applied Mathematics
   with Oceanology and Computer
   Programming
Vidyasagar University
Midnapore, West Bengal, India

**Yasuhiro Saito**
Department of Maritime Safety
   Technology
Japan Coast Guard Academy
Kure, Japan

**Adesh Kumar Sharma**
Computer Centre & Dairy Economics,
   Statistics & Management Division
ICAR-National Dairy Research Institute
Karnal, Haryana, India

**S. Sonker**
Department of Mathematics
National Institute of Technology
 Kurukshetra
Kurukshetra, Haryana, India

**O. Singh**
Department of Operational Research
University of Delhi
Delhi, India

**S.B. Singh**
Department of Mathematics, Statistics &
 Computer Science
G. B. Pant University of Agriculture &
 Technology
Pantnagar, Uttarakhand, India

**Pankaj Kumar Srivastava**
Department of Mathematics
Jaypee Institute of Information Technology
Noida, Uttar Pradesh, India

**Maharshi Subhash**
Department of Science and Methods in
 Engineering
University of Modena and Reggio Emilia
Reggio Emilia, Italy

**Michele Trancossi**
Department of Science and Methods in
 Engineering
University of Modena and Reggio Emilia
Reggio Emilia, Italy

**Pandu R. Vundavilli**
School of Mechanical Sciences
IIT Bhubaneswar
Bhubaneswar, Odisha, India

*chapter one*

# Application of the Laplace transform in problems of studying the dynamic properties of a material system and in engineering technologies

**Lubov Mironova**
*Russian University of Transport (MIIT)*

**Leonid Kondratenko**
*Moscow Aviation Institute (State National Research University)*

## Contents

This chapter is written by engineers for engineers. The authors try to convey to the reader the simplicity and accessibility of the methods in a concise form with the illustration of the calculation schemes. For a more extensive study of the stated problems of mathematical modeling, at the end of the chapter are given the literature sources, from which the reader can obtain the necessary additional explanations. The list of authors includes well-known scientists in the field of mathematics and mechanics – G. Doetsch, A.I. Lur'e, L.I. Sedov, V.A. Ivanov, and B.K. Chemodanov. In compiling the theoretical material, we refer to the authors mentioned. This chapter reflects the experience of lecturing on mathematical methods of modeling, as well as the personal participation of the authors in the work in this technical field.

The material presented can be of interest to students, graduate students and other specialists.

## 1.1   Designation

$j$ — the imaginary unit; $e$—the base of natural logarithms;
$\alpha = \sigma + j\omega$—the complex number;
Re—real part, Im—the imaginary part of the complex number;
$s$—a complex variable; $s = x + jy$, $x = \mathrm{Re}\ s$, $y = \mathrm{Im}\ s$;
$L$—the transformation (Laplace transform);
$F(s)$—function of complex variable $s$ (Laplace representation);
$f(t)$—function of the real variable $t$ (the original);
$L[f(t)]$—direct Laplace transform;
$L^{-1}[F(s)]$—inverse Laplace transform; and
$\rightarrow$—the sign of the correspondence of the transformation:
for the direct transformation $-f(t) \rightarrow F(s)$; for the inverse transformation $-F(s) \leftarrow f(t)$.

In many formulas, fractional numbers are not represented by a standard record but by a slash or by multiplication of the factor in the degree ($n$), irrational numbers are expressed as a number in fractional power. For a correct understanding of these symbols, examples are given:

$$a/bc = \frac{a}{bc}, \quad a+b/c = a+\frac{b}{c}, \quad a/(b+c) = \frac{a}{b+c}, \quad a(bc+d)^{-1} = \frac{a}{bc+d}, \quad a^{1/2} = \sqrt{a}, \quad b^{-1/3} = \frac{1}{\sqrt[3]{b}}.$$

## 1.2   Laplace transform and operations mapping

The Laplace transform is a powerful mathematical method for solving differential, difference, and integral equations. By means of these equations, one can describe any physical (technological) process and conduct mathematical modeling of the behavior of the object and of the reaction of the environment under the influence of force or other factors, investigate the dynamic properties of the element of construction, and much more.

In many engineering problems, it is important to investigate a function $f(t)$, where real variable $t$ is time. Such problems in mechanics relate to dynamic problems.

The simplest and most economical solution of such problems is possible with the help of methods of the theory of operational calculus [1].

An important role in applied mathematical analysis is played by the Laplace integral

$$I = \int_0^\infty f(t)e^{-st}\, dt. \tag{1.1}$$

Here, $s$ is a complex variable; $s = x + jy$; $t > 0$.

To calculate the Laplace integral, the definition and behavior of the function $f(t)$ for a negative value of the argument $t < 0$ are immaterial. Therefore, we consider a subclass of piecewise continuous functions $f(t)$ of the real variable $t$ defined for $t > 0$ and assumed to be zero for $t < 0$ (Figure 1.1).

This class of functions is characterized by a certain order of growth, such that for any $t > 0$ the module $f(t)$ grows more slowly than some exponential function [1].

Expression (1.1) in the operational calculus and its applications is applied in the following form:

$$F(s) = \int_0^\infty f(t)e^{-st}\, dt. \tag{1.2}$$

Here, the function $F(s)$ is a function of the complex variable $s = x + jy$.

In the new expression (1.2), the function of the complex variable $s$ is put in correspondence with the function of the real variable $t$. Such a correspondence is called the Laplace transform. Symbolically, the Laplace transform is written in form

$$L\big[f(t)\big] = F(s) = \int_0^\infty f(t)e^{-st}\, dt. \tag{1.3}$$

The record $L[f(t)]$ means $L$-transformation. The Laplace transform connects the single-valued function $F(s)$ of the complex variable $s$ (image) with the corresponding function $f(t)$ of the real variable $t$ (the original). A brief description of the essence of the Laplace transform and the correspondence table of operations can be found in Ref. [2].

As can be seen from (1.2), this transformation consists of multiplying the function $f(t)$ by the exponential function $e^{-s}$ and integrating the product of these functions with respect to the argument $t$ in the range from 0 to $\infty$.

From the image of $F(s)$, if it exists, one can always find the original $f(t)$. Such a transition is called the *inverse Laplace transform*, symbolically denoted by $L^{-1}$, and corresponds to

$$f(t) = L^{-1}\big[F(s)\big] = \frac{1}{2\pi} \int_{x-j\infty}^{x+j\infty} F(s)e^{st}\, ds, \quad t > 0. \tag{1.4}$$

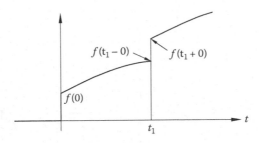

*Figure 1.1* A piecewise-continuous function: $t_1$ is a point of discontinuity of the first kind; $f(+0) = \lim_{t \to +0} f(t); f(-0) = 0; f(t_1 + 0) - f(t_1 - 0) -$ function jump $f(t)$.

The right-hand side of Equation (1.4) is called the *inverse of the Laplace integral* and is a complex integral.

If relation (1.4) is rewritten in the form

$$\int_{-\infty}^{+\infty} e^{jyt}\left(x+jy\right)dy = 2\pi e^{-xt} f(t) \quad \text{for } t>0;$$

$$\int_{-\infty}^{+\infty} e^{jyt}\left(x+jy\right)dy = 0 \quad \text{for } t<0, \tag{1.5}$$

then the formulas (1.2) and (1.5) have a physical meaning. For a constant value of $x$ in the complex variable $s = x + jy$, the function $F(x + jy)$ is the spectral density of the damped time function $e^{-st}f(t)$ for which the variable $y$ is the circular frequency. Such a change in the variable $s$ in the complex plane corresponds to the displacement of the point along the vertical line with the abscissa $x$ (Figure 1.2).

From the mathematical point of view, multiplying the function $f(t)$ by $e^{-st}$ makes the improper integral of the right-hand side of expression (1.2) a convergent in the half-plane Re $s > x_0$ (Figure 1.3).

The function $f(t)$ can be an original only if the following conditions are satisfied:

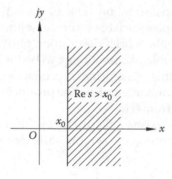

**Figure 1.2** Changing the complex variable.

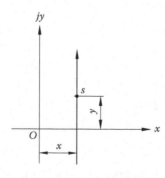

**Figure 1.3** The domain of convergence of the Laplace integral.

1. The function $f(t)$ is continuous for all values $t \geq 0$. Continuity can be violated only at points of discontinuity of the first kind. The number of these points must be finite in any interval of limited length (Figure 1.1).
2. The function $f(t) = 0$ for the values $t < 0$.
3. The function $f(t)$ has a limited order of increasing (i.e., one can find constant numbers).
4. $M > 0$ and $x_0 \geq 0$ such that

$$f(t) < Me^{x_0 t}, \quad t > 0.$$

Here, $x_0$ is the exponent of the growth of the function $f(t)$.

The set of all $f(t)$ is called the space of originals, and the set of all $F(s)$ is the image space.

An important property of the Laplace transform is the following: the images obtained as a result of the L-transformation are analytic functions. Such functions (as complex functions) can be differentiated as many times as desired.

The true meaning of the Laplace transform is that this transformation has the character of a single-valued mapping of the function from the space of originals in the image space, in which the operations performed on the image function are much simpler and more obvious. The transition from the image to the original makes it possible to obtain the desired solutions in a simpler way.

By virtue of the brevity of the following presentation, we give without proofs theorems characterizing the important properties of the Laplace transform. A complete exposition and proofs of the theorems are given in Refs. [1,3,4].

**Theorem 1.1:** If the function $f(t)$ is an original, then this function is Laplace transformed and the image of the given function $F(s)$ is defined in the half-plane Re $s > x_0$, where $x_0$ is the growth index of the function $f(t)$.

Under the condition Re $s > x_0$, the integral (1.2) is an absolutely convergent integral. The number $x_0$ is called the *abscissa of absolute convergence of the integral* (1.2).

**Theorem 1.2:** The image $F(s)$ of the original $f(t)$ in the half-plane, for which Re $s > x_0$, where $x_0$ is the growth index of the original, is an analytic function.

The continuity of the function $F(s)$ follows from the proof of Theorem 1.1.

This important property makes it possible to use powerful methods of the theory of functions of a complex variable in calculations, because in the practical application of Laplace, calculations are performed not over given functions, but over their images.

The analytic expression of the original through the image is formulated by Theorem 1.3.

**Theorem 1.3:** The original $f(t)$ at points of continuity is defined by

$$f(t) = \frac{1}{2\pi} \int_{x-j\infty}^{x+j\infty} F(s)e^{st}\, ds, \tag{1.6}$$

where $F(s)$ is the Laplace representation of the original $f(t)$, and the integral on the right-hand side of this equation is understood in the sense of the principal value

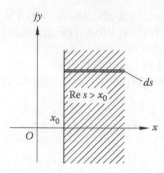

*Figure 1.4* The domain of inversion of the Laplace integral: $ds = jdy$.

$$\int_{x-j\infty}^{x+j\infty} F(s)e^{st}\,ds = \lim_{y\to\infty}\int_{x-jy}^{x+jy} F(s)e^{st}\,ds,$$

and this is taken along a straight line parallel to the imaginary axis and located in the half-plane Re $s > x_0$ (Figure 1.4).

Formula (1.6) is called the *Laplace inversion formula* and establishes a connection between the image of $F(s)$ and the single-valued corresponding original $f(t)$. The process of obtaining the original from a given image is written by the expression (1.4).

The formula (1.6) defines the original only at the points of its continuity. However, for the piecewise-continuous functions $f(t)$, illustrated in Figure 1.1, the limit of the right-hand side of (1.6) at the points of discontinuity of the first kind exists and is defined by

$$\lim_{y\to\infty}\frac{1}{2\pi j}\int_{x-j\infty}^{x+j\infty} F(s)e^{st}\,ds = \frac{1}{2}\left[f(t+0)+f(t-0)\right].$$

From this follows another important property—the single value of the Laplace transform.

The original always corresponds to a single image, since the values of the original at points of discontinuity do not change the view of the image. At the same time, the same image can be associated with a set of originals, the values of which differ from each other only at points of discontinuity [4].

*Corollary of Theorem 1.3:* If the original is a differentiable function everywhere in the interval $0 < t < \infty$, then the original with respect to the given image is uniquely determined.

It should be noted that not all analytical functions can be images. In particular, periodic functions of the form $e^{as}$, cos $s$, sin $s$, are not images, and not all functions can be originals $(1/t,\ \text{tg}\ \omega t, e^{t^2})$. The proof of the theorem, which determines sufficient conditions when the function $F(s)$ is an image, is find in Refs. [4,5].

The Laplace transform is the result of the extension of the Fourier transform to functions that satisfy the Dirichlet conditions in the interval $0 < t < \infty$ but do not satisfy the condition of absolute integrability in this interval. The connection between the Fourier transform and the Laplace transform is clearly presented in Ref. [3]. The Fourier and Laplace transforms are widely used in the theory of automatic regulation.

The next important property of the Laplace transform is the linearity of the transformation, which is formulated by a theorem that establishes the "original-image" correspondence.

**Theorem 1.4:** If the functions $f_1(t), f_2(t), ..., f_n(t)$ are originals, and the images of these functions are, respectively, $F_1(s), F_2(s), ..., F_n(s)$, and if $\lambda_1, \lambda_2, ..., \lambda_n$ are quantities that do not depend on $t$ and $s$, then the following equalities hold:

$$L\left[\sum_{k=1}^{n} \lambda_k f_k(t)\right] = \sum_{k=1}^{n} \lambda_k F_k(s); \tag{1.7}$$

$$L^{-1}\left[\sum_{k=1}^{n} \lambda_k F_k(s)\right] = \sum_{k=1}^{n} \lambda_k f_k(t). \tag{1.8}$$

The linearity property allows, in the practical application of the Laplace transform, performance of calculations not on the given functions, but on their images, applying the table of correspondences between the originals and the images. In this case, you need to know not only the images of individual functions, but also the rules for displaying operations performed on such functions. Therefore, following we formulate other properties of transformation (differentiability, integrability, etc.) in the form of rules, and we call such rules later when solving some mathematical problems.

## 1.3 Linear substitutions

We give the rules for linear transformation of an argument in the original or image. For simplicity of clarity, instead of the symbolic designation of the transformation, we introduce the arrows indicating the direct and inverse Laplace transformations.

**Rule I. Theorem 1.5:** *Similarity theorem.* Multiplying the argument of the original (image) by a certain positive number results in the division of the image (the original) and argument of the image (the original) into the same positive number

$$f(at) \rightarrow \frac{1}{a}F\left(\frac{s}{a}\right), \quad F(as) \leftarrow \frac{1}{a}f\left(\frac{t}{a}\right) \quad a > 0. \tag{1.9}$$

This operation characterizes the change in the scale of the independent variable.

**Rule II. Theorem 1.6:** *First displacement theorem (the lag theorem).* If the function $f(t)$ is an original and $F(s)$ is its image, then the image of the displaced original $f(t - a)$, where $a$ is a real number, is determined by the expression

$$f(t-a) \rightarrow e^{-as}F(s); \quad a > 0. \tag{1.10}$$

For the inverse transformation is valid

$$e^{-as}F(s) = 0, \quad \text{for } t < a;$$

$$e^{-as}F(s) \leftarrow f(t), \quad \text{for } t > a.$$  (1.11)

Since $t < a$, the argument $t - a$ is negative, then the function $f(t - a)$ is equal to zero. The graph of this function is obtained from the graph of the function $f(t)$ by shifting its graph to the right by a distance $a$ (Figure 1.5a, b).

The displacement theorem has a wide application in the theory of automatic regulation, as well as in the study of processes described by piecewise-continuous and periodic functions.

**Rule III. Theorem 1.7:** *Second bias theorem (the bias theorem).* If the function $f(t)$ is the original and $F(s)$ is the image, then the image of the displaced original $f(t + a)$, where $a$ is a real number, is determined by the expression

$$f(t+a) \rightarrow e^{as}\left(F(s) - \int_0^a e^{-st} f(t)dt\right); \quad a > 0.$$  (1.12)

The essence of this theorem is that the image of $F(s)$ cannot be linearly transformed into the original of the function $f(t + a)$, since the right-hand side of Equation (1.12) has a finite Laplace integral. Its calculation is carried out in the interval of variation of the real variable $0 \le t < a$.

This rule is the opposite of the second rule. The graph of the function $f(t)$ is shifted to the left by a distance $a$ (Figure 1.5a, c).

The bias theorem determines the ratio of the image and the original in the case when the complex variable is displaced by $a$ [1]:

$$F(s+a) \leftarrow e^{-at}f(t) + a\int_0^t e^{-at}f(t)dt.$$  (1.13)

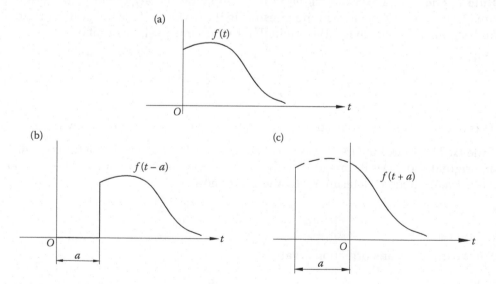

*Figure 1.5* Offset function to the right and left.

Rule III plays an important role in solving difference equations [3].

**Rule IV. Theorem 1.8:** *Damping theorem.* If the function *f(t)* is the original and *F(s)* is the image, and if $\alpha$ is any complex number, then we can write

$$e^{\alpha t} f(t) \rightarrow F(s - \alpha) \qquad (1.14)$$

or

$$e^{-\alpha t} f(t) \rightarrow F(s + \alpha). \qquad (1.15)$$

It follows that if the correspondence *f(t)* $\rightarrow$ *F(s)* holds in the half-plane.
Re $s = x > x_0$, then the correspondence (1.14) is meaningful for Re $(s + \alpha) > x_0$, i.e.,
Re $s > x_0$ - Re $\alpha$. The actual attenuation of the original occurs only if $\alpha$ is a positive real number.

## 1.4   Differentiation and integration

The following theorems establish two important properties of the Laplace transform.

**Theorem 1.9:** The differentiation theorem for the original.
If the function *f(t)* and its derivative are *f'(t)* originals and *F(s)* is the image of the original *f(t)*, then is justly the equality

$$L\left[f'(t)\right] = sF(s) - f(+0), \quad \text{where } f(+0) = \lim_{t \to 0} f(t). \qquad (1.16)$$

If we set the initial value *f* (+0) = 0, then from formula (1.16) we obtain

$$L\left[f'(t)\right] = sF(s).$$

Hence, we formulate the following rule.

**Rule V.** The operation of differentiating the original corresponds to the operation of multiplying the image of this original by the complex number *s*.

$$f'(t) \rightarrow sF(s). \qquad (1.17)$$

If the derivatives of higher orders $f^{(2)}(t), f^{(3)}(t), \ldots, f^{(n)}(t)$ are originals, then the following relations hold:

$$f^{(2)}(t) \rightarrow s^2 F(s) - f(+0) - f'(+0);$$

$$f^{(3)}(t) \rightarrow s^3 F(s) - f(+0)s^2 - f'(+0)s - f^{(2)}(+0);$$

and

$$f^{(n)}(t) \rightarrow s^n F(s) - \sum_{k+1}^{n} s^{n-k} f^{(k-1)}(+0). \qquad (1.18)$$

The essence of Rule V is as follows. The differentiation, which in the origin space is a transcendent process [3], is replaced in the image space by multiplying the image by the degree of argument $s$ with the simultaneous addition of the polynomial whose coefficients are the initial values of the original.

Rule V assumes that the derivative of the highest order $f^{(n)}(t)$ exists at each point $t > 0$ and has an image. This rule is especially valuable in solving differential equations.

In the operational calculus, instead of the Laplace integral (1.3), we prefer to consider the function

$$F(s) = s\, L[f(t)]; \quad F(s) = s\int_0^\infty e^{-st} f(t)\,dt \tag{1.19}$$

or

$$\frac{F(s)}{s} = \int_0^\infty e^{-st} f(t)\,dt. \tag{1.20}$$

Taking into account (1.19), we give the most important ratio of operational calculus

$$f'(t) \to s\left[F(s) - f(0)\right]; \quad f''(t) \to s^2\left[F(s) - f(0) - \frac{f'(0)}{s}\right];$$

$$f^{(n)}(t) \to s^n\left[F(s) - f(0) - \frac{f'(0)}{s} - \cdots - \frac{f^{(n-1)}(0)}{s^{(n-1)}}\right]. \tag{1.21}$$

There are no contradictions between the formulas for differentiating the original (1.18) and the expressions (1.21). According to the rule for calculating the integral, these expressions differ only in the integration constants. Only in (1.18), these constants are real, and in (1.21) are only complex quantities.

The essence of the function (1.19) lies in the fact that in transformations we work in image space only with analytic functions and their initial values. This important technique is widely used in mechanics and other technical applications. Then, using a concrete example, we show the advantages of applying formulas (1.21).

**Rule VI. Theorem 1.10:** *Differentiation theorem for an image.*
If the function $f(t)$ is an original and $F(s)$ is an image, then is justly the equality

$$L[tf(t)] = -\frac{d}{ds} F(s). \tag{1.22}$$

Since the image of $F(s)$ is always an analytic function and possesses all derivatives, on the basis of (1.22) one can obtain derivatives of any order.

Thus, the operation of image differentiation with respect to $s$ corresponds to the operation of multiplying the original by an independent variable $t$ taken with the opposite sign:

$$-tf(t) \to F'(s); \quad t^2 f(t) \to F^{(2)}(s); \quad (-1)^n t^n f(t) \to F^{(n)}(s). \tag{1.23}$$

**Rule VII. Theorem 1.11:** If the function $f(t)$ is an original, and $F(s)$ is the image, then the integral is also an original and is justly equality

$$L\left[f^{-1}(t)\right] = \frac{F(s)}{s} + \frac{f^{-1}(+0)}{s}. \tag{1.24}$$

Here, $f^{-1}(t) = \int f(t)\,dt = \int_0^t f(\tau)\,d\tau + f^{-1}(+0)$, where $f^{-1}(+0)$ is the integration constant.

Hence, under the condition that $f^{-1}(+0) = 0$, Rule VII is formulated.

The operation of integrating the original corresponds to the operation of dividing the image of this original by the complex number $s$:

$$\int_0^t f(t)\,dt \rightarrow \frac{1}{s}F(s). \tag{1.25}$$

Theorem 1.11 extends to integrals of higher orders.

Let $f^{(-k)}(t) = \int f(t)\,dt \int f(t)(dt)^2 \dots \int f(t)(dt)^k$, then

$$L\left[f^{(-2)}(t)\right] = \frac{F(s)}{s^2} + \frac{f^{-1}(+0)}{s^2} + \frac{f^{(-2)}(+0)}{s};$$

$$L\left[f^{(-3)}(t)\right] = \frac{F(s)}{s^3} + \frac{f^{-1}(+0)}{s^3} + \frac{f^{(-2)}(+0)}{s^2} + \frac{f^{(-3)}(+0)}{s};$$

and

$$L\left[f^{(-n)}(t)\right] = \frac{F(s)}{s^n} + \sum_{k=1}^{n} \frac{f^{(-k)}(+0)}{s^{nk+1}}. \tag{1.26}$$

The rule of integration for an image is rarely used in practice, so we do not give it here.

The expressions for the derivative and integral representations are of primary importance in operational calculus; therefore, the number $s$ acquires the character of the operator [1].

## 1.5   *Multiplication and curtailing*

Consider operations on combinations of several functions.

**Rule VIII. Theorem 1.12:** The image of the sum of a finite number of originals is equal to the sum of the images of these originals:

$$\left[f_1(t) + f_2(t)\right] \rightarrow F_1(s) + F_2(s). \tag{1.27}$$

**Rule IX. Theorem 1.13:** *Convolution theorem.* If the functions $f_1(t)$ and $f_2(t)$ are originals and their images are, respectively, $F_1(s)$ and $F_2(s)$, then is justly the equality

$$L\left[\int_0^t f_1(t-\tau)f_2(\tau)d\tau\right] = F_1(s)F_2(s). \tag{1.28}$$

Here the integral combination of functions is called the *convolution of the originals* $f_1(t)$ and $f_2(t)$ and is denoted by

$$f_1 f_2 = \int_0^t f_1(t-\tau)f_2\tau d\tau. \tag{1.29}$$

Rule IX establishes a correspondence between the convolution of originals and the product of images

$$f_1(t)f_2(t) \rightarrow F_1(s)F_2(s).$$

The inverse transformation is formulated as follows. The product of two images is an image and is equivalent to convolution of the originals:

$$F_1 F_2 \rightarrow \int_0^t f_1(t-\tau)f_2\tau d\tau, \quad t > 0. \tag{1.30}$$

Rule IX has found wide application in technical applications.

**Rule X. Theorem 1.14:** *Theorem of complex convolution.* If the functions $f_1(t)$ and $f_2(t)$ are originals and their images are, respectively, $F_1(s)$ and $F_2(s)$, then the product of the functions $f_1(t)$ and $f_2(t)$ is also an original and is justly the equality

$$L\left[\int_0^t f_1(t)f_2(t)\right] = \frac{1}{2\pi j}\int_{x-j\infty}^{x+j\infty} F_1(s-w)F_2(w)dw. \tag{1.31}$$

Here $w$ is a complex number.

Expression (1.31) is valid only in the case when the abscissas $x$ along which integration is performed will be chosen so that the variables entering into the functions $F_1(s)$ and $F_2(s)$ move in the half-plane of absolute convergence of the integrals $L[f_1]$ and $L[f_2]$.

Rule X establishes a correspondence between the product of two originals and a complex convolution of images

$$f_1(t)f_2(t) \rightarrow \frac{1}{2\pi j}\int_{x-j\infty}^{x+j\infty} F_1(w)F_2(s-w)dw, \quad x_1 \le x < \operatorname{Re} s - x_2; \tag{1.32}$$

$$f_1(t)f_2(t) \rightarrow \frac{1}{2\pi j}\int_{x-j\infty}^{x+j\infty} F_1(s-w)F_2(w)dw, \quad x_2 \le x < \operatorname{Re} s - x_1. \tag{1.33}$$

The domains of absolute convergence of these integrals are illustrated in Figure 1.6.

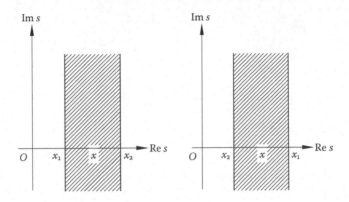

**Figure 1.6** Area of absolute convergence.

The equality of Parseval is widely used in technical applications:

$$\int\limits_{0}^{\infty}\left|f(t)\right|^2 dt = \frac{1}{2\pi}\int\limits_{-\infty}^{+\infty}\left|F(jy)\right|^2 dy. \tag{1.34}$$

The reader will find a more detailed exposition of this question in Ref. [3].

The left-hand-side integral (1.34) is called the *quadratic quality criterion*. In optimization processes, minimizing this integral is the defining characteristic.

## 1.6 The image of a unit function and some other simple functions

We consider the cases of Laplace transform of the simplest functions, which have found wide application in many scientific disciplines.

**The first case.** The function $f(t)$ is the original. This function takes the unit value $t > 0$ and is zero for all $t < 0$. This function is often called either a *single step function*, or a *single jump function* or simply a *single jump*, respectively, and various notations are used. For example, a unit function is widely used of the mathematical apparatus of theories of automatic control, signal processing and other technical applications. In the operational calculus, this function is called the *Heaviside unit function* and has the form

$$u(t) = 0, \quad \text{for } t < 0,$$
$$u(t) = 1, \quad \text{for } t > 0. \tag{1.35}$$

The graph of the unit function is shown in Figure 1.7a.

As can be seen from Equation (1.35) and Figure 1.7a, the unit function is undefined at the point $t = 0$. The choice of one or another value of the unit function for $t = 0$ is related to the features of a particular task.

**The second case.** A single jump occurs at time $t = a > 0$. Such a process is described by the unit function $u\,(t-a)$, which is called the *shifted unit function* (Figure 1.7b). Here it is fair:

$$u(t-a) = 0, \quad \text{for } t < a,$$
$$u(t-a) = 1, \quad \text{for } t > a. \tag{1.36}$$

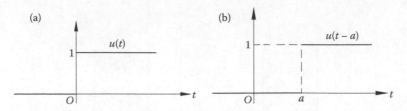

*Figure 1.7* Function of a unit step.

We find the image of the unit functions (1.35) and (1.36).
    For the function (1.35):

$$L[u(t)] = \int_0^\infty u(t)e^{-st}\,dt = \int_0^\infty e^{-st}\,dt = -\frac{e^{-st}}{s}\Big|_0^\infty = \frac{1}{s} \text{ or } u(t) \to 1/s. \tag{1.37}$$

For the function (1.36):

$$L[u(t-a)] = \int_0^\infty u(t-a)e^{-st}\,dt = \int_0^\infty e^{-st}\,dt = \frac{e^{-as}}{s} \text{ or } u(t-a) \to e^{-as}/s. \tag{1.38}$$

In mechanics, the initial values of the function are used in solving problems (i.e., values of the function for $t = 0$). We denote it as $u_0(t)$ and the shifted function, respectively, $u_0(t - \tau)$. Taking (1.18) into account, we can write for them

$$u_0(t) \to 1; \tag{1.39}$$

$$u_0(t - \tau) \to e^{-s\tau}. \tag{1.40}$$

The physical meaning of the initial function is that at time $t = 0$ it takes the value of the constant $C$. It follows from (1.39) that any constant $C$ is an image of the same constant. In the case of the Laplace transform, we always mean that the "constant" (initial function) is a function of $t$, vanishes for $t < 0$ and is equal to $C$ for $t > 0$ [1].

   *The third case.* Special functions. To the category of special functions is the *Dirac delta function*, also called the *first-order impulsive unit function*. The delta function is defined by

$$\delta(t) = 0, \quad \text{for } t \neq 0,$$
$$\delta(t) = \infty, \quad \text{for } t = 0. \tag{1.41}$$

In this case is integral

$$\int_{-\infty}^{+\infty} \delta(t)\,dt = 1. \tag{1.42}$$

Conditions (1.41) and (1.42) are incompatible from the point of view of classical mathematical analysis, and therefore the delta function does not belong to the "function" in the usual sense. However, in the class of generalized functions, the delta function occupies an equal place [2]. The notion of a "delta function" turns out to be significant when extending the operation for differentiating discontinuous functions. For example, a sequence of functions

$$f_\delta(t,a) = \frac{u_0(t) - u_0(a)}{a},$$

characterizing pulses of height $1/a$ and duration $a$ (Figure 1.8), for $a \to 0$ converges to a delta function. For example, the function

$$u_1(t,a) = 1/a[u_0(t) - u_0(t-a)] = 1/a, \quad t < a,$$

$$u_1(t,a) = 1/a[u_0(t) - u_0(t-a)] = 0, \quad t > a \tag{1.43}$$

has a physical meaning in mechanics, as a force of constant magnitude, acting for a period of time $a$. The momentum of this force for a time interval of action is equal to one, regardless of the value of $a$. Such a function is called a delta function of the first order. This function is zero for all $t$ except $t = 0$, when becomes infinite, so that $\lim_{a \to 0} a u_1(t, a) = 1$.

The shifted Dirac delta function $\delta(t - \tau)$ is defined by

$$d(t - \tau) = 0 \quad \text{for } t \neq \tau,$$

$$d(t - \tau) = \infty \quad \text{for } t = \tau. \tag{1.44}$$

Similarly, the displaced unit impulsive force will be denoted as $u_1(t - \tau)$, and

$$u_1(t - \tau) = 0 \quad \text{for } t \neq \tau,$$

$$u_1(t - \tau) = \infty \quad \text{for } t = \tau. \tag{1.45}$$

This force is interpreted as a force, instantaneously communicating at a time $t = \tau$ for the point of unit mass a speed, this equal to one.

For Equations (1.44) and (1.45) we have the relations

$$u_1(t) \to s; \quad u_1(t - \tau) \to e^{-s\tau} s. \tag{1.46}$$

The reader will find delta functions of the second order (Figure 1.9) in Refs. [1,2]. We give only the correlation of the originals and images

$$u_2(t) \to s^2; \quad u_2(t - \tau) \to s^2 e^{-s\tau}. \tag{1.47}$$

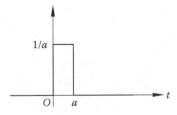

*Figure 1.8* The function $u_1(t, a)$.

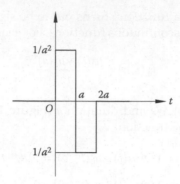

*Figure 1.9* The function $u_2(t, a)$.

**The fourth case.** The time function is $e^\alpha$, where $\alpha$ is an arbitrary complex or real number. We represent such a function in the form

$$u(t)e^{\alpha t} = 0 \qquad \text{for } t < 0,$$
$$u(t)e^{\alpha \tau} = 1e^{\alpha t} \qquad \text{for } t > 0. \tag{1.48}$$

We take the function $u(t)$ as a unit function, which for convenience of calculations we additionally represent in form of the multiplier $\mathbf{1}(t)$.

The image of the function (1.48) will be

$$L\left[\mathbf{1}(t)e^{-\alpha t}\right] = \int_0^\infty e^{\alpha t}e^{-st}\,dt = -\frac{e^{-(s-\alpha)t}}{s}\Big|_0^\infty = \frac{1}{s-\alpha}. \tag{1.49}$$

If $\alpha$ is a complex number, then, depending on the values that the real and imaginary parts take, function (1.48) characterizes the types of vibrations and motions. Expression (1.48) takes on an explicit physical meaning. The reader will find a detailed exposition of this question in Ref. [3].

The graph of the function (1.48), where $\alpha$ is the real negative number ($\alpha < 0$), is shown in Figure 1.10.

**The fifth case.** The function of time is $t$ and the ratio

$$f(t) = 0 \quad \text{for } t < 0,$$
$$f(t) = t \quad \text{for } t > 0 \tag{1.50}$$

is fair.

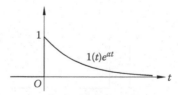

*Figure 1.10* A truncated exponential with real $\alpha < 0$.

We find the image of this function using integration by parts, we obtain for Re $s > 0$:

$$L[t] = \int_0^\infty t e^{-st}\, dt = -\frac{t e^{-st}}{s}\Big|_0^\infty + \frac{1}{s}\int_0^\infty e^{-st}\, dt = -\frac{e^{-st}}{s^2}\Big|_0^\infty = \frac{1}{s^2}. \tag{1.51}$$

*The sixth case.* The function of time is $t^n$ and ratio

$$f(t) = 0 \quad \text{for } t < 0,$$
$$f(t) = t^n \quad \text{for } t > 0 \tag{1.52}$$

is fair.

Using the previous method and repeated integration, we obtain an image for the function (1.52) in the following form:

$$L\left[t^n\right] = \frac{n!}{s^{n+1}}. \tag{1.53}$$

*The seventh case.* We represent the image $(1/s)\cdot F(s)$ in the form

$$\frac{1}{s^n} F(s) = \frac{1}{s} F(s) \frac{1}{s^{n-1}}. \tag{1.54}$$

Using Rule IX (the convolution theorem), we obtain an image of an integer power of the variable $t$:

$$\frac{F(s)}{s^n} \rightarrow \frac{1}{(n-1)!} \int_0^t f(\tau)(t-\tau)^{n-1} d\tau. \tag{1.55}$$

The mechanism of finding the image according to the given original, if this original exists, reduces to calculating the integral (1.2). For the simplest functions, such an operation does not present mathematical difficulties.

Therefore, the results of the transformations are given in tables of correspondence between originals and images [1–4]. We give some correspondences in Tables 1.1 and 1.2.

*Table 1.1* Table of originals and images of these originals

| Number | The original | The image |
| --- | --- | --- |
| 1 | $\mathbf{1}(t)$ | $1/s$ |
| 2 | $e^{-\alpha t}$ | $1/(s+\alpha)$ |
| 3 | $t$ | $1/s^2$ |
| 4 | $t^n$ | $n!/s^{n+1}$ |

*Table 1.2* Table of images and originals of these images

| Number | The image | The original |
|--------|-----------|--------------|
| 1 | $1$ | $\delta(t)$ |
| 2 | $1/s$ | $1$ |
| 3 | $1/s^2$ | $t$ |
| 4 | $n!/s^n$ | $t^n$ |
| 5 | $1/(s-a)$ | $e^{at}$ |
| 6 | $1/s\,(s-a)$ | $\dfrac{1}{a}(e^{at}-1)$ |
| 7 | $s^2[X(s)-x_0-x_0'/s]$ | $x''(t)$ |
| 8 | $\omega/(s^2+\omega^2)$ | $\sin \omega t$ |
| 9 | $s/(s^2+\omega^2)$ | $\cos \omega t$ |

## 1.7  Examples of solving some problems of mechanics

Many problems of mechanics reduce to the solution of differential equations of various orders, in the course of which a general solution is first sought, and then, substituting the initial values of the functions and their derivatives, find a particular solution. In mechanics, to determine the solution uniquely, additional initial or boundary conditions are given. The number of such conditions must coincide with the order of the differential equation or system. Depending on the method of specifying additional conditions in differential equations, the following problems are distinguished:

- The Cauchy problem, when all additional conditions are given in one point (as a rule in the starting point) of interval
- Boundary value problem, when additional conditions are indicated by the values of the function and its derivatives at the boundary of the interval—at the beginning and at the end of the integration

As is known, the solution of such problems is connected with the problem of integrating partial differential equations under given boundary conditions [6].

We show the advantages of the Laplace transform in solving a differential inhomogeneous first-order equation with constant coefficients.

**Example:** Let some process be described by the following differential equation, which we call the initial equation:

$$y' + c_0 y = f(t). \tag{1.56}$$

Equation (1.56) is represented in the original space. In the image space, this equation corresponds to the depicting equation, which has the form

$$L[y'] + c_0 L[y] = L[f(t)]. \tag{1.57}$$

The symbol $L$ [...] denotes the transformation of the original equation by multiplying both parts by $e^{-st}$ and integrating from 0 to $\infty$.

Applying Rule V, we write in the images of Equation (1.57):

$$sY(s) - y(+0) + c_0 Y(s) = F(s). \tag{1.58}$$

Thus, we obtained a linear algebraic equation with the initial value of the function $y(t)$ corresponding to the value $y(+0)$ for the initial point $t = 0$ (Theorem 1.9).

The solution of this equation is quite simple:

$$Y(s)\,[(s+c_0)] = F(s) + y(+0),$$

and finally we have

$$Y(s) = F(s)\frac{1}{s+c_0} + y(+0)\frac{1}{s+c_0}. \qquad (1.59)$$

To the resulting image of $Y(s)$ we find the corresponding original, using the inversion formula (1.4) and Table 1.1 (item 8):

$$y(t) = f(t)e^{-c_0 t} + y(+0)e^{-c_0 t}. \qquad (1.60)$$

We note that the first term on the right-hand side of (1.59) is the product of two images, which, according to Rule IX, corresponds to the convolution of two originals, the first term in (1.60). Finally, we get

$$y(t) = \int_0^t f(\tau)e^{-c_0(t-\tau)} = e^{-c_0 t}\int_0^t f(\tau)e^{c_0 \tau}\,d\tau + y(+0)e^{-c_0 t}. \qquad (1.61)$$

We note that (1.61) is a solution of the differential equation (1.56) for a given initial value of the function $y(t)$. This solution can be obtained even easier. We outline the course of the solution. Let $f(t) = 1(t)$ and $y(+0) = 0$. Using Tables 1.1 (item 1) and 1.2 (item 6), and passing from the image to the original, we obtain the solution (1.56) in the following form:

$$\mathbf{1}(t) = 1/s;\, Y(s) = \frac{1}{s(s+c_0)};\, Y(s) \leftarrow y(t);\, y(t) = \frac{1}{c_0}(1 - e^{-c_0 t}).$$

The solution of a homogeneous second-order differential equation with constant coefficients is given on a concrete example.

Example of the second. Find the solution of equation

$$\frac{d^2 x}{dt^2} + 6\frac{dx}{dt} + 5x = 0$$

with the initial conditions at $t = 0$, $x(0) = 0$, $x'(0) = 1$.

We write the depicting equation

$$L[x''] + L[6x'] + L[5x] = 0.$$

According to Rule V we have

$$L[x''] = s^2 X(s) - sx(0) - x'(0) = s^2 X(s) - 1;$$

$$L[5x'] = 6L[x'] = 6sX(s) - 6s\,x(0) = 6s\,X(s);\, L[5x] = 5X(s)$$

or

$$s^2X(s) - 1 + 6sX(s) + 5X(s) = 0, \text{ or } X(s)[(s^2 + 6s + 5)] = 1.$$

From which

$$X(s) = 1/(s^2 + 6s + 5) = k_1/(s+1) + k_2/(s+5).$$

Here $k_1 = 1$, $k_2 = -1$; the denominator of the fraction has roots $s_1 = -1$, $s_2 = -5$.

Now it is time to move from the image to the original. The inverse Laplace transform gives us the following solution:

$$x(t) = L^{-1}[X(s)] = L^{-1}[1/(s+1)] - L^{-1}[1/(s+5)] = e^{-t} - e^{-5t}.$$

A detailed solution of differential equations of any order may be found in Ref. [3].

We give examples of the solutions of certain problems of mechanics with the aid of the Laplace transform.

*The first task.* The motion of a material point of mass $m$ under the action of a force that depends on time. The differential equation of motion of a point of mass $m$ has the form

$$mx''(t) = f(t). \tag{1.62}$$

*Decision.* We write the depicting equation

$$L[mx''(t)] = L[f(t)] \text{ or } ms^2X(s) - msx(0) - mx'(0) = F(s). \tag{1.63}$$

We rewrite (1.63) in the form

$$ms^2X(s) = msx_0 + mx_0' + F(s). \tag{1.64}$$

Here $x_0$ and $x_0'$ are the initial values of the function $f(x)$ and its first derivative for $t = 0$. Then

$$X(s) = \frac{1}{s}x_0 + \frac{1}{s^2}x_0' + \frac{1}{m}\frac{F(s)}{s^2}. \tag{1.65}$$

Turning to the original, using Table 1.2 (items 2, 3) and (1.53), we obtain the solution of Equation (1.62) in the form

$$x = x_0 + x_0't + \frac{1}{m}\int_0^t f(\tau)(t-\tau)d\tau. \tag{1.66}$$

The same result can be obtained using the formulas (1.20), Table 1.2 (paragraph 1.7). Assuming the initial conditions in the form

$$x = x_0, x' = v_0 \text{ or } t = 0, \tag{1.67}$$

we obtain the following depicting equation

$$ms^2 X(s) = ms^2 x_0 + msv_0 + F(s). \qquad (1.68)$$

Or

$$X(s) = x_0 + sv_0 + (1/ms^2)F(s). \qquad (1.69)$$

Analyzing the two depicting Equations (1.64) and (1.68), we draw the following conclusions: the original of the second derivative $x''(t)$ corresponds to only one image $s^2 X(s)$; the initial values of the functions for (1.64) are given by real quantities; and in Equation (1.68), they automatically become complex numbers.

Passing from the image (1.69) to the original, we immediately obtain Equation (1.66). The depicting Equation (1.68) can also be obtained using the impulsive functions (1.39), (1.46). This can be done if we assume that for zero initial values of the coordinate and velocity to a point of mass $m$, at the time $t = 0$, a pulse is applied, $mx_0'u_1(t)$. This action imparts a velocity $x_0'$ to the point, as well as two oppositely directed impulsive shocks, which impart an instantaneous displacement of $x_0$ the point.

Let us show this by the example of motion of a point of unit mass according to the law of motion

$$x''(t) = f(t). \qquad (1.70)$$

Let's write down the depicting equation, applying single impulsive functions,

$$s^2 X(s) = u_0(t)F(s) + u_1(t)v_0 + u_2(t)x_0. \qquad (1.71)$$

Taking into account the expressions (1.39), (1.47), we obtain

$$s^2 X(s) = 1F(s) + sv_0 + s^2 x_0. \qquad (1.72)$$

Moving from the image to the original, we have

$$x = \int_0^t f(\tau)(t-\tau)d\tau + v_0 t + x_0. \qquad (1.73)$$

**The second task.** Vibrations of the simplest vibrator. Let the load $P = mg$ suddenly be suspended at the end of the stressed spring. We neglect the weight of the cargo. At the same time the cargo is given a deviation by the value displacement of the spring $x_0$ and a deviation the speed of $x_0'$. It is necessary to find a change in the elongation of the spring $x(t)$ under a given force (Figure 1.11).

*Decision.* Due to the fact that the spring is stiff, during the action of the force $P$ and the reported stroke length, this spring will change until the whole system reaches equilibrium and the spring reaches at rest. Therefore, the motion of the particles of the spring can be regarded as longitudinal oscillations, which are made at some point in time.

Moreover, these oscillations will first be forced oscillation, and then at the initial moment of equilibrium, the oscillations take on the character of free oscillations. We use the method of introducing single impulsive functions, following the example considered above. We write the equation of cargo movement

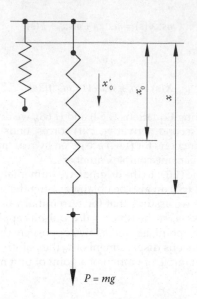

**Figure 1.11** The calculation scheme.

$$mx'' + cx = mgu_0(t) + mx_0' u_1(t)v_0 + mx' u_2(t)x_0,$$ (1.74)

where $c$ is the spring stiffness, and $x$ is elongation.

The equation in the images (1.74) is written in the form

$$(ms^2 + c)X(s) = mg + mx_0's + mx_0 s^2.$$ (1.75)

Then

$$X(s) = \frac{g}{s^2 + k^2} + \frac{x_0's}{s^2 + k^2} + \frac{x_0 s^2}{s^2 + k^2}.$$ (1.76)

Here $k = \sqrt{\dfrac{c}{m}}$ is the frequency of free oscillations.

We transform the first term on the right-hand side of (1.76) as follows: $\dfrac{g}{s^2 + k^2} = g \dfrac{1}{s^2 + k^2}$.

The second factor of the resulting expression is considered as

$$\frac{1}{s^2 + k^2} = \frac{1}{k^2}\left(1 - \frac{s^2}{s^2 + k^2}\right) = \frac{1}{k^2}\left(\frac{1}{s} - \frac{s}{s^2 + k^2}\right).$$

Applying the analogous method for transforming the remaining terms (1.76) and passing to the origin space (Table 1.2, items 8, 9), we finally obtain the well-known solution

$$x(t) = \frac{g}{k^2}\left(1 - \cos kt\right) + \frac{x_0'}{k}\sin kt + x_0 \cos kt.$$

The reader may find an extensive exposition of this material in the literature [1].

## 1.8 Laplace transform in problems of studying oscillation of rods

Consider elementary movements of an elastic cylindrical rod of mass $m$. Mass forces, distributed mass pairs, and surface pairs are absent. We write the equations of displacements of particles of an elementary volume of a rod for the following cases:

1. The rod is given only translational motion (Figure 1.12)

$$\frac{\partial^2 u}{\partial t^2} = c_1 \left( \frac{\partial^2 u}{\partial x^2} \right),$$  (1.77)

where $c_1$ is the coefficient characterizing the properties of the rod material.

Taking $c_1 = E/\rho$, we finally obtain

$$\frac{\partial^2 u}{\partial t^2} = \frac{E}{\rho} \frac{\partial^2 u}{\partial x^2}.$$  (1.78)

Here, $E$ is the modulus of elasticity of the material; $\rho$ is the density of the material.

2. Only twisting movements appear in the rod (the rod is only exposed to the torsion pulse), assuming that the set of planar cross sections of the rod rotate sequentially at a distance $dx$ from each other (Figure 1.13)

$$\frac{\partial^2 \varphi}{\partial t^2} = c_2 \left( \frac{\partial^2 \varphi}{\partial x^2} \right).$$  (1.79)

Taking $c_2 = G/\rho$, we finally obtain

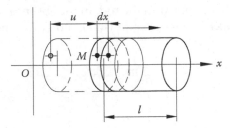

*Figure 1.12* Linear motion of the rod.

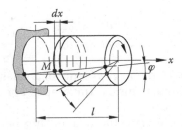

*Figure 1.13* Twisting motion of the rod.

$$\frac{\partial^2 \varphi}{\partial t^2} = \frac{G}{\rho} \frac{\partial^2 \varphi}{\partial x^2}. \tag{1.80}$$

Here $G$ is the shear modulus of the material.

In the case of wave processes propagating in an elastic rod, longitudinal and transverse oscillations with allowance for (1.77) and (1.79) can be described by the formulas [7]

$$\frac{\partial^2 u}{\partial x^2} = \frac{1}{a_1^2} \frac{\partial^2 u}{\partial t^2}, a_1 = \sqrt{\frac{1-\mu}{(1+\mu)(1-2\mu)}}; \tag{1.81}$$

$$\frac{\partial^2 \varphi}{\partial x^2} = \frac{1}{a_2^2} \frac{\partial^2 \varphi}{\partial t^2}, a_2 = \sqrt{\frac{G}{\rho}}. \tag{1.82}$$

Here $\mu$ is Poisson's ratio; $G = E/[2(1 + \mu)]$.

## 1.9   Relationship between the velocities of the particles of an elementary volume of a cylindrical rod with stresses

If the elastic body is subject to wave perturbations, then the displacement of particles of the elementary volume of the rod is accompanied by the appearance of stresses in it. We also note that the elastic wave is two independently propagating waves: longitudinal and transverse waves. In the case of propagation in the elastic body of a longitudinal wave, the movement of particles is effected by the action of a normal perturbing force in the direction of propagation of the wave itself (i.e., along the longitudinal axis of the rod). This means that only normal stresses $\sigma$ arise in the rod. In the propagation of a transverse wave, the displacement of particles of elementary sections occurs under the action of a shearing force in a plane orthogonal to the direction of propagation of the wave.

Thus, tangential stresses $\tau$ appear in the rod. In the absence of mass forces, distributed mass and surface pairs, taking into account (1.77) and (1.79), the following relationships are valid [8]:

$$\rho \frac{\partial \upsilon}{\partial t} = -\frac{\partial \sigma}{\partial x}; \tag{1.83}$$

$$\frac{1}{E} \frac{\partial \sigma}{\partial t} = -\frac{\partial \upsilon}{\partial x}; \tag{1.84}$$

$$r\rho \frac{\partial \Omega}{\partial t} = -\frac{\partial \tau}{\partial x}; \tag{1.85}$$

$$\frac{1}{G\rho} \frac{\partial \tau}{\partial t} = -\frac{\partial \Omega}{\partial x}. \tag{1.86}$$

Here $\upsilon$ is the travel speed of an elementary volume of an elastic rod along axis, $\upsilon = \partial u/\partial t$; $r$ is the radius of the rod; and $\Omega$ is the angular travel speed of the particles of the elastic rod in the plane of the section, $\Omega = \partial \varphi/\partial t$.

Equations (1.83) and (1.84) describe the relationship between the velocities of longitudinal displacement of plane sections of the elementary volume of an elastic rod and changes in normal stresses with the gradients of changes of these variables along the length of the rod.

Equations (1.85) and (1.86) describe the relationship between the shear rate of flat sections of the elementary volume of an elastic rod and the rate of change of the maximum tangential stresses with the gradients of the variations of these variables along the length of the rod.

For a short rod, Equations (1.83–1.86) can be written in ordinary derivatives. A detailed exposition of this question is given in Ref. [8].

This approach, proposed by L. Kondratenko, allowed for development of a new method for studying the dynamics of rotating and longitudinally moving elements of the construction. The method makes it possible to estimate the magnitude and voltage oscillations in the structural elements, as well as the speed of movement of the functional element in engineering technologies.

Let us explain the essence of Kondratenko's method.

## 1.10   An inertial disk rotating at the end of the rod

As a model, let us consider a structure consisting of a cylindrical rod, at the end of which a disk of mass $m$ with a flywheel moment of inertia $J$ is fixed. The rod is imparted rotary motion in the absence of mass forces, distributed mass, and surface pairs. Rotation of the disc is impeded by the moment of resistance $M_r$ (Figure 1.14). Such a model can be taken as an imitation model of the machine spindle operation during hole machining. Rotational motion to the rod is transmitted at point 1 (Figure 1.14). The driven link is a disk fixed to the end of the rod (position 2). We denote the angular velocities of these elements as $\Omega_1$ and $\Omega_2$, respectively. Obviously, the values of the velocities will be different in magnitude.

Integrating Equation (1.86) with respect to the coordinate $x$ and then differentiating it with respect to $t$, we obtain the relation for the torsional vibrations of the rod

$$\vartheta_k \frac{d\tau}{dt} = \Omega_1 - \Omega_2, \tag{1.87}$$

where $\vartheta_k$ is the coefficient of torsional elasticity, $\vartheta_k = \dfrac{l}{\rho G W}$; $\Omega_1$, $\Omega_2$ is the angular velocity of rotation of the rod section and the cross section of the rod near the disk, respectively;

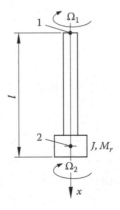

*Figure 1.14* The design scheme.

and $W$ is the geometric moment of resistance of the cross section of the rod near the disk, $W = \pi d^3/16$.

We integrate (1.87). We obtain the dependence of the tangential stresses on the variation of the twist angle. We integrate (1.87). We obtain the dependence of the tangential stresses on the variation of the twist angle

$$v_{1k}\tau = \varphi_1 - \varphi_2, \vartheta_{1k} = \frac{l}{\rho G}. \tag{1.88}$$

The tangential stresses developed in the rod overcome the resistance of the forces.

$M_r = M_{r0} + h_k\,\Omega_2$, as well as rising inertia forces $M_d = Jd\Omega_2/dt$. Here $h_x$ is the loss factor proportional to the angular velocity of rotation of the disk. Taking into account that $\tau = M/W$, we finally obtain

$$\tau(t)W = M_{r0}(t) + h_x\Omega_2(t) + J\frac{d\Omega_2}{dt}. \tag{1.89}$$

## 1.11   Equations of torsional oscillations of a disk

For Equations (1.85) and (1.86), we write the partial differential equations, assuming that the density and the shear modulus of the rod material are equal and constant along the length:

$$r\rho s\Omega(s) = -\frac{d\tau(s)}{dx}; \tag{1.90}$$

$$\frac{s\tau(s)}{rG} = -\frac{d\Omega(s)}{dx}. \tag{1.91}$$

Differentiating (1.90) with respect to the coordinate $x$, eliminating the derivative $d\Omega(s)/dx$ with the help of (1.91), and introducing the new variable $\theta_k(s) = \pm s\,(G^{-1}\rho)^{1/2}$, we obtain a new differential equation second-order

$$\frac{\partial^2\tau(s)}{\partial x^2} - \theta_k^2(s)\tau(s) = 0. \tag{1.92}$$

The solution of this equation has the form

$$\tau(s,x) = C_1\exp[\theta_k(s)x] + C_2\exp[-\theta_k(s)x]. \tag{1.93}$$

The constants of the integration $C_1$, $C_2$ are determined by the boundary conditions, for $x = 0$,

$$\tau(s,x) = t_1(s,0); \frac{\partial\tau(s,x)}{\partial x} = -\frac{\theta_k^2(s)G}{s}\cdot\Omega_1(s,0). \tag{1.94}$$

The final solution (1.90) and (1.91) will be

$$\Omega(s,x) = \Omega_1(s,0)ch[\theta_k(s)x] - \frac{1}{G\theta_k(s)}s\tau_1(s,0)sh[\theta_k(s)x]; \tag{1.95}$$

$$\tau(s,x) = \tau_1(s,0)ch\big[\theta_k(s)x\big] - \frac{1}{s}rG\theta_k\Omega_1(s,0)sh\big[\theta_k(s)x\big]. \tag{1.96}$$

The coefficient $\theta_k(s)$ is the symbolical coefficient of wave propagation. The solution obtained in the images makes it possible to calculate the frequency characteristics of the driven link depending on the change in the speed of the leading link, taking into account the emerging reactive force factors of the medium. The reader may find an extensive exposition of this question in the literature [8].

## 1.12  Equations of longitudinal oscillations of a disk

Consider a sufficiently short rod of length $l$, at the end of which a disk with mass $m$ is fixed, counting, as before, the element 1 by the leading link, and the disk by the driven link (Figure 1.15). Let's give the leading link a linear motion. This model corresponds to the mechanism of the pusher, where the driving link is the pusher. The motion of the driven link is impeded by the force $F$. There are no distributed surface forces by the length of the pusher. The oscillations of the disk do not affect the speed of the pusher.

For such a system, the relations (1.82) and (1.83) are valid. We rewrite these expressions in ordinary derivatives:

$$\rho\frac{dv}{dt} = -\frac{d\sigma}{dx}; \tag{1.97}$$

$$\frac{1}{E}\frac{d\sigma}{dt} = -\frac{dv}{dx}. \tag{1.98}$$

We perform actions similar to Section 1.10. Integrating (1.97) with respect to the coordinate $x$ and then differentiating with respect to $t$, we obtain the relation for the longitudinal vibrations of the rod

$$\vartheta'_{n0}\frac{d\sigma}{dt} = v_1 - v_2, \tag{1.99}$$

where $\vartheta'_{n0}$ is the coefficient characterizing the longitudinal elasticity, $\vartheta'_{n0} = l/E$; $v_1$, $v_2$ are the linear velocities of the displacement of the points of the rod and disk sections, respectively; and $E$ is the modulus of elasticity of the material. Assuming that the modulus of elasticity is the same and constant, we integrate (1.99), and we obtain the dependence of the normal stresses on the displacements of the points of the rod:

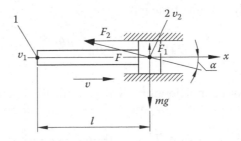

*Figure 1.15* Linear motion of the rod.

$$\vartheta'_{n0}\sigma = (v_1 - v_2)t. \tag{1.100}$$

Or

$$\frac{l}{E}\sigma = x_1 - x_2, \quad \frac{l}{E}\sigma = \Delta u. \tag{1.101}$$

Expression (1.101) is the well-known Hooke's law.

The normal stresses developed in the rod overcome the resistance arising on the disk (slave link), which is the resultant force of its two components $F_2 = F_f = k \sin \alpha$. Note that the force $F_2$ is the friction force, which is determined by the expression $F_2 = F_f = k_f \sin \alpha$. Here $k_f$ is the coefficient of friction. The force $F_1$ is the inertial component, and $F_1 = -F_i = m dv_2/dt$.

Without taking into account the direction of the speed of motion and taking $F = \sigma f$, where $f$ is the cross-sectional area of the body, we write the following relation [8]:

$$\beta\sigma f = F_0(t) + k_1 v_1(t) + h v_2 + m\frac{dv_2}{dt}. \tag{1.102}$$

Here $\beta$ is the proportionality coefficient, which depends on the coefficient of friction caused by the contact pressure and the direction of motion of the driven link, $\beta = 1 - c \operatorname{sgn} v_2$; $k_1$, $h$ are the coefficients of friction loss proportional to the speeds of the $\beta$ driving and driven links. Solving with the help of the symbolical method jointly (1.99) and (1.102), taking into account the direction of motion of the driven link, we finally obtain

$$v_1(t)\big(1 - c\operatorname{sgn}v_2 - k_1 v_{n0}p\big) - \vartheta_{n0}pF_0(t) = v_2(t)\big(1 - c\operatorname{sgn}v_2 + hv_{n0}p + m\vartheta_{n0}p^2\big). \tag{1.103}$$

Here $\vartheta_{n0}$ is the elasticity of the mechanical system, $\vartheta_{n0} = l/fE$; $p$ is a differential operator, $p \equiv d/dt$.

Further transformations will be based on the energy approach of the deformation of an elastic body and the rheological representation of the transfer of dynamic energy through a metallic body (the Zener model) [9].

It is known that the realization of the principle of continuity of deformations in an elastic body corresponds to the minimum value of the potential deformation energy accumulated by the body [10] (i.e., in deformation processes, the stored energy in the body is spent on performing work to restore the body shape to its original state after the load is removed).

Taking into account the phenomenological Zener model, we write the differential equation characterizing the redistribution of stresses and deformations in the body under the static load of the body in some time [8]

$$\sigma + \frac{\eta}{E_2}\frac{d\sigma}{dt} = E_1\theta + \eta\frac{d\theta}{dt}. \tag{1.104}$$

Here $\eta$ is the coefficient of proportionality, which characterizes the viscosity of the body; $E_1$, $E_2$ are elastic constants of isothermal and isobaric deformation processes; and $\theta$ is linear deformation of the body.

If we take $\theta' = 0$, then expression (1.104) is transformed into equation

$$\sigma + \frac{\eta}{E_2}\frac{d\sigma}{dt} = E_1\theta_0, \quad \tau_\varepsilon = \frac{\eta}{E_1} \tag{1.105}$$

with the decision

$$\sigma(t) = E_1\theta_0 + (\sigma_0 - E_1\theta_0)\exp(-t/\tau_\varepsilon). \tag{1.106}$$

Here $\tau_\varepsilon$ is the relaxation time under the condition of constant deformation.
    If we take $\sigma' = 0$, then the solution of Equation (1.104) is

$$\theta(t) = E_1\sigma_0 + \left(\theta_0 - \frac{\sigma_0}{E_1}\right)\exp\left(-\frac{t}{\tau_\sigma}\right); \tau_\sigma = \frac{\eta}{E_1}, \tag{1.107}$$

where $\tau_\sigma$ is the time of retardation (lag).
    We transform Equation (1.104), going over to the operator form:

$$p\sigma = E_2 p\theta - \frac{1}{\tau_\varepsilon}(\sigma - E_1\theta), \quad p \equiv \frac{d}{dt}. \tag{1.108}$$

We perform one more transformation

$$\sigma\left(p + \frac{1}{\tau_\varepsilon}\right) = \theta E_2\left(p + \frac{1}{k\tau_\varepsilon}\right); \quad k = \frac{E_2}{E_1}. \tag{1.109}$$

Passing under zero initial conditions to Laplace transforms, we rewrite Equation (1.109) in the images

$$\sigma(s)\left(s + \frac{1}{\tau_\varepsilon}\right) = \theta(s)E_2\left(s + \frac{1}{k\tau_\varepsilon}\right). \tag{1.110}$$

Taking into account the stepwise deformation, the Laplace transform of the stress change function (1.110) is written in the form

$$\sigma(s) = \theta_0 E_2 \frac{s + \dfrac{1}{k\tau_\varepsilon}}{\left(s + \dfrac{1}{\tau_\varepsilon}\right)s}. \tag{1.111}$$

The original is determined by means of residues with respect to the poles

$$\sigma(t) = \theta_0 E_2\left[\frac{1}{k} + \left(1 - \frac{\tau_\varepsilon}{k\tau_\varepsilon}\right)\right]\exp\left(-\frac{t}{\tau_\varepsilon}\right) = \theta_0 E_1\left[1 + \left(\frac{E_2}{E_1} - 1\right)\exp\left(-\frac{t}{\tau_\varepsilon}\right)\right]. \tag{1.112}$$

Assuming that the relaxation constant is zero within the elastic range, when the stresses do not exceed the yield strength, without taking into account the direction of motion in accordance with (1.102) and (1.103), we write the following equations of motion of the material object:

$$\sigma f(1-c) = F_0(t) + hv_2 + m\frac{dv_2}{dt} \tag{1.113}$$

and

$$v_1(t)(1-c) - \vartheta_{n0}pF_0(t) = v_2(t)\left(1 - c + h\vartheta_{n0}p + m\vartheta_{n0}p^2\right). \tag{1.114}$$

In this case, the oscillations of the velocity of motion of a material object can be described by the following equation [8]:

$$v_2(t) = \frac{v_1(t) - a\vartheta_{n0}pF_0(t)}{1 + ah\vartheta_{n0}p + am\vartheta_{n0}p^2}, \quad a = \frac{1}{1-c}. \tag{1.115}$$

For the difference in the velocities of motion for $v_1(t) = $ const, the following relation is valid:

$$\Delta v(t) = \frac{-a\vartheta_{n0}pF_0(t)}{1 + ah\vartheta_{n0}p + am\vartheta_{n0}p^2}. \tag{1.116}$$

Integrating (1.116) with respect to $t$ and passing to the differential form, we obtain the equation of displacement of the material point relative to the leading member:

$$\Delta u(t) + ah\vartheta_{n0}\frac{d\Delta u}{dt} + am\vartheta_{n0}\frac{d^2\Delta u}{dt^2} = -a\vartheta_{n0}\frac{dF_0}{dt}. \tag{1.117}$$

From the solution of the system of equations (1.113) and (1.114), the stresses in the rod are determined by the equality

$$\sigma(t) = \frac{a}{f}\left[F_0(t) + hv_2(t)(1 + Tp)\right], \tag{1.118}$$

where $T = m/h$ is the inertial time constant.

A detailed exposition of this question can be found in Ref. [8].

The equations obtained make it possible to apply the motion transfer scheme and, on its basis, to investigate the longitudinal and torsional oscillations of the moving technological object fixed to the end of the rod (the input and output links of the system).

## 1.13   Application of the Laplace transform in engineering technology

The presented method allows mathematical modeling of many technological operations of mechanical engineering to be carried out and features of the dynamic phenomena at work of the equipment and interaction of the tool to be revealed with detail.

### 1.13.1   Method of studying oscillations of the velocities of motion and stresses in mechanisms containing rod systems

To study the dynamic characteristics of the process of the functioning of mechanisms containing rod systems, let us take a generalized model of the inertial disk rotating at the end of the rod. Such a model can be taken as an imitation model when studying the process of hole processing. The structural scheme for the transfer of rotational motion is shown in Figure 1.16 [8].

Accepted designations: $\vartheta_k$—coefficient of torsional elasticity; $p$—differentiation operator ($p = d/dt$); $\tau$—tangential stresses; $\tau_\varepsilon$—relaxation constant; $k_\varepsilon$—ratio of the adiabatic and isothermal elasticity modulus of the rod material; $M_r$—resultant moment of the resistance forces; $l$—length of the rod with the disk; $h_k$—coefficient of friction loss proportional to the rotational speed; 1, 2—leading and driven links.

The transfer function of the effect of the oscillations of the torque on the rotational speed of the disk is the relation

$$W_\Omega(s) = \frac{\Omega_2(s)}{M_r(s)}. \tag{1.119}$$

Here $\Omega_2$ is the angular velocity of rotation of the disk, and $M_r$ is the moment of resistance of forces.

The transfer function (1.119) is a Laplace transform of the impulse response $k(t)$ [4]. To determine $k(t)$, it is necessary to find the roots of the characteristic equation.

The proposed mathematical model allows us to investigate the dynamics of the rotating parts of the structural element, and also obtain equations describing in the rod the relationship between the angular acceleration of elementary sections and the gradient of the tangential stress and the rate of change of this voltage with the angular velocity gradient.

### 1.13.2   Features of functioning of a drive with a long force line

To study the dynamic phenomena of the drive with a long force line, we use a generalized model of an inertial disk rotating at the end of the rod.

The solution of differential equations (1.85) and (1.86) in the originals gives information about oscillations in a mechanism with a short rod (i.e., in an elastic system with lumped parameters).

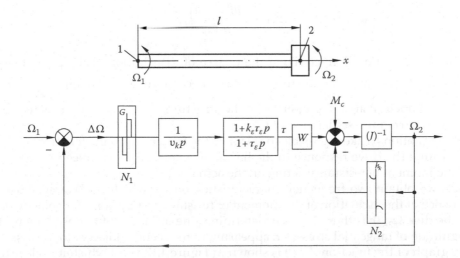

*Figure 1.16* The structural scheme for the transfer of rotational motion.

In the presence of long lines (an elastic system with distributed parameters), it is expedient to carry out an investigation of such processes in the complex domain by means of a one-dimensional Laplace transform.

Based on the proposed scheme of motion transfer (Figure 1.14), the transfer function can be described by expression

$$W_\Omega(s) = \frac{\Omega_2(s)}{M_r(s)} = \frac{\vartheta_k s(s+\alpha_1)}{[\alpha_2 + s(1+\vartheta_k h_k \alpha_1) + s^2\vartheta_k(h_k + J\alpha_1) + s^3\vartheta_k J]}. \tag{1.120}$$

Here $\alpha_1$, $\alpha_2$ are quantities that take into account the peculiarities of the rotation of the disk in the interaction medium (contact interaction, etc.).

### 1.13.3 Investigation of dynamic features of the system in the technologies of deephole machining

When modeling dynamic processes in the technologies of deephole machining, we use expression (1.120).

In view of the brevity of the presentation at the condition of the matched load, when input to the system the moment of momentum, which is completely consumed in the load, we give the final equations in the following form:

$$\Omega_2(s)\,[1 + h_k\vartheta_k(s)s + J\vartheta_k(s)s^2] = \Omega_1(s)\,ch^{-1}[\theta_k(s)l] - M_r(s)s\vartheta_k(s);$$

$$\tau_c(s) = \psi(s)\cdot\left[M_r + \frac{\Omega_1(s)(h_k + Js)}{ch[\theta_k(s)l]}\right];$$

$$\psi(s) = \frac{1}{W[1 + h_k\vartheta_k(s) + J\vartheta_k(s)s^2]};$$

$$\vartheta_k(s) = \frac{l}{GrW}Z_k(s);$$

$$Z_k(s) = \frac{th[\theta_k(s)l]}{\theta_k(s)l}.$$

Here $Z_k(s)$ is a function characterizing the degree of distribution of the parameters. For $\Omega_1 = 0$ and the Laplacian $s = j\omega$, the function $Z_k(j\omega)$ becomes real, i.e. $Z_k(j\omega) = \frac{tg\,\alpha_k}{\alpha_k}$; $\alpha_k$ is the parameter characterizing the properties of the structure, $\alpha_k = l\omega\,(\rho G^{-1})^{0.5}$; $\omega$ is the circular frequency of harmonic oscillations; $j = (-1)^{1/2}$.

These equations make it possible to calculate the frequency characteristics of the drive and determine the drive response to the harmonic variation in the speed of the driving link or the moment of resistance acting on the actuator.

Thus, we obtain two frequency characteristics, one of which is $-W_M(j\omega)$, illustrating the influence of the oscillation of the moment of resistance on the angular velocity of rotation of the disk $\Omega_2$; the other $-W_{M\tau}(j\omega)$, determines the influence of the oscillations $M_r$ on the magnitude of tangential stresses $\tau$, appearing in the section adjacent to the disk.

The graph of the function $Z_k(\alpha_k)$ is shown in Figure 1.17, from which it is clear that as the parameter tends to zero, the function $Z_k$ tends to unity $\alpha_\kappa \to 0$, $Z_\kappa \to 1$.

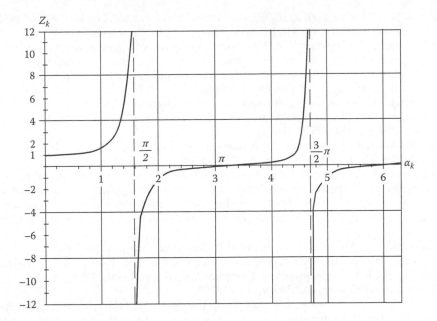

*Figure 1.17* Change in the function $Z_k$ as a function of the dimensionless parameter $\alpha_k$.

In cases of variation of $\alpha_k$ in the intervals $\pi/2 + k\pi > \alpha_k > \pi + k\pi$, the function $Z_\kappa$ takes negative values $Z_k < 0$.

The graph in Figure 1.17 clearly defines the zones of stable and unstable operation of the mechanism.

A detailed exposition of this material and questions of mathematical modeling of the rolling of the tube can be found in the literature [11–16].

## References

1. Lur'e A.I. *[Operacionnoe ischislenie i ego prilozheniya k zadacham mekhaniki]* Operational Calculus and Its Applications to Problems of Mechanics. GITTL, Moscow, 1950. (In Russ.).
2. Korn G., Korn T. *Mathematical Handbook for Scientists and Engineers Definitions, Theorems and Formulas for Reference and Review.* McGraw-Hill Book Company, New York, San Francisco, Toronto, London, Sydney, 1968.
3. Doetsch G. *Anleitung zum praktiscen gebrauch der Laplace-transformation.* R. Oldenbourg, München, 1961.
4. Ivanov V.A., Chemodanov B.K., Medvedev V.S. *[Matematicheskie osnovy teorii avtomaticheskogo regulirovaniya]* Mathematical Foundations of the Theory of Automatic Control. High School, Moscow, 1971. (In Russ.).
5. Lavrent'ev M.A., SHabat B.G. *[Metody teorii funkcij kompleksnogo peremennogo]* Methods of the Theory of Functions of a Complex Variable. The Science, Moscow, 1965. (In Russ.).
6. Mironova L.I. *[Komp'yuternye tekhnologii v reshenii zadach teorii uprugosti]* Computer Technologies in Solving Problems in the Theory of Elasticity. Palmarium Academic Publishing, ISBN-13: 978-3-659-72395-7; ISBN-10: 3659723959. (In Russ.).
7. Sedov L.I. *[Mekhanika sploshnoj sredy]* Continuum Mechanics. Nedra, Moscow. T.1, T.2, 1970. (In Russ.).
8. Kondratenko L. A. *[Raschet kolebanij detalyah i uzlah mashin]* Calculation of Velocity Variations and Stresses in Machine Assemblies and Components. Sputnik, Moscow, 2008. (In Russ.).
9. Eirich Frederik R. (Edit). *Rheologiy.* Academic Press, Inc., New York, VI, 1965.

10. Bezuhov N.I. *[Osnovy teorii uprugosti, plastichnosti i polzuchesti] Fundamentals of the Theory of Elasticity, Plasticity and Creep.* High School, Moscow, 1968. (In Russ.).
11. Kondratenko L.A., Terekhov V.M., Mironova L.I. [Ob odnom metode issledovaniya krutil'nyh kolebanij sterzhnya i ego primenenii v tekhnologiyah mashinostroeniya] About one method of research torsional vibrations of the core and this application in technologies of mechanical engineering. *Engineering & Automation Problems.* 2017, vol. 1, pp. 133–137. (In Russ.).
12. Kondratenko L., Terekhov V., Mironova L. The aspects of roll-forming process dynamics. Vibroengineering PROCEDIA. At the 22nd International Conference on Vibroengineering, Moscow, 2016, pp. 460–465. (In Russ.).
13. Kondratenko L.A. [Mekhanika rolikovogo val'cevaniya teploobmennyh trub] Mechanics Roller Rolling Heat Exchange Tubes. Sputnik, Moscow, 2015. (In Russ.).
14. Kondratenko L.A., Terekhov V.M., Mironova L.I. [K voprosu o vliyanii dinamiki rolikovogo val'cevaniya na kachestvo izgotovleniya teploobmennyh apparatov v atomnyh ehnergeticheskih ustanovkah] On the effect of the dynamics of the roller rolling on the quality of manufacture of heat exchangers of nuclear power units. *Heavy Engineering Construction.* 2016, vol. 3, pp. 10–14. (In Russ.).
15. Kondratenko L., Mironova L., Terekhov V. Investigation of vibrations during deepholes machining. 25th International Conference Vibroengineering, Liberec, Czech Republic. JVE International LTD. Vibroengineering Procedia. 2017, Vol. 11. ISSN 2345-0533, pp. 7–11. Crossref DOI link: https://doi.org/10.21595/vp.2017.18285.
16. Kondratenko L., Mironova L., Terekhov V. On the question of the relationship between longitudinal and torsional vibrations in the manufacture of holes in the details. 26th Conference in St. Petersburg, Russia. JVE International LTD. Vibroengineering Procedia. 2017, Vol. 12. ISSN 2345-0533, pp. 6–11. Crossref DOI link: https://doi.org/10.21595/vp.2017.18461.

# chapter two

# Fourier series and its applications in engineering

*Smita Sonker and Alka Munjal*
National Institute of Technology Kurukshetra

## Contents

## 2.1   Introduction

Mathematics has its roots embedded within various streams of engineering and sciences. The concepts of the famous Fourier series were originated from the field of physics. The following two physical problems are the reasons for the origin of Fourier series:

    i. Heat conduction in solid
   ii. The motion of a vibrating string

Jean Baptiste Joseph Fourier (1768–1830) was the first physicist, mathematician, and engineer who developed the concepts of Fourier analysis in dealing with the problems of vibrations and heat transfer. He claimed that any continuous or discontinuous function of $t$ could also be expressed as a linear combination of $\cos(t)$ and $\sin(t)$ functions.

In the mathematical analysis, we do not usually get a full decomposition into the simpler things, but an approximation of a complex system is usually achieved by a more elementary system. When we truncate the Taylor series expansion of a function, we approximate the function by using the polynomial.

The form of a Taylor series is as follows (*infinite series*):

$$f(t) = \sum_{n=0}^{\infty} a_n t^n,$$

where $a_0, a_1, a_2, \ldots$ are called the constant coefficients of the infinite series. A Taylor series does not include terms with negative powers. The quality of the approximation depends on the number of terms taken under consideration. Of course, for a function to have a Taylor series, it must (among other things) be infinitely differentiable in some interval, and this is a very restrictive condition.

The Fourier series, which is a sum of sines and cosines, can be used for the approximation of any periodic function. Sines and cosines serve as much more versatile "prime elements" than powers of $t$. Sines and cosines can be used to approximate not only non-analytic functions, but they even do a good job in the wilderness of the discontinuous functions.

## 2.2   Periodic functions

A function satisfying the identity $l(t) = l(t + T)$ for all $t$, where $T > 0$, is called periodic or $T$-periodic as shown in Figure 2.1.

For a $T$-periodic function

$$l(t) = l(t+T) = l(t+2T) = l(t+3T) = \cdots = l(t+nT).$$

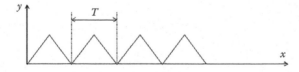

*Figure 2.1* A periodic function with $T$ period.

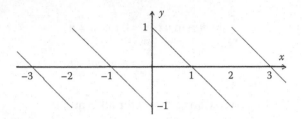

*Figure 2.2* A periodic function with $T$ period.

Here, $nT$ is also a period for any integer $n > 0$, and $T$ is called a fundamental period. Any interval of length $T$ is the same as the definite integral of a $T$-periodic function. The following example uses this property to integrate a 2-periodic function as shown in Figure 2.2.

> **Example 2.1:** Let there exist a 2-periodic function $f$ and $I$ be a positive integer. If $f(x) = -x + 1$ on the interval $0 \le x \le 2$, compute $\int_{-N}^{N} f^2(x)\,dx$.

**Solution:**

$$\int_{-I}^{I} f^2(x)\,dx = \int_{-I}^{I+2} f^2(x)\,dx + \cdots + \int_{I-2}^{I} f^2(x)\,dx$$

$$\int_{-I}^{I} f^2(x)\,dx = I \int_{I-2}^{I} f^2(x)\,dx = I \int_{0}^{2} (-x+1)^2\,dx = I\left\{ -\frac{1}{3}(-x+1)^3 \right\}\Big|_{0}^{2}$$

$$\int_{-I}^{I} f^2(x)\,dx = -\frac{I}{3}[-1-1] = \frac{2}{3}I.$$

The most important periodic functions are those in the $2\pi$-period of the trigonometric system

$$1, \cos t, \cos 2t, \cos 3t, \ldots, \cos mt, \ldots$$

$$\sin t, \sin 2t, \sin 3t, \ldots, \sin mt, \ldots$$

## 2.3   *Orthogonality of sine and cosine functions*

Two functions $f$ and $g$ are orthogonal over the interval $[a, b]$, if

$$\int_{a}^{b} f(x)g(x)\,dx = 0.$$

Examples of orthogonal functions:

$$\int_{-\pi}^{\pi} \cos mx \cos nx\,dx = 0 \text{ for } m \ne n$$

$$= \pi \text{ for } m = n$$

$$\int_{-\pi}^{\pi} \sin mx \sin nx \, dx = 0 \text{ for } m \neq n$$

$$= \pi \text{ for } m = n$$

$$\int_{-\pi}^{\pi} \cos mx \sin nx \, dx = 0 \text{ for all } m \text{ and } n.$$

Certain sequences of sin $nt$ and cos $nt$ functions are orthogonal on certain intervals. The resulting expansions,

$$f = \sum_{i=1}^{\infty} c_i \varphi_i$$

using the sin $nt$ and cos $nt$ become the Fourier series expansions of the function $f$. First, we just consider the functions $\varphi_n(t) = \cos nt$. These are orthogonal on the interval $0 < t < \pi$.

**Example 2.2:** The functions $\varphi_0(x) = 1$, $\varphi_1(t) = \cos t$, $\varphi_2(t) = \cos 2t$, $\varphi_3(t) = \cos 3t$, ..., $\varphi_n(t) = \cos nt$, ... are orthogonal on the interval $0 < t < \pi$. Furthermore, $|\varphi_0|^2 = \pi$ and $|\varphi_n|^2 = \dfrac{\pi}{2}$ for $n = 1, 2, ...$.

**Proof:** Using $(\varphi_n(t), \varphi_m(t)) = \displaystyle\int_0^{\pi} \cos(nt)\cos(mt)\,dt$

$$= \int_0^{\pi} \left[ \frac{1}{2}\cos(n+m)t + \frac{1}{2}\cos(n-m)t \right] dt$$

$$= \left[ \frac{1}{2(n+m)}\sin(n+m)t + \frac{1}{2(n-m)}\sin(n-m)t \right]_0^{\pi} = 0,$$

so, the $\varphi_n$ are orthogonal.

The fact that $|\varphi_0|^2 = \pi$ is an easy verification:

$$|\varphi_n|^2 = \int_0^{\pi} \cos^2(nt)\,dt = \int_0^{\pi} \frac{1}{2}[1+\cos 2nt]\,dt$$

$$= \frac{1}{2}\left[ t + \frac{1}{2n}\sin 2nt \right]_0^{\pi} = \frac{\pi}{2}.$$

Next, we just consider the functions $\psi_n(t) = \sin nt$. These are also orthogonal on the interval $0 < t < \pi$. The resulting expansion is called as the Fourier sine series expansion of $f$.

**Example 2.3:** The functions $\psi_1(t) = \sin t$, $\psi_2(t) = \sin 2t$, $\psi_3(t) = \sin 3t$, ..., $\psi_n(t) = \sin nt$, ... are orthogonal on the interval $0 < t < \pi$. Furthermore, $|\psi_n|^2 = \dfrac{\pi}{2}$ for $n = 1, 2, ...$

**Proof:**

$$(y_n(t), y_m(t)) = \int\limits_0^\pi \sin(nt)\sin(mt)\,dt = \int\limits_0^\pi \left[\frac{1}{2}\cos(n+m)t - \frac{1}{2}\cos(n-m)t\right]dt$$

$$= \left[\frac{1}{2(n+m)}\sin(n+m)t + \frac{1}{2(n-m)}\sin(n-m)t\right]_0^\pi = 0,$$

so, the $\psi_n$ are orthogonal.

$$|\psi_n|^2 = \int\limits_0^\pi \sin^2(nt)\,dt = \int\limits_0^\pi \frac{1}{2}[1-\cos 2nt]\,dt$$

$$= \frac{1}{2}\left[t - \frac{1}{2n}\sin 2nt\right]_0^\pi = \frac{\pi}{2}.$$

Finally, we consider the functions $\varphi_n(t) = \cos nt$ and $\psi_n(t) = \sin nt$. These are orthogonal on the interval $-\pi < t < \pi$.

**Example 2.4:** The functions $\varphi_0(t) = 1$, $\varphi_1(t) = \cos t$, $\varphi_2(t) = \cos 2t$, $\varphi_3(t) = \cos 3t$, ..., $\varphi_n(t) = \cos nt$, ... and $\psi_1(t) = \sin t$, $\psi_2(t) = \sin 2t$, $\psi_3(t) = \sin 3t$, ..., $\psi_n(t) = \sin nt$, ... are orthogonal on the interval $-\pi < t < \pi$. Furthermore, $|\varphi_0|^2 = 2\pi$ and $|\varphi_n|^2 = |\psi_n|^2 = \pi$ for $n = 1, 2, ...$

**Proof:** The fact that

$$(\varphi_n(t), \varphi_m(t)) = 0$$

$$\text{and}\quad (\psi_n(t), \psi_m(t)) = 0$$

is shown in Examples 2.2 and 2.3. For $(\varphi_n(t), \psi_m(t))$, the third identity is used:

$$(\varphi_n(t), \psi_m(t)) = \int\limits_{-\pi}^\pi \cos(nt)\,\sin(mt)\,dt$$

$$= \int\limits_{-\pi}^\pi \left[\frac{1}{2}\sin(m+n)t - \frac{1}{2}\sin(m-n)t\right]dt$$

$$= \left[-\frac{1}{2(n+m)}\cos(n+m)t + \frac{1}{2(m-n)}\cos(n-m)t\right]_{-\pi}^\pi = 0.$$

Then $|\varphi_0|^2 = 2\pi$ is an easy verification and $|\varphi_n|^2 = |\psi_n|^2 = \pi$ is shown in the same way (see Examples 2.2 and 2.3).

## 2.4   Fourier series

Fourier series are special representation of the functions (signals) of the form

$$f(x) = a_0 + \sum_{n=1}^\infty (a_n \cos(nx) + b_n \sin(nx)),$$

where $a_0, a_1, a_2, \ldots, b_1, b_2, \ldots$ are Fourier coefficients.

The coefficient $a_0$ is determined by integrating both sides over the interval $[-\pi, \pi]$:

$$\int_{-\pi}^{\pi} f(x)dx = \int_{-\pi}^{\pi} a_0\,dx + \sum_{n=1}^{\infty} \int_{-\pi}^{\pi} (a_n \cos(nx) + b_n \sin(nx))dx.$$

Since, $\quad \int_{-\pi}^{\pi} \cos nx\,dx = \int_{-\pi}^{\pi} \sin nx\,dx = 0$ for $n = 1, 2, \ldots$

$$\int_{-\pi}^{\pi} f(x)dx = \int_{-\pi}^{\pi} a_0\,dx = 2\pi a_0 \Rightarrow a_0 = \frac{1}{2\pi} \int_{-\pi}^{\pi} f(x)dx.$$

The coefficient $a_n$ is determined by multiplying both sides with $\cos mx$ and integrating the resulting equation over the interval $[-\pi, \pi]$:

$$\int_{-\pi}^{\pi} f(x)\cos(mx)dx = \int_{-\pi}^{\pi} a_0 \cos(mx)dx + \sum_{n=1}^{\infty} \int_{-\pi}^{\pi} a_n \cos(nx)\cos(mx)dx$$

$$+ \sum_{n=1}^{\infty} \int_{-\pi}^{\pi} b_n \sin(nx)\cos(mx)dx.$$

Since $\int_{-\pi}^{\pi} \cos mx\,dx = 0,\ \int_{-\pi}^{\pi} \cos mx \sin nx\,dx = 0$ for all $m$ and $\int_{-\pi}^{\pi} \cos mx \cos nx\,dx = 0$ for $m \neq n$:

$$\int_{-\pi}^{\pi} f(x)\cos(mx)dx = a_n \int_{-\pi}^{\pi} (\cos nx)^2\,dx = \pi a_n$$

$$a_n = \frac{1}{\pi} \int_{-\pi}^{\pi} f(x)\cos(mx)dx = \frac{1}{\pi} \int_{0}^{2\pi} f(x)\cos(mx)dx.$$

Similarly, the coefficient $b_n$ is determined by multiplying both sides with $\sin mx$ and integrating the resulting equation over the interval $[-\pi, \pi]$:

$$b_n = \frac{1}{\pi} \int_{-\pi}^{\pi} f(x)\sin(mx)dx = \frac{1}{\pi} \int_{0}^{2\pi} f(x)\sin(mx)dx.$$

Trigonometric Fourier series associated with $f$ is

$$s_n(f;x) = \frac{a_0}{2} + \sum_{k=1}^{n} (a_k \cos kx + b_k \sin kx), \quad \forall n \geq 1 \text{ with } s_0(f;x) = \frac{a_0}{2},$$

denotes the $(n + 1)$th partial sums, called trigonometric polynomials of degree (or order) $n$, of the Fourier series of $f$. The conjugate Fourier series of the series of $f$ is defined by

$$\sum_{n=1}^{\infty}(b_n\cos nx - a_n\sin nx) = \sum_{n=0}^{\infty}v_n,$$

and its $n$th partial sum is given by

$$\tilde{s}_n(f;x) = \sum_{k=1}^{n}(b_k\cos kx - a_k\sin kx), \ \forall n \geq 1 \text{ and } \tilde{s}_n(f;x) = 0,$$

where $a_k = \dfrac{1}{\pi}\displaystyle\int_{-\pi}^{\pi} f(x)\cos kx\, dx,$ $k = 0,1,2,\dots,$ and $b_k = \dfrac{1}{\pi}\displaystyle\int_{-\pi}^{\pi} f(x)\sin kx\, dx,$ $k = 1,2,3,\dots$ are

called the Fourier coefficients of $f$. The sequence of partial sums of series $\sum_{k=0}^{\infty} u_k(x)$, given by

$s_n(f;x) = \dfrac{a_0}{2} + \sum_{k=1}^{n}(a_k\cos kx + b_k\sin kx),$ is a trigonometric polynomial of order $n$.

## 2.5   *Dirichlet's theorem*

The Fourier series of a piecewise smooth integrable function $f$ converges at each point $x$ to

$$\frac{f(x^+) + f(x^-)}{2}.$$

Hence, the Fourier series converges to $f(x)$ at points of continuity and to the average of the limiting values at a jump discontinuity.

## 2.6   *Riemann–Lebesgue lemma*

Fourier coefficient $a_n$ and $b_n$ of any function tends to zero as $n$ tends to infinity—that is,

$$\lim_{n\to\infty}\frac{1}{\pi}\int_{-\pi}^{\pi} f(x)\cos kx\, dx = 0$$

and

$$\lim_{n\to\infty}\frac{1}{\pi}\int_{-\pi}^{\pi} f(x)\sin kx\, dx = 0.$$

Validation of the asymptotic approximations for integrals can be done by the Riemann–Lebesgue lemma. The method of steepest descent (rigorous treatments) and stationary phase method are based on the Riemann–Lebesgue lemma.

## 2.7   *Term-wise differentiation*

A continuous, piecewise smooth $2\pi$-periodic function is $f$ on all of R with Fourier series

$$\frac{a_0}{2} + \sum_{n\in N} a_n\cos(nt) + \sum_{n\in N} b_n\sin(nt).$$

If $\dot{f}$ is piecewise smooth, then the series can be differentiated term by term to yield the following point-wise convergent series at every point $t$:

$$\frac{f'(t^+) + f'(t^-)}{2} = \sum_{n \in N}(nb_n\cos(nt) - na_n\sin(nt)).$$

## 2.8 Convergence of Fourier series

The smoothness of the integrable function $f$ is represented as the convergence of a Fourier series. If $f$ is continuous and piecewise smooth, series converges uniformly and absolutely to $f$ and otherwise, its Fourier series may diverge. The Fourier coefficients $a_n$ and $b_n$ of a continuous and $k$-times differentiable function approach 0 faster than $1/n_k$.

**Example 2.5:** Let $f(x) = x$ be $2\pi$-periodic function on $[-\pi, \pi]$, and its second partial sum is

$$S_2 = -\sin(2x) + 2\sin(x).$$

Figure 2.3 is a graph comparing the approximation to $f$; the Fourier series.

For the approximations of a function, trigonometric Fourier series expression is very useful due to infinitely differentiable and term-wise integrability. Because of infinite differentiability and fewer limitations, Fourier series are more applicable than Taylor series.

In approximating the function $f$, the Taylor series expansion must be centered at a certain point with the converging point $x$ in a neighborhood of a certain radius around that point. Note that the Fourier series does not have this kind of limitation (centered at any specific point). Furthermore, differentiability of the function is a necessary condition to have a Taylor series expansion, whereas a merely integrable function can be approximated by the Fourier series. Some concepts like "small order" and "big 'oh'" are required to understand the convergence of the Fourier series, because a divergent Fourier series will be of no use when approximating a function.

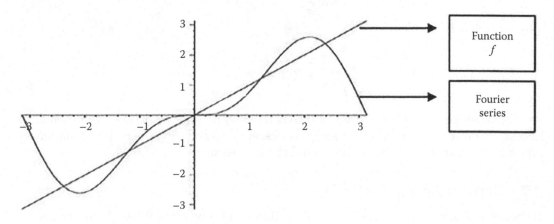

*Figure 2.3* $f(x)$ vs $S_2(f(x))$.

## 2.9 Small order

Function $g(n)$ is of a smaller order than function $h(n)$, or $g(n)$ approaches to 0 faster than $h(n)$, if

$$\lim_{n \to \infty} \frac{g(n)}{h(n)} = 0,$$

and we write $g = o(h)$.

> **Example 2.6:** Fourier coefficients of a piecewise smooth continuous periodic functions are of smaller order than $1/n$.

## 2.10 Big "oh" for functions

For real-valued functions $f$ and $g$, if

$$\left| \frac{f(t)}{g(t)} \right| \text{ is bounded as } t \to a,$$

where $a$ could be $\pm\infty$, then $f = O(g)$, or $f$ is at most of order $g$.

> **Example 2.7:** $\sin t = O(1)$ as $t \to a$ for any $a$.

## 2.11 Fourier analysis and Fourier transform

1. Fourier analysis is the study of the Fourier transform, Fourier series, and related concepts.
2. The Fourier transform, Fourier series, and several related concepts are just special cases of constructions from representation theory (writing a conjugacy-invariant function on a group as a linear combination of characters).
3. Fourier analysis is not just a special case of representation theory—not even close.
4. These might at first sound contradictory, but they really are not. Of course there is some subtlety in "related concepts," but that is not really the fundamental problem. Consider the following similar set of true statements:
   a. Number theory is the study of the integers (and related concepts).
   b. The integers are just a special case of some construction from category theory (the initial object in some category of rings).
   c. The integers are also just a special case of some construction from group theory (the endomorphism algebra of the free Abelian group on one generator).
   d. The integers are also a special case from set theory, model theory, etc.
   e. But number theory is not a special case of any of those fields.

Even worse, number theory is not even the only field of mathematics devoted to studying the integers—much of combinatorics is as well.

The fundamental problem is not that number theorists bring in additional concepts like number fields, Galois groups, and modular forms. They do, but the issue arises even when working with purely elementary statements like the four-square theorem. What does this mean in the case of Fourier analysis? One question you might study in Fourier analysis is whether the Fourier transform exists from one space of functions (may be an $L_p$ space) to another. Now to define the Fourier transform on this space (i.e., to uniquely characterize it) one just needs to know what it is on some dense set—the smooth compactly supported functions probably work. The Fourier transform on smooth compactly supported functions is not very hard to set up, and it is a special case of a construction from representation theory, as well as being a special case of a construction from integral calculus, and probably many other fields.

In some sense, because of this uniqueness property, everything about the Fourier transform on $R$ in all its incarnations is determined by just this restriction to smooth compactly supported functions, which is almost a purely algebraic object as one needs very little analysis to define it. Furthermore, the styles of argument and thought typical in representation theory are not so helpful for analyzing about $L_p$ norms. All of these concepts could fall under the umbrella of Fourier analysis. In the Fourier series approximation, the periodic functions are represented as a sum of simple waves of sine and cosine. Extension of the Fourier series is Fourier transform. Fourier transform is used when the time period of given function is lengthened and approach to infinity.

## 2.12 Fourier transform

Form of Fourier integral representation of $f(x)$ is

$$f(x) = \frac{1}{2\pi} \int\limits_{-\infty}^{\infty} \int\limits_{-\infty}^{\infty} f(t)e^{is(t-x)} \, dt \, ds$$

$$= \frac{1}{2\pi} \int\limits_{-\infty}^{\infty} e^{-isx} \left( \int\limits_{-\infty}^{\infty} f(t)e^{ist} \, dt \right) ds.$$

$$\text{If} \quad F(s) = \int\limits_{-\infty}^{\infty} f(t)e^{ist} \, dt,$$

$$\text{then} \quad f(x) = \frac{1}{2\pi} \int\limits_{-\infty}^{\infty} F(s)e^{-isx} \, ds.$$

The function $F(s)$ is called as the **Fourier transform** of the function $f(x)$, and the function $f(x)$ itself is called as the **inverse Fourier transform** of $F(s)$. Linearity, similarity theorem (change of scale property), shifting, and modulation properties are followed by Fourier transform. Fourier transform is applicable for dealing with the boundary value problems occurred in mathematical, physical, and engineering sciences like in heat conduction, vibration of strings, and so on. In two-dimensional problems, it is sometimes required to apply the transform twice and the required solution is obtained by double inversion.

## 2.13 Gibbs phenomenon

**J. Willard Gibbs,** an American physicist, studied the peculiar manner of Fourier series. He stated that near the discontinuity manifested, due to lack of development in the approximations, there is a continual presence of the overshoot or undershoot. After his name, this phenomenon is called as **Gibbs phenomenon** (ringing artifacts) in which the Fourier series of a periodic function $f$ containing a piecewise continuously differentiability behaves at jump discontinuous points. Near the jump point, due to large oscillations of the $n$th partial sum of the Fourier series of the function $f$, the value of maximum (minimum) of the partial sum of the function $f$ might increase above (below) that of the function itself. There would be always a minimum error due to an overshoot or undershoot near the discontinuous point even if the approximation is found by the maximum number of the terms of the series (Fourier series). Hence, the value of the size of the overshoot is independent of the number of terms used in the Fourier approximation. Value of the size of overshoot (or undershoot) always is about 9% of the value of the size of the jump at the discontinuous points.

The reason of the overshoot is to approximate a discontinuous function $f$ with the help of a partial (i.e., finite) sum of continuous functions (linear combination of the sine and cosine functions). Any finite sum of continuous functions is always a continuous function; therefore, it is not possible to approximate the discontinuity of the function within any arbitrary accuracy. But, the sum of the infinite number of terms of continuous functions can represent the discontinuous function. In this way, the exhibition of the Gibbs phenomenon can be ignored.

## 2.13.1   *Gibbs phenomenon with an example*

The purpose of this example is to deal with the following terms:

   i. Function $f(t)$
  ii. The sequence of partial sums $S_n(t)$
 iii. Nörlund mean $N_p$ of partial sums of a Fourier series
  iv. The sequence of averages (i.e., $\sigma_1^n(t)$-means or (C, 1) mean)

Let the Nörlund mean of partial sums of a Fourier series be denoted by $N_p$. The behavior of the Nörlund mean is better than the sequence of partial sums ($t$). Similarly the $\sigma_1^n(t)$-means or (C, 1) mean also behave better than the ($t$) for the following function:

$$f(t) = \begin{cases} -1, & -\pi \le t < 0 \\ 1, & 0 \le t \le \pi \end{cases}.$$

For all real values of $t$,

$$f(t + 2\pi) = f(t).$$

Trigonometric Fourier series is given by

$$\frac{2}{\pi} \sum_{n=1}^{\infty} \frac{1 - (-1)^n}{n} \sin nt, \quad -\pi \le t \le \pi.$$

The $n$th Cesàro sum ($\sigma^1(x)$) for the trigonometric Fourier series is given by

$$\sigma_n^1(t) = \frac{2}{\pi} \sum_{k=1}^{n} \left(1 - \frac{k}{n}\right)\left(\frac{1 - (-1)^k}{k}\right) \sin(kt),$$

where $\delta = 1$.

For trigonometric Fourier series $s_n(t)$ denoted the $n$th partial sum which is given by

$$s_n(t) = \frac{4}{\pi}\left( \sin(t) + \frac{\sin(3t)}{3} + \frac{\sin(5t)}{5} + \cdots + \frac{\sin(nt)}{n} \right).$$

The $(C, 1)$ method is weaker than Nörlund method $N_p$ if $N_p$ has increasing weights $\{p_n\}$ (Theorem 20 of Hardy's "Divergent Series").

The Nörlund mean $N_p$ of the function $f(t)$ for $p_n = n + 1$ is given by

$$t_n^N(f;t) = \frac{2}{(n+1)(n+2)} \sum_{k=0}^{n} (n-k+1)s_k(f;t).$$

In the interval $[-\pi, \pi]$, one can observe that $s_n(t)$ converges to $f(t)$, but its converging rate is very slow. The converging rate of $\sigma_n^1(t)$ and $t_n^N(f;t)$ toward $f(t)$ is higher than the converging rate of $s_n(t)$ toward $f(t)$. Near the points of discontinuities ($-\pi$, 0, and $\pi$), as $n$ increases, the peaks of $s_5$ and $s_{10}$ move closer to the line passing through points of discontinuity (Gibbs phenomenon), but for $n = 5$, 10 the peaks of the graph of $\sigma_n^1(t)$ and $t_n^N(f;t)$ go flatter. Hence, an overshoot or undershoot of the peculiarity of the Fourier series and other series of the eigen-functions at simple discontinuous points is a Gibbs phenomenon; that is, near the point of discontinuity the converging rate of the trigonometric Fourier series is very slow.

For the case of the various summable means (by using various summability methods for approximation) of the trigonometric Fourier series of the function $f(t)$, overshoot or undershoot the Gibbs phenomenon and one can observe that the effect of the summability method is very smooth. Hence, $t_n^N(f;t)$ and $\sigma_n^1(t)$ is the better approximant than ($t$). One can observe that $N_p$ method is stronger than $(C, 1)$ method.

The graph in Figure 2.4 implies that except the point $t_0 = 0$ (point of discontinuity of $f(t)$), the sequence converges to $f(t)$. Gibbs focused on this point and around this point the behavior of the Fourier partial sums.

In the continuous area ($-\pi < t < 0$ and $0 < t < \pi$), the graph tends to look more like that of the original one by increasing the number of Fourier coefficients. But the amplitude of the wiggles remains constant near the discontinuous point (around the origin). Hence, the

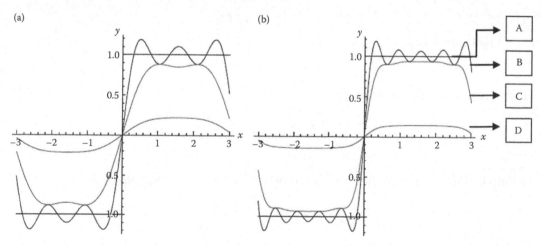

***Figure 2.4*** For $n = 5$ and $n = 10$, $f(t)$ (A), $s_n(t)$ (B), $\sigma_n^1(t)$ (C), and $t_n^N(f;t)$ (D).

partial sums of the trigonometric Fourier series will not smoothly converge to the mean value at the points of discontinuity.

### 2.13.2   Results related to Gibbs phenomenon

1. Gibbs phenomenon occurs for a function that is piecewise smooth with a jump discontinuous point at zero.
2. Gibbs phenomenon occurs at each of jump discontinuities for a piecewise smooth function with a finite number of discontinuous points.
3. Gibbs phenomenon occurs at all jumps discontinuous points of wavelet approximations of the function.
4. If a summation method (convolution of a positive kernel with the function $f$) is used for approximation of a discontinuous function $f$, then Gibbs phenomenon does not occur at the jump discontinuous points.
5. Gibbs phenomenon also does not occur in the case of approximation by wavelets that involve positive delta sequences. This result also used to show that in case of Haar wavelets, Gibbs phenomenon does not occur.

## 2.14   Trigonometric Fourier approximation

The computation of error function $E_n(f)$, defined by $E_n(f) = \min\|T_n(t) - f(t)\|$,   $n > 0$ is called the error estimation of function through trigonometric Fourier series using summability techniques. The trigonometric polynomial $T_n(t)$ is known as the Fourier approximant of $f$. This method is called the Fourier approximation method. The summability methods are used to find the Fourier approximation of the function.

## 2.15   Summability

In 1890, Cesàro deals with the sum of some divergent series and defined **Cesàro summation (summability methods)**.

There exist three types of summability:

   i. Ordinary summability
  ii. Absolute summability
 iii. Strong summability

### 2.15.1   Ordinary summability

Let $\Sigma u_n$ be an infinite series of real numbers with sequence of partial sums $\{s_n\}$. Let $T \equiv (a_{n,k})$ be a real or complex constants infinite matrix and $t_n$ given by

$$t_n = \sum_{k=0}^{n} a_{n,k} s_k, \quad n = 0, 1, 2, \ldots,$$

defines the matrix transform of the sequence $\{s_n\}_{n=1}^{\infty}$. Here the column vector of the $t_n$ is the product of the matrix $T$ with the column vector of the $s_n$. The sequence $\{s_n\}$ or the series $\Sigma u_n$ is said to be matrix summable to $s$, if $\lim_{n \to \infty} t_n = s$.

## 2.15.2  *Absolute summability*

The series $\sum_{n=0}^{\infty} a_n$ with partial sum sequence $\{s_n\}$ is absolute summable to $s$, i.e., if

$$\lim_{n \to \infty} t_n = s$$

and if $\{t_n\}$ (sequence of the mean) is of bounded variation, i.e.,

$$\sum_{n=1}^{\infty} |t_n - t_{n-1}| < \infty.$$

**Absolute summability of index $q$:** The infinite series $\sum_{n=0}^{\infty} a_n$ with the sequence of the partial sum $\{s_n\}$ is absolute summable with the index $q$ to $s$, i.e., if

$$t_n \to s, \quad \text{as } n \to \infty,$$

and $$\sum_{k=1}^{n} k^{q-1} |t_k - t_{k-1}|^q < \infty, \quad \text{as } n \to \infty.$$

It is denoted by $|A, q|$.

## 2.15.3  *Strong summability*

The infinite series $\sum_{n=0}^{\infty} a_n$ with the sequence of the partial sum $\{s_n\}$ is strong summable with index $q$ to $s$, i.e., if

$$\sum_{k=1}^{n} k^q |t_k - t_{k-1}|^q = O(n), \quad \text{as } n \to \infty,$$

and $t_n \to s$, as $n \to \infty$.
   It is denoted by $[A, q]$.
   The following inclusion relations hold,

$$|A, q| \subset [A, q] \subset (A).$$

## 2.16  *Methods for summability*

These are some summability methods:

   i. $(C, 1)$ means when $a_{n,k} = \dfrac{1}{n+1}, \quad 0 \le k \le n.$

   ii. Harmonic means when $a_{n,k} = \dfrac{1}{(n-k+1)\log n}, \quad 0 \le k \le n.$

   iii. Cesàro $(C, \delta)$ means when $a_{n,k} = \dfrac{E_{n-k}^{\delta-1}}{E_n^{\delta}}, \quad 0 \le k \le n.$

   iv. Nörlund means when $a_{n,k} = \dfrac{p_{n-k}}{P_n}, \quad 0 \le k \le n.$

v. Riesz means when $a_{n,k} = \dfrac{P_k}{P_n}$, $0 \le k \le n$.

vi. General Nörlund $(N, p, q)$ means when $a_{n,k} = \dfrac{p_{n-k} q_k}{R_n}$,

where $R_n = \displaystyle\sum_{k=0}^{n} p_k q_{n-k}$.

vii. Deferred Cesàro means: Agnew defined the deferred Cesàro mean of the sequence $x = (x_k)$ by

$$(D_{p,q})_n = \frac{1}{q(n) - p(n)} \sum_{k=p(n)+1}^{q(n)} x_k$$

$$q(n) < p(n) \text{ and } \lim_{n \to \infty} q(n) = \infty,$$

where $q(n)$ and $p(n)$ are positive natural numbers sequences.

## 2.17 Regularity condition

The summability matrix $T$ is regular, if

$$\lim_{n \to \infty} s_n = s \Rightarrow \lim_{n \to \infty} t_n = s.$$

Toeplitz and Silverman (1913) obtained necessary and sufficient conditions for the regularity of matrix $T$.

1. $\displaystyle\sum_{k=0}^{n} |a_{n,k}| \le M$, where $M$ (finite constant) is independent of $n$.

2. $\lim_{n \to \infty} a_{n,k} = 0$, $\quad \forall k$.

3. $\lim_{n \to \infty} \displaystyle\sum_{k=0}^{n} a_{n,k} = 1$.

## 2.18 Norm

A function that assigns strictly the positive length or size in a vector space is known as the norm.

A function $p: V \to R$ is a norm on $V$ if it satisfies the following properties ($\forall a \in R$ and $u, v \in V$):

i. $p(a\,v) = |a|\ p(v)$.
ii. $p(u + v) \le p(u) + p(v)$.
iii. If $p(v) = 0$ then $v$ is the zero vector.
iv. $p(v) \ge 0$ (non-negativity).

In the analysis of the Fourier series, importance of the $L_p$ norm cannot be ignored as it is an essential tool. The condition $p \to \infty$ will give the value of essential upper bound of the $L_p$

norm and $L_p$ behavior represents the Lipschitz behavior at $p \to \infty$. Hence, by replacing the power function with the help of more general classes of functions, the results of Fourier series can be generalized.

1. **$L_0$-Norm:** The first norm is a $L_0$-norm. $L_0$-norm of $x$ is $\|x\|_0 = \sqrt[0]{\sum_i x_i^0}$, that is a total number of nonzero elements in a vector.
2. **$L_1$-Norm (Absolute value norm):** It is the sum of absolute difference between two vectors or matrices, $\|x\|_1 = \sum_i |x_i - x_{i-1}|$.
3. **$L_2$-Norm (Euclidean norm):** It is a sum of squared difference $\|x_1 - x_2\|_2 = \sqrt{\sum_i (x_{1_i} - x_{2_i})^2}$

   and has wide applicability in the signal processing field for mean-squared error (MSE) measurement.
4. **$L_p$-Norm:** It is given by $\|x_1 - x_2\|_p = \sqrt[p]{\sum_i (x_{1_i} - x_{2_i})^p}$, $1 \le p < \infty$.
5. **$L_\infty$-Norm:** It is the maximum entries' magnitude of the vector $\|x\|_\infty = \max(|x_i|)$.

## 2.19   Modulus of continuity

The modulus of continuity $\omega(f, \delta)$ of a continuous function $f$ in $[a, b]$ is defined by

$$\omega(f,\delta) = \sup_{|y-x| \le \delta} \{|f(x) - f(y)|, \quad x,y \in [a,b]\}.$$

Let $f \in L_p[a, b]$, $p \ge 1$, then a function $\omega_p(f, \delta)$ is called the integral modulus of continuity and defined by

$$\omega_p(f,\delta) = \sup_{0 < |t| \le \delta} \left\{ \int_a^b |f(x+t) - f(x)|^p \, dx \right\}^{\frac{1}{p}}.$$

## 2.20   Lipschitz condition

Let $f(x)$ be defined on an interval $I$ and suppose we can find two positive constants $M$ and $\alpha$ such that

$$|f(x_1) - f(x_2)| \le M|x_1 - x_2|^\alpha$$

for all $x_1, x_2 \in I$. Then $f$ is said to satisfy a Lipschitz condition of order $\alpha$.

## 2.21   Various Lipschitz classes

**Lip $\alpha$:** A function $f \in$ Lip $\alpha$, for $0 < \alpha \le 1$, if $|f(x+t) - f(x)| \le O(t^\alpha)$.
   **Lip $(\alpha, p)$:** A function $f \in$ Lip $(\alpha, p)$, for $p \ge 1$, $0 < \alpha \le 1$, if

$$\left\{ \int_a^b |f(x+t) - f(x)|^p \, dx \right\}^{1/p} \le O(t^\alpha).$$

**Lip ($\xi(t)$, $p$):** For a positive increasing function $\xi(t)$ and an integer $p \geq 1$, a signal $f \in$ Lip ($\xi(t)$, $p$), if

$$\left\{ \int_a^b |f(x+t) - f(x)|^p \, dx \right\}^{1/p} \leq O(\xi(t)).$$

**$W(L_p, \xi(t))$:** For a given positive increasing function $\xi(t)$, an integer $p \geq 1$ and $\beta \geq 0$, $f$ belongs to weighted class $W(L_p, \xi(t))$, if

$$\left\{ \int_a^b |\{f(x+t) - f(x)\} \sin^\beta x|^p \, dx \right\}^{\frac{1}{p}} = O(\xi(t)).$$

**Note:** If $\beta = 0$, $W(L_p, \xi(t))$ coincides with the class Lip ($\xi(t)$, $p$); if $\xi(t) = t^\alpha$, Lip ($\xi(t)$, $p$) reduces to Lip ($\alpha$, $p$) and if $p \to \infty$, then Lip ($\alpha$, $p$) reduces to Lip $\alpha$:

$$\text{Lip } \alpha \subseteq \text{Lip } (\alpha, p) \subseteq \text{Lip } (\xi(t), p) \subseteq W(L_p, \xi(t)).$$

**$L_p$-space:** The set of Lebesgue integrable functions $f$: $E[a, b] \to R$, such that $\int_a^b |f(x)|^p \, dx \leq \infty$, for $1 \leq p \leq \infty$ is denoted by $L_p$ or $L^p$. If $E = [a, b]$ is the interval of finite length, then we write $L_p$ $[a, b]$. The $L_p$ ($E$)-space ($p \geq 1$) is a Banach space under the norms defined by

$$\|f\|_p = \left\{ \int_a^b |f(x)|^p \, dx \right\}^{\frac{1}{p}}, \quad \text{for } 1 \leq p < \infty \text{ and } \|f\|_\infty = \sup\{|f(x)| : x \in [a,b]\}.$$

## 2.22  Degree of approximation

A major portion of the study of theory of signals (functions) is concerned with the connections between the structural properties of a function and its degree of approximation. The objective is to relate the smoothness (by trigonometric Fourier approximation) of the function to the rate of decrement in the degree of approximation to zero. This chapter is to discuss the trigonometric approximation of the function (signal) and the concept needed to find the approximation degree using Fourier series. Trigonometric approximation is the most classical setting where the results are the most penetrating and satisfying. One of the basic problems in the theory of Fourier series is to examine the approximation degree using certain methods. In this sense, one of the important results is encountered. **Quade** [1] solved a problem related with approximation by trigonometric polynomial by using Nörlund summability in $L_p$ norm.

**Theorem 2.1 [1]:** Let $f \in \text{Lip}(\alpha, p)$, $0 < \alpha \leq 1$. Then

$$\|f - \sigma_n(f)\|_p = O(n^{-\alpha})$$

for either

  i. $p > 1$ and $0 < \alpha \leq 1$ or
  ii. $p = 1$ and $0 < \alpha < 1$.

And if $p = \alpha = 1$, then

$$\|f - \sigma_n(f)\|_1 = O(n^{-1}\log(n+1)).$$

**Chandra** [2] improved the result [1] and proved the following:

**Theorem 2.2** [2]: Let $f \in \text{Lip}(\alpha, p)$ and let $(p_n)$ be positive such that

$$(n+1)p_n = O(P_n).$$

If either

  i. $p > 1$ and $0 < \alpha \leq 1$ and
  ii. $(p_n)$ is monotonic

or

  i. $p = 1$ and $0 < \alpha < 1$ and
  ii. $(p_n)$ is nondecreasing.

Then

$$\|f - N_n(f)\|_p = O(n^{-\alpha}).$$

These are very important basic results of the degree of approximation and are a motivation for researchers working in this area. After this, several mathematicians studied the degree of approximation by using different summability techniques of a signal that belongs to various classes like Chandra [3,4], Khan [5,6], Mursaleen and Mohiuddine [7], Mishra and Mishra [8], Chen and Hong [9], Mishra et al. [10–12], Chen and Jeng [13], Mishra [14,15], and Alexits [16]. Bor [17–21] gave a number of theorems dealing with summability factors of the series and provided many applications. Recently, Sonker and Munjal [22–29] gave a number of theorems exploring the applications of summability and absolute summability of the Fourier and infinite series.

Many engineering problems can be solved using the summability methods. $(C, 1)$ and $(C, 2)$ can be used for increasing the rate of convergences of Gibbs phenomenon. For getting the information of the system and any process, analysis of signals or time functions can be done, and it is of great importance. Psarakis and Moustakides [30] presented a method for designing the finite impulse response (FIR) digital filters.

## 2.23   Fourier series and music

In mathematical physics, the wave equations can be represented in the form of the trigonometric Fourier series. Due to this applicability, the synthesis and analysis of the music can be understood easily. The human eardrums vibrate due to the presence of the variations in air and they hear a sound. Vibration due to music is created by the following cases:

  i. The piano string is struck.
  ii. A guitar string is plucked.
  iii. The bow is drawn across a violin string, etc.

This process transmits the vibrations of the music to the air and amplified the vibrations of the air. The human eardrums feel the air pressure fluctuations and the human brain converts them into electrical signals.

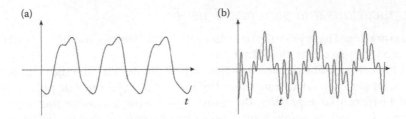

**Figure 2.5** Waveforms (a) flute and (b) violin.

For the study of the two different music instruments **(a) flute** and **(b) violin**, the graphs are plotted in Figure 2.5. For the sustained note D (294 vibrations per second), the graphs of **waveforms** show the difference between flute and violin. The flute waveform is simpler than that of the violin.

Fourier series approximation of this music is expressed as

$$P(t) = a_0 + a_1 \cos\left(\frac{\pi t}{L}\right) + b_1 \sin\left(\frac{\pi t}{L}\right) + a_2 \cos\left(\frac{2\pi t}{L}\right) + b_2 \sin\left(\frac{2\pi t}{L}\right) + \cdots$$

The sum of simple pure sounds is used for the expression of the Fourier coefficients, which have different values corresponding to the different musical instruments.

The $n$th term is called **$n$th harmonic** of $P$,

$$a_n \cos\left(\frac{n\pi t}{L}\right) + b_n \sin\left(\frac{n\pi t}{L}\right).$$

**Amplitude** is

$$A_n = \sqrt{a_n^2 + b_n^2}$$

and the **energy** of the $n$th harmonic is its square,

$$A_n^2 = a_n^2 + b_n^2.$$

It can be observed again that flute waveform is quite simple in comparison to the violin waveform. In violin, the highest harmonics are very strong but the energy of the flute keeps decreasing very fast.

Hence, trigonometric Fourier series is very useful in expressing the sounds of musical instruments. Complex musical sounds can be made of a combination of various pure sounds.

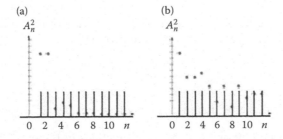

**Figure 2.6** Energy spe°ctra (a) flute and (b) violin.

## 2.24   *Applications and significant uses*

1. **Approximation Theory:** With the help of the summability methods, Fourier approximation of the given function can be found.
2. **Signal Processing:** Fourier analysis has significant uses for the signal processing. We use Fourier series to write a signal in the form of a trigonometric polynomial.
3. **Partial Differential Equation:** By using the variable separable method, the partial differential equations of the higher order can be solved. But for better understanding of behavior of the solution of the differential equation, Fourier series and error of Fourier approximation is used. With the help of Fourier series a prediction can be taken about the dynamic nature of the solution.
4. **Control Theory:** The nonharmonic Fourier series is used to deal with control theory.
5. **In Signal Processing (Speech Processing, Digital Image Processing and Audio Signal Processing):** Fourier approximation has a number of direct applications in rectification of signals in FIR filter and infinite impulse response (IIR) filter. These filters have numerous applications in image processing, speech processing, and medical signal processing, communications, sonar, radar, etc.
6. **Medical Signal Processing:** It is very difficult to deal with a large number of long sequences as DNA sequences. Multiple alignments by fast Fourier transform (MAFFT) is the fastest and most accurate program for the "lining up" of DNA sequences to reveal mutations, additions, and deletions between them in computational biology/bioinformatics. Ramanujan-Fourier series is used for comparative analysis of DNA sequences.
7. **Digital Image Processing:** For the reconstitution of a picture, approximation theory is used. It is also applicable for removing speckles from photographs with the help of the filter (summability methods). For the separation of the speckles and photo, the highest-frequency components can be dropped. Fourier approximation has a great mathematical importance and significance for $L_2$ and $L_\infty$ of the signals. In Ref. [31], importance of the filter in $L_p$ space is explained for the theory of signals. It is also useful for investigating the perturbations of matrix valued functions and bounds of the lattice norms.
8. **Data Analysis:** A necessary and sufficient condition for a system to be bounded input bounded output (BIBO) stable is that the impulse response be absolutely summable, i.e.,

$$\text{BIBO stable} \Leftrightarrow \sum_{n=-\infty}^{\infty} |h(n)| < \infty.$$

Summability techniques are trained to minimize the error. With the use of summability technique, the output of the signals (found by Fourier approximation) can be made stable, bounded, and used to predict the behavior of the input data, the initial situation, and the changes in the complete process.

## *References*

1. E. S. Quade, "Trigonometric approximation in mean," *Duke Mathematical Journal*, vol. 3, pp. 529–542, 1937.
2. P. Chandra, "Trigonometric approximation of functions in L$_p$-norm," *Journal of Mathematical Analysis and Applications*, vol. 275, pp. 13–26, 2002.

3. P. Chandra, "On the degree of approximation of a class of functions by means of Fourier series," *Acta Mathematica Hungarica*, vol. 52, no. 3–4, pp. 199–205, 1988.

4. P. Chandra, "A note on the degree of approximation of continuous functions," *Acta Mathematica Hungarica*, vol. 62, no. 1–2, pp. 21–23, 1993.

5. H. H. Khan, "On the degree of approximation of a functions belonging to the class *Lip*(α, *p*)," *Indian Journal of Pure and Applied Mathematics*, vol. 5, pp. 132–136, 1974.

6. H. H. Khan, Approximation of Classes of Functions [Ph.D. thesis], AMU, Aligarh, India, 1974.

7. M. Mursaleen and S. A. Mohiuddine, *Convergence Methods for Double Sequences and Applications*, Springer, ISBN: 978-81-322-1611-7, 2014. doi: 10.13140/2.1.3899.5526.

8. V. N. Mishra and L. N. Mishra, "Trigonometric approximation in *L*(*p* ≥ 1)spaces," *International Journal of Contemporary Mathematical Sciences*, vol. 7, pp. 909–918, 2012.

9. J. T. Chen and H.-K. Hong, "Review of dual boundary element methods with emphasis on hyper singular integrals and divergent series," *Applied Mechanics Reviews*, vol. 52, no. 1, pp. 17–32, 1999.

10. V. N. Mishra, K. Khatri, and L. N. Mishra, "Product ($N, p_n$) ($C$, 1) summability of a sequence of Fourier coefficients," *Mathematical Sciences*, vol. 6, p. 38, 2012.

11. V. N. Mishra, K. Khatri, and L. N. Mishra, "Product summability transform of conjugate series of Fourier series," *International Journal of Mathematics and Mathematical Sciences*, vol. 2012, Article ID 298923, 13 pages, 2012.

12. V. N. Mishra, H. H. Khan, K. Khatri, and L. N. Mishra, "On approximation of conjugate of signals (Functions) belonging to the generalized weighted *W*(*Lr*, ξ(*t*)), (*r* ≥ 1)-class by product summability means of conjugate series of Fourier series," *International Journal of Mathematical Analysis*, vol. 6, no. 35, pp. 1703–1715, 2012.

13. J. T. Chen and Y. S. Jeng, "Dual series representation and its applications to a string subjected to support motions," *Advances in Engineering Software*, vol. 27, no. 3, pp. 227–238, 1996.

14. V. N. Mishra, "On the degree of approximation of signals (Functions) belonging to the weighted (*Lp*, (*t*)), (*p* ≥ 1)-class by almost matrix summability method of its conjugate Fourier series," *International Journal of Applied Mathematics and Mechanics*, vol. 5, no. 7, pp. 16–27, 2009.

15. V. N. Mishra, "On the degree of approximation of signals (functions) belonging to generalized weighted (*L_p*, (*t*)), (*p* ≥ 1)-class by product summability method," *Journal of International Academy of Physical Sciences*, vol. 14, no. 4, pp. 413–423, 2010.

16. G. Alexits, *Convergence Problems of Orthogonal Series*, Pergamon Elmsford, New York, 1961.

17. H. Bor, "Factors for generalized absolute Cesàro summability," *Mathematical and Computer Modelling*, vol. 53, no. 5, pp. 1150–1153, 2011.

18. H. Bor, "An application of almost increasing sequences," *Applied Mathematics Letters*, vol. 24, no. 3, pp. 298–301, 2011.

19. H. Bor, "Generalized absolute Cesàro summability factors," *Bulletin of Mathematical Analysis and Applications*, vol. 8, no. 1, pp. 6–10, 2016.

20. H. Bor, "On absolute summability factors," *Proceedings of the American Mathematical Society*, vol. 118, no. 1, pp. 71–75, 1986.

21. H. Bor, "Almost increasing sequences and their new applications II," *Filomat*, vol. 28, no. 3, pp. 435–439, 2014.

22. S. Sonker and A. Munjal, "Sufficient conditions for triple matrices to be bounded," *Nonlinear Studies*, vol. 23, no. 4, pp. 533–542, 2016.

23. S. Sonker and A. Munjal, "Absolute summability factor $\varphi - |C, 1; \delta|_k$ of infinite series," *International Journal of Mathematical Analysis*, vol. 10, no. 23, pp. 1129–1136, 2016.

24. S. Sonker and A. Munjal, "Approximation of the function *f* belong to *Lip* (α, *p*) using infinite matrices of Cesàro sub-method," *Nonlinear Studies*, vol. 24, no. 1, pp. 113–125, 2017.

25. S. Sonker and A. Munjal, "A note on boundness conditions of absolute summability $\varphi - |A|_k$ factors," *International Conference on Advances in Science and Technology*, vol. 67, pp. 208–210, 2017. Proceedings ICAST-2017 Type A, ISBN: 9789386171429.

26. S. Sonker and A. Munjal, "Absolute summability $\varphi - |C, \alpha, \beta; \delta|_k$ of infinite series," *Journal of Inequalities and Applications*, vol. 168, pp. 1–7, 2017. doi: 10.1186/s13660-017-1445-5.

27. S. Sonker and A. Munjal, "Absolute summability factor $|N, p_n|_k$ of improper integrals," *International Journal of Engineering and Technology*, vol. 9, no. 3S, pp. 457–462, 2017.

28. S. Sonker and A. Munjal, "Absolute Nörlund summability $|N; p_n|_k$ of improper integrals," *National Conference on Recent Advances in Mechanical Engineering (NCRAME-2017)*, vol. II, no. 90, pp. 413–415, ISBN: 978-93-86256-89-8, 2017.

29. S. Sonker, Xh. Z. Krasniqi, and A. Munjal, "A note on absolute Cesàro $\varphi - |C, \ 1;\delta;1|_k$ summability factor," *International Journal of Analysis and Applications*, vol. 15, no. 1, pp. 108–113, 2017.

30. E. Z. Psarakis and G. V. Moustakides, "An $L_2$-based method for the design of 1-D zero phase FIR digital filters," *IEEE Transactions on Circuits and Systems I: Fundamental Theory Applications*, vol. 44, no. 7, pp. 591–601, 1997.

31. M. I. Gil', "Estimates for entries of matrix valued functions of infinite matrices," *Mathematical Physics Analysis and Geometry*, vol. 11, no. 2, pp. 175–186, 2008.

*chapter three*

# Soft computing techniques and applications

**Pankaj Kumar Srivastava and Dinesh Bisht**
*Jaypee Institute of Information Technology*

**Mangey Ram**
*Graphic Era (Deemed to be University)*

## Contents

## 3.1   Introduction: Soft computing

In this highly incorporated era, we need a tool that can be applicable to the problems aris-ing in different fields. In this sense soft computing is an emerging computing approach for upcoming years. Soft computing gives the flexibility to model the problem according to given constraints. It helps to find quick solutions to the problems arising in various dis-ciplines. These methods mimic human behavior. The roots of soft computing are found in fuzzy logic, data analysis, and intelligent systems. The main objective of soft computing is to develop intelligent machines to provide solutions to real-world problems, which are not modeled or too difficult to model mathematically.

Its goal is to utilize the capacity for imprecision, uncertainty, approximation, and par-tial truth to achieve resemblance with decision-making of human like. The development history of soft computing can be described as follows:

1943: Neural Network by McCulloch and Pitts
1960: Evolutionary Computing by Rechenberg
1962: Evolutionary Programming by Fogel
1965: Evolutionary Strategies by Rechenberg
1965: Fuzzy logic by Zadeh
1970: Genetic Algorithms by Holland
1981: Soft Computing by Zadeh
1992: Genetic Programming by Koza

## 3.2   Fuzzy logic

Life is full of uncertainties; it can be vague or imprecise. To deal with such uncertainties probability theory used to be a tool for mathematicians, which is based on classical set theory. Zadeh, in 1965 [1], challenged that there are some uncertainties which are out of the scope of probability theory. For example a company owner needs an honest person for his company. Now there are available choices that can be extremely honest, very hon-est, honest some of the time, and dishonest; which cannot be defined using classical logic because in this logic there are only two choices—honest and dishonest. Zadeh named this new concept as fuzzy set theory based on membership functions. Classical set theory is about yes or no concepts, whereas fuzzy set theory includes gray part also. Fuzzy set the-ory deals with appropriate reasoning in linguistic terms. Logic that deals mathematically with imprecise information usually employed by humans is fuzzy logic. A multivalued logic extends Boolean logic usually employed in computer science. Fuzzy logic is based on the concept of logic having multidimensions that provide intermediate values to be defined between conventional opposite evaluations like true or false, high or low, heat or cold, etc. Fuzzy concept, introduced by Zadeh in 1960, resembles uncertainty to generate decisions by human reasoning [1–3]. Fuzzy logic is a variety of multivalued logic which is consequent of fuzzy set theory. This logic deals with interpretation of those that are near rather than strict fuzzy logic system works with vague concepts as well. In fuzzy logic, the membership of precision of a statement ranges between 0 and 1, while in classical it is 0 or 1. Plato was the first person who believed that there was something beyond true and false. But it was Lukasiewicz who proposed an organized option to the bi-valued logic of Aristotle [4–7]. The existence of Greeks is still a standard example of fuzziness and precision at the same time [8].

Many decisions are based on beliefs. Occasionally, beliefs concerning uncertain events are expressed in numerical form as odds or subjective probabilities. These beliefs are usually expressed in statements such as

"I think that… "
"chances are… "
"it is unlikely that… "
and so forth.

The fuzzy expression contains a fuzzy proposition with its truth value in the interval [0,1]. It represents a mapping from [0,1] to [0,1] such as

$$g : [0, 1] \rightarrow [0, 1].$$

The generalization of the domain into $n$ dimension converts it as

$$G : [0, 1]^n \rightarrow [0, 1].$$

In view of this we may define it as a logical expression satisfying the following:

i. Truth values 0 and 1 and variable $x_i \in [0,1]$, $i = 1,2,3,\ldots,n$ are fuzzy.
ii. If $u$ is a fuzzy expression, then $\bar{u}$ is also fuzzy.
iii. If $u$ and $v$ are fuzzy expression, then $u \wedge v$ and $u \vee v$ are fuzzy.

### 3.2.1 Evolution of fuzzy logic

1965: From crisp sets to fuzzy sets
1973: From fuzzy sets to granulated fuzzy sets (linguistic variable)
1999: From measurements to perceptions

## 3.3 Fuzzy sets

Fuzzy set $F(m)$ is represented by a pair of two components: first is the member $m$ and the second is its membership grade $\mu_F(m)$ which maps any element $m$ of universe of discourse $M$ to the membership space [0,1], as given below:

$$F(m) = \{(m, \mu_F(m)), \quad m \in M. \tag{3.1}$$

### 3.3.1 Equal fuzzy sets

Two fuzzy sets $F_1$ and $F_2$ are said to be equal if all the members of $F_1$ belong to $F_2$ with the same membership grade as in $F_1$.

## 3.3.2  Membership function

A function that describes the membership grades of elements in a fuzzy set is said to be a membership function. A membership function can be discrete or continuous. It needs a uniform membership function representation for efficiency. Some well-known membership functions are as discussed in the following sections.

### 3.3.2.1  Z-Shaped membership function

Z-Shaped membership function is given by

$$Z(x;p,q) = \begin{cases} 1 - 2\left(\dfrac{x-p}{q-p}\right)^2 ; & \text{if } p < x \geq (p+q)/2 \\[2ex] 2\left(\dfrac{x-p}{q-p}\right)^2 ; & \text{if } (p+q)/2 < x \leq q \\[2ex] 1; & \text{if } x \leq p \\[1ex] 0; & \text{otherwise} \end{cases} \tag{3.2}$$

### 3.3.2.2  Triangular membership function

Triangular membership function is given by

$$T(x;p,q,r) = \begin{cases} \dfrac{x-p}{q-p} ; & \text{if } p < x \geq q \\[2ex] \dfrac{r-x}{r-q} ; & \text{if } q < x \leq r \\[1ex] 0; & \text{otherwise} \end{cases} \tag{3.3}$$

### 3.3.2.3  Trapezoidal membership function

Trapezoidal membership function is represented by

$$T(x;p,q,r,s) = \begin{cases} \dfrac{x-p}{q-p} ; & \text{if } p < x \geq q \\[2ex] 1; & \text{if } q < x \leq r \\[1ex] \dfrac{s-x}{s-r} ; & \text{if } r < x \leq s \\[1ex] 0; & \text{otherwise} \end{cases} \tag{3.4}$$

### 3.3.2.4  Gaussian membership function

Representation of the Gaussian membership function is as follows:

$$G(x;\sigma,m) = \frac{1}{e^{\frac{1}{2}\left(\frac{x-m}{\sigma}\right)^2}}. \tag{3.5}$$

## 3.4 Fuzzy rule base system

To understand the fuzzy rule base system, let us take a statement, "If there is *high* traffic jam and *heavy* rain then I may get a *little* late." Here high, heavy and little are fuzzy sets related to variables traffic jam, rainfall, and late, respectively. Mathematically it can be represented as follows: IF ($x$ is $F_1$ and $y$ is $F_2$) THEN ($z$ is $F_3$), where $F_1$, $F_2$, and $F_3$ are fuzzy sets, and $x$, $y$, and $z$ are variables. A collection of all such rule for a particular system is known as rule base [9].

## 3.5 Fuzzy defuzzification

It is quite difficult to take decision on the bases of fuzzy output; in that case this fuzzy output is converted into crisp value. This process of converting fuzzy output into crisp output is known as defuzzification [10]. Different methods are available in the literature; some widely used methods are discussed in the following sections.

### 3.5.1 Center of area (CoA) method

This method is also known by the names *centroid method* and *center of gravity* method. CoA is the most popular method of defuzzification among researchers. This method is based on CoA taken by the fuzzy set. Its defuzzified value is calculated after considering the entire possibility distribution and is given by

$$m^* = \begin{cases} \dfrac{\int (\mu(m) \times m)\, dm}{\int \mu(m)\, dm}; & \text{for continuous membership value of } m \\[4mm] \dfrac{\sum \mu(m) \times m}{\sum \mu(m)}; & \text{for discrete membership value of } m \end{cases} \tag{3.6}$$

### 3.5.2 Max-membership function

This method is also called the height method and is applicable to peaked output functions. Expression for this method is given by

$$\mu(m^*) \geq \mu(m) \quad \forall\, m \in M \,(\text{Universe of discourse}). \tag{3.7}$$

### 3.5.3 Weighted average method

This method is applied when output is symmetrical and is given by

$$m^* = \frac{\sum \mu(\bar{m}) \times \bar{m}}{\sum \mu(\bar{m})}, \tag{3.8}$$

where $\bar{m}$ is the centroid of each symmetric membership function. This method is computationally efficient but less popular.

### 3.5.4   Mean-max method

This method is similar to the max-membership method; the only difference is that the locations of maximum membership can be more than one. Expression is given by

$$m^* = \frac{m_1 + m_2}{2},\tag{3.9}$$

where $m_1 + m_2$ are the mean of maximum interval.

### 3.5.5   Center of sums

This method is based on the algebraic sum of fuzzy subsets. This method is very fast in terms of calculations. The defuzzified value is give by

$$m^* = \frac{\displaystyle\sum_{i=1}^{N} m_i \sum_{k=1}^{n} \mu(m_i)}{\displaystyle\sum_{i=1}^{N}\sum_{k=1}^{n} \mu(m_i)}.\tag{3.10}$$

## 3.6   Comparison of crisp to fuzzy

*Table 3.1* Comparison between crisp sets and fuzzy sets

|  | Crisp | Fuzzy |
| --- | --- | --- |
| Response | Is rain water colorless?<br>Yes or No | Is Narendra Honest?<br>Extremely dishonest (0)<br>Honest sometimes (0.3)<br>Very honest (0.7)<br>Extremely honest (1) |
| Operation | 1. Operation of union<br>2. Operation of intersection<br>3. Complement Law<br>4. Operation of difference | 1. Operation of union<br>2. Operation of intersection<br>3. Complement Law<br>4. Equality<br>5. Differences<br>6. Disjunctive sum |
| Properties | 1. Commutative Law<br>2. Associative Law<br>3. Distributive Law<br>4. Idempotent Law<br>5. Identity Law<br>6. Law of Absorption Law<br>7. Transitive Law<br>8. Involution Law<br>9. De-Morgan's Law<br>10. Excluded Middle Law<br>11. Contradiction Law | 1. Commutative Law<br>2. Associative Law<br>3. Distributive Law<br>4. Idempotent<br>5. Identity Law<br>6. Absorption Law<br>7. Transitive Law<br>8. Involution Law<br>9. De-Morgan's Law |

## 3.7 Examples of uses of fuzzy logic

Examples of uses of fuzzy logic are

- Foam detection
- Imbalance compensation
- Water level adjustment
- Washing machine
- Food cookers
- Taking blood pressure
- Determination of "socioeconomic class"
- Cars

## 3.8 Artificial neural networks

A neural network is a complex network of neurons. Neurons are responsible for processing and storing the information. This stored information can be used for the future. An artificial neural network (ANN) is motivated by the functioning of the human brain. The human brain and ANN both acquire knowledge from experience; here we can term it as *learning*. Neurons are used to store this learning output.

In this modern era of computers, we are having computers perform mathematical problems, but in some cases, computers are unable to give entirely accurate results (e.g., the distorted image of any animal or word can be easily recognized by the human brain but computers fail to do so). The human brain can easily understand words spoken in a noisy environment. This motivates study of the human brain. A fundamental unit of the brain is called *nerve cells* or *neurons*. Neurons in the human brain vary in their shape and size. Neurons are made up of soma which itself consist of the cell nucleus. Neurons are biochemical units, which process the electrical signals. Dendrites linked with cell body, receive signals from other neurons. Whereas the axon long fiber going out from the cell body processes the output of the neuron. Eventually, the axon also branches into strands and substrands called *synapses*. Synapses release chemical elements to dendrites, which increases or decreases the electrical potential of neurons. If the total input established at the cell body ranges to a threshold value, neurons get fired. After this fire, the neuron cannot fire again for some fraction of time. This time duration is known as the *absolute refractory period*. The dendrites' work is to receive signals from other neurons, whereas the axon transmits triggered activity to other neurons. A receptor neuron receives material from muscles. The size of synapses depends on the process of learning. McCulloch and Pitts in 1943 introduced this concept of neural networks. McCulloch a psychiatrist and Pitts a mathematician took a long time to come up with this conclusion. They claimed that any complex problem can be easily solved by this method. Although this presented model of McCulloch–Pitts looked simple, it was a revolutionary innovation in the field of artificial intelligence [11]. Figure 3.1 demonstrates the basic principle of ANNs.

The progress of neural networks includes development of learning rules for neurons. The first learning rule is given by Hebb (1949). This rule states that the strength of two neurons increased by activations of neurons at the same time [12]. Linearly unseparable functions could not be solved using perceptron [13,14]. Due to this drawback ANN did not gain popularity until the discovery of the backpropagation algorithm [15–17]. The backpropagation algorithm is based on multilayer networks. The backpropagation algorithm was developed by Paul Werbos in 1974 and rediscovered independently by Rumelhart and

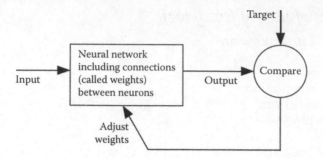

***Figure 3.1*** Basic principle of artificial neural networks.

Parker. Discovery of backpropagation is a benchmark in this field. This learning algorithm has been used in feed-forward multilayer neural networks. The backpropagation algorithm gives flexibility to train ANNs in parallel. This training improves the efficiency of multilayer perceptron. The gradient descent supervised learning method minimizes the error of the network [18–20].

### 3.8.1   Artificial neurons

McCulloch–Pitts gave the concept of the simplest neuron; it is also known as a threshold logic unit (TLU). In this model inputs and outputs are taken as binary values [11]. Input gets activated with the help of other neurons. Further synaptic weights are added and compared with the threshold value. If this value is more than threshold, then this particular neuron gets fired (i.e., it gets activated). If it is less than threshold, the neuron will not be activated.

### 3.8.2   Firing rule

One of the main concepts of ANN is the firing rule. A firing rule decides whether the neuron will get activated or not with any input pattern. Hamming distance technique is one of the basic and simple firing techniques. This technique is widely used due to its simple calculations. There are two types of training patterns for a node: the first are responsible for the firing of any neuron, called 1-taught set of inputs. The second type of training patterns that oppose this change are called 0-taught set of inputs.

Let $K = (k_1, k_2, \ldots, k_n)$ and $L = (l_1, l_2, \ldots, l_n)$, then hamming distance is given by

$$H = \sum_{i}^{n} |k_i - l_i|.$$

### 3.8.3   Different types of neural networks

On the basis of the unique working of the human brain, different types of ANNs have been proposed. All of these ANNs have some similarities to the functioning of the human brain, so that they can be used for many complex problems (e.g., pattern recognition and classification).

### 3.8.3.1 Feedback ANN

The feedback network reverses the information. These are applicable to the error corrections of the internal system.

### 3.8.3.2 Feed-forward ANN

It is a neural network containing

  i. An input layer
 ii. An output layer
iii. One or more layers of neurons

Decision is based on group behavior of the connected neurons.

### 3.8.3.3 Classification-prediction ANN

Classification-prediction ANN identifies particular patterns and converts these into specific groups.

## 3.9 Training of neural networks

In training of ANN weights, biases or any other parameter, get adjusted according to data set. Training is a kind of curve fitting where parameters are optimized to give best results. Learning can be of two types: supervised or unsupervised.

### 3.9.1 Supervised training

Inputs and outputs both are given in supervised training. These inputs are processed into a network and the outcome is compared with the actual output. Hence, the difference of actual output and desired output is adjusted by adjusting weights. This process is repeated until optimized error is achieved. These data which are used in adjustment of weights are called training data. This learning is similar to class learning where a teacher is always there to correct mistakes of students, thus it is sometimes referred to as supervised learning. If the input data lack some precise knowledge, then training may not be possible for the particular network. For a good learning an appropriate number of data set is required, otherwise the networks may not converge. In standard conditions we divide the data set into three parts: one for training, one for testing, and the last part for validation. For an unbeaten network it is necessary to examine all the basic parts again and again (e.g., number of hidden layers, connection weights, etc.). The adaption of feedback is done by most the popular backpropagation technique. At last when the network is trained appropriately, the final weights can be frozen.

### 3.9.2 Unsupervised training

In this type of learning only inputs are given to the network, not the outputs. The decision of selecting features is done by grouping the input data. Hence, this learning algorithm is known as *adaption*. In real life there are several examples where exact training of data is not possible (e.g., in an army, the situation where an army faces new weapons). Tuevo Kohonen designed a self-organizing neural network that can learn without desired output [21]. As this learning algorithm resembles a class of students where students are learning themselves, without the help of a teacher, it is sometimes referred to as learning without a teacher.

### 3.9.3 Reinforced training

This type of learning comes under supervised learning with condition. Here the teacher works as a guide to tell either the output is correct or not but will not give the actual output to the network. This learning algorithm is not popular among researchers.

## 3.10 Adaptive neuro fuzzy inference system

Adaptive neuro fuzzy inference system (ANFIS) is a combination of neural networks and fuzzy logic to utilize their advantages. Neural network and fuzzy logic have their own strengths and weaknesses. In this combination, neural network and fuzzy logic complement each other. In an ANFIS neural networks play the role of supporting tool to fuzzy system [22]. The ANFIS makes it possible to plan complete neural network knowledge to develop a fuzzy inference system. This strong combination helps overcome the shortcomings of both techniques. To generate fuzzy rules for a fuzzy inference system, ANFIS can easily learn from a sufficiently large data set [23–29].

## 3.11 Genetic algorithms

Optimization is a process for finding the best out of available solutions. Optimization is needed everywhere, even in our daily routine, when we decide our to-do list or prioritize our tasks for the day [30]. There are many traditional methods available in the literature to solve the optimization problem. These traditional methods have the following drawbacks:

1. Traditional methods may stick up at local optimum.
2. Gradient-based methods are not applicable in case of discontinuous objective function.
3. Parallel computing is difficult for these methods.
4. Only unimodal problems can be handled.
5. Different methods are needed for different kinds of problems.

The above drawbacks of traditional methods motivate the search for a robust and efficient method. Biological systems are flexible, robust, efficient, and self-guided. Genetic algorithms are inspired by Darwin's theory about evolution: "survival of the fittest." The method was developed by John Holland (1975) [30–32].

## 3.12 Working of genetic algorithm

The following steps are involved in genetic algorithms (Figure 3.2):

**Step 1—Generation of population:** Depending on the complexity of the problem, an initial population generated. This generation is purely random. The most commonly used encoding scheme is binary encoding. In binary encoding, the solutions are represented in the form of a binary string. The length of each chromosome (solution in form of bit) is dependent on the accuracy we required.

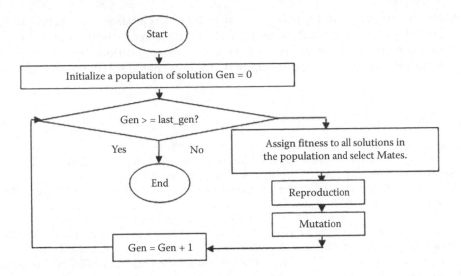

*Figure 3.2* Working cycle of a genetic algorithm.

**Step 2—Fitness evolution:** To calculate the fitness value of each solution, the decoding is done to get the real value of the solution. This value of the variable is then substituted to the given objective function to compare fitness of solution.

**Step 3—Reproduction:** Good strings ("fittest") in a population are selected and assigned a large number of copies to form a mating pool.

**Step 4—Crossover:** In this step parents exchange properties.

**Step 5—Mutation:** The concept of biological mutation is also preserved here. A sudden change in population is done to take the solution out of local optima.

## 3.13   Applications of soft computing

- A temperature control system tuned through soft computing is applicable to change the temperature and humidity according to the climate conditions.
- Washing machines adjust operations for the best wash results on the basis of dirt conditions. A combination of neural network and fuzzy logic can also be implemented to get better wash results [33,34].
- Soft computing is also applicable in aircraft deciding vehicles.
- A Coke oven gas cooling plant is a first-rate application of the soft computing methodology.
- These days control of automatic exposure in video cameras is also done through soft computing.
- Wastewater treatment process control, reverse osmosis plants, quantitative pattern analysis for industrial quality assurance, control of constraint satisfaction problems in structural design and control of water purification plants are a few global applications of soft computing.
- Soft computing finds its measure application in medical diagnostic support systems like in use of anesthesia, logical findings in Alzheimer's patients, and diagnosis of diabetes and prostate cancer.

- Automatic underground train operation, railway acceleration, train schedule control, transportation, braking and stopping, and adaptive filter for nonlinear channel equalization control of broadband noise are some other applications of soft computing.
- The motion planning of mobile robots is also done using soft computing techniques.

## References

1. L. A. Zadeh, "Fuzzy sets," *Inf. Control*, vol. 8, no. 3, pp. 338–353, 1965.
2. L. A. Zadeh, "Outline of a new approach to the analysis of complex systems and decision processes," *IEEE Trans. Syst. Man Cybern.*, vol. 3, no. 1, pp. 28–44, 1973.
3. M. Gr. Voskoglou, "Measuring the uncertainty of human reasoning," *Am. J. Appl. Math. Stat.*, vol. 2, no. 1, pp. 1–6, 2013.
4. C. Lejewski, "Jan Lukasiewicz," *Encycl. Philos.*, vol. 5, pp. 104–107, 1967.
5. D. C. S. Bisht, M. Raju, and M. Joshi, "Simulation of water table elevation fluctuation using fuzzy-logic and ANFIS," *Comput. Model. New Tech.*, vol. 13, no. 2, pp. 16–23, 2009.
6. A. Gupta, and N. Singhal, "Advice generation using fuzzy logic in OMR Pheonix technique," *Int. J. Comput. Appl.*, vol. 52, no. 16, pp. 6–10, 2012.
7. E. Egrioglu, U. Yolcu, C. H. Aladag, and C. Kocak, "An ARMA type fuzzy time series forecasting method based on particle swarm optimization," *Math. Probl. Eng.*, vol. 2013, pp. 1–12, 2013.
8. A. Reigber, My life with Kostas. Unpublished report, Neverending Story Press, 1999.
9. J. M. Mendel, *Uncertain Rule-Based Fuzzy Logic Systems: Introduction and New Directions.* Prentice Hall PTR, Upper Saddle River, NJ, 2001.
10. R. R. Yager, and D. P. Filev, *Essentials of Fuzzy Modeling and Control.* John Wiley & Sons, New York, 388 pp, 1994.
11. W. S. McCulloch, and W. Pitts, "A logical calculus of the ideas immanent in nervous activity," *Bull. Math. Biophys.*, vol. 5, no. 4, pp. 115–133, 1943.
12. D. O. Hebb, "Organization of behavior. New York: Wiley, 1949, pp. 335," *J. Clin. Psychol.*, vol. 6, no. 3, pp. 307–307, 1950.
13. M. Dougherty, "A review of neural networks applied to transport," *Transp. Res. Part C Emerg. Technol.*, vol. 3, no. 4, pp. 247–260, 1995.
14. A. L. Glass, and K. J. Holyoak, "Alternative conceptions of semantic theory," *Cognition*, vol. 3, no. 4, pp. 313–339, 1974.
15. J. A. Anderson, "A simple neural network generating an interactive memory," *Math. Biosci.*, vol. 14, no. 3–4, pp. 197–220, 1972.
16. L. Glass, and R. E. Young, "Structure and dynamics of neural network oscillators," *Brain Res.*, vol. 179, no. 2, pp. 207–218, 1979.
17. K. Fukushima, "Cognitron: A self-organizing multilayered neural network," *Biol. Cybern.*, vol. 20, no. 3–4, pp. 121–136, 1975.
18. I. Jung, L. Koo, and G.-N. Wang, "Two states mapping based neural network model for decreasing of prediction residual error," *Int. J. Ind. Manuf. Eng.*, vol. 1, no. 7, pp. 322–328, 2007.
19. K. L. Priddy, and P. E. Keller, Artificial Neural Networks: An Introduction, vol. 68. SPIE Press, Bellingham, WA, 2005.
20. S. Sapna, A. Tamilarasi, and M. P. Kumar, "Backpropagation learning algorithm based on Levenberg Marquardt Algorithm," *Comp. Sci. Inf. Technol. CS IT*, vol. 2, pp. 393–398, 2012.
21. T. Kohonen, ed., *Self-Organizing Maps.* Springer-Verlag, New York, Secaucus, NJ, 1997.
22. B. Dixon, "Applicability of neuro-fuzzy techniques in predicting ground-water vulnerability: A GIS-based sensitivity analysis," *J. Hydrol.*, vol. 309, no. 1, pp. 17–38, 2005.
23. U. Nauck, and R. Kruse, "Design and implementation of a neuro-fuzzy data analysis tool in Java," *Manual Technical University of Braunschweig Germany*, 1999.
24. E. Khan, Neural fuzzy based intelligent systems and applications In: Jain, LC and NM Martin eds., *Fusion of Neural Networks, Fuzzy Sets and Genetic Algorithms: Industrial Applications.* CRC Press, Washington, DC, 1999.

25. J.-S. Jang, "ANFIS: Adaptive-network-based fuzzy inference system," *IEEE Trans. Syst. Man Cybern.*, vol. 23, no. 3, pp. 665–685, 1993.
26. J. M. Keller, R. Krishnapuram, and F.-H. Rhee, "Evidence aggregation networks for fuzzy logic inference," *IEEE Trans. Neural Netw.*, vol. 3, no. 5, pp. 761–769, 1992.
27. I. N. Aghdam, B. Pradhan, and M. Panahi, "Landslide susceptibility assessment using a novel hybrid model of statistical bivariate methods (FR and WOE) and adaptive neuro-fuzzy inference system (ANFIS) at southern Zagros Mountains in Iran," *Environ. Earth Sci.*, vol. 76, no. 6, p. 237, 2017.
28. A. M. Ahmed, and S. M. A. Shah, "Application of adaptive neuro-fuzzy inference system (ANFIS) to estimate the biochemical oxygen demand (BOD) of Surma River," *J. King Saud Univ.-Eng. Sci.*, vol. 29, no. 3, pp. 237–243, 2017.
29. A. Karkevandi-Talkhooncheh, S. Hajirezaie, A. Hemmati-Sarapardeh, M. M. Husein, K. Karan, and M. Sharifi, "Application of adaptive neuro fuzzy interface system optimized with evolutionary algorithms for modeling $CO_2$–crude oil minimum miscibility pressure," *Fuel*, vol. 205, pp. 34–45, 2017.
30. A. Boultif, A. Kabouche, and S. Ladjel, "Application of genetic algorithms (GA) and threshold acceptance (TA) to a ternary liquid–liquid equilibrium system," *Int. Rev. Model. Simul. IREMOS*, vol. 9, no. 1, pp. 29–36, 2016.
31. I. Cruz-Vega, C. A. R. García, P. G. Gil, J. M. R. Cortés, and J. de J. R. Magdaleno, "Genetic algorithms based on a granular surrogate model and fuzzy aptitude functions," in *Evolutionary Computation (CEC), 2016 IEEE Congress on*, 2016, pp. 2122–2128.
32. R. L. Haupt, and S. E. Haupt, *Practical Genetic Algorithms*. John Wiley & Sons, New York, 2004.
33. D. C. S. Bisht, P. K. Srivastava, and M. Ram, "Role of fuzzy logic in flexible manufacturing system," *Diagnostic Techniques in Industrial Engineering*. Springer, Cham, pp. 233–243, 2018.
34. N. Mathur, P. K. Srivastava, and A. Paul, "Algorithms for solving fuzzy transportation problem," *Int. J. Math. Oper. Res.*, vol. 12, no. 2, pp. 190–219, 2018.

# chapter four

# New approach for solving multi-objective transportation problem

**Gurupada Maity and Sankar Kumar Roy**
*Vidyasagar University*

## Contents

## 4.1 Introduction

Operations research (OR) is a discipline that encompasses a wide range of methods in solving real-life decision-making problems. The mathematical methods are applied in pursuit of improving decision-making and efficiency in the areas of mathematical optimization, econometric methods, simulation, neural networks, decision analysis, and the analytic hierarchy process. The study of OR arose during World War II. During this time, OR was considered as a scientific way for providing respective departments with a quantitative basis for making decisions corresponding to the operations under the entire system. The term "optimization" is the root of the study in OR. The optimization is used in different areas of study, like mathematical optimization, engineering optimization, economics and business, information technology, etc.

In an optimization problem (OP), we basically treat the objective function, either maximization or minimization, with or without some prescribed set of constraints.

Requirements in real-life decision-making situations enlarge the area of mathematical OPs in different fields of multi-objective optimization (MOO) problems.

The transportation problem (TP), a special kind of decision-making problem, may be considered as the central nervous system to keep the balance in economical infrastructure from ancient times to today. TP can be treated as a special case of a linear programming problem (LPP). A graphical network of TP is shown in Figure 4.1.

The classical sense of TP fixes how many units of goods are to be transported from each node of origin to various nodes of destinations, satisfying availabilities of sources, and demands of destinations, while minimizing the total cost of transportation along with minimizing the costs per unit of commodities for the purchasers.

TP was originally developed by Hitchcock [1], and later the mathematical model was presented by Koopmans [2]. A number of research works have been invented by several authors in the thrust area of TP, such as that by Ebrahimnejad [3], Kaur and Kumar [4], Mahapatra et al. [5], Maity and Roy [6,7], Maity et al. [8], Midya and Roy [9,10], Roy [11], Roy and Maity [12], and Roy et al. [13]. The single-objective TP is not sufficient to tackle real-life decision-making problems in competitive market scenarios. To tackle all the real-life situations on TP, we have to introduce here multi-objective TP. Several approaches for solving managerial problems involving multiple conflicting objective functions are introduced by Charnes and Cooper [14]. Maity and Roy [15] solved a multi-objective TP by fuzzy programming. A study on a multi-objective TP in a fuzzy environment to obtain a compromise solution was introduced by Waiel [16]. Roy et al. [17] presented a study on a multi-objective transportation problem (MOTP) using a conic scalarization approach. Kumar et al. [18,19] solved different types of optimization problems using particle swarm optimization. Recently, Pant et al. [20,21] applied the particle swarm optimization technique to solve different types of optimization problems.

Considering the situations of real-life decision-making problems, we design this chapter on TPs in multi-objective ground where the objective functions are conflicting. Many situations occurred where the solution of a MOTP is found as a compromise solution, but the solution often depends on the weights of objective functions proposed by the decision maker (DM). Then in the MOTP, the compromise solution satisfies the goals corresponding to the objective functions which play an effective role in solving it.

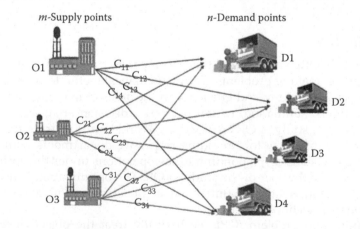

*Figure 4.1* Graphical network of TP.

Several approaches are available for solving a MOTP, such as fuzzy programming, goal programming, revised multi-choice goal programming, etc. Zimmermann [22] introduced the concept of solving a MOO problem using fuzzy programming. Basically, in a fuzzy programming approach, a MOTP is converted to a single-objective optimization problem and then its solution is treated as a compromise solution.

Goal programming (GP), an analytical tool, is introduced to address the decision-making problem involving objective functions that are conflicting and noncommensurable to each other, and targets have been assigned as goals to the objective functions. The DM is interested in maximizing the aspiration level of the corresponding goals. The concept of GP was introduced by Charnes and Cooper [14] and further developed by researchers such as Hannan [23], Ignizio [24], Tamiz et al. [25], Romero [26], Liao [27], Tabrizi et al. [28], and many others. However, the resources ambiguity and the incomplete information make it almost impossible to set the specific aspiration levels (goals) and select a better decision by DM. To tackle this situation, Chang [29] presented the multi-choice goal programming (MCGP) approach to solve the multi-objective decision-making (MODM) problem. During this time, again Chang [30] proposed RMCGP which is the revised form of MCGP for solving the MODM.

In this chapter, we introduce a new approach for solving the MOTP. Especially, we intend to solve the MOTP using the Vogel approximation method (VAM). The usefulness of the algorithm is tested through a numerical example.

The rest of this chapter is organized in the following way: Section 4.2 describes the preliminaries of the proposed chapter. Section 4.3 contains the mathematical model of TP and MOTP. The solution procedure is presented in Section 4.4 which contains five subsections. Fuzzy programming, goal programming, and revised multi-choice goal programming are briefly presented in Sections 4.4.1–4.4.3, respectively. An algorithm for solving the proposed MOTP by VAM is introduced in Section 4.4.4. Section 4.4.5 contains the merits and demerits of the proposed approaches for solving MOTP. A numerical example is taken into consideration to justify our study; and comparison among the obtained solutions from the approaches is carried out in Sections 4.5 and 4.6, respectively. The chapter ends with the conclusion and an outlook of the study in Section 4.8.

## 4.2 Preliminaries

In an optimization problem, there are mainly two perspectives, namely, formulation of the model and then finding its solution. Here, we present some useful definitions in connection of the study.

**Definition 4.1:** Optimization is a mathematical discipline which is concerned with finding maximum or minimum of objective functions with or without constraints. In the study of optimization, basically we need to optimize a real function $f(x_1, x_2, \ldots, x_n)$, of $n$ variables $x_1, x_2, \ldots, x_n$ with or without constraints.

In an OP, for modeling a physical system, if there be only one objective function, and the task is to obtain the optimal solution, then it is referred to as a single-objective optimization problem. The general form of single-objective optimization problem can be depicted as follows:

$$\text{minimize or maximize} \quad f(x_1, x_2, \ldots, x_n) \tag{4.1}$$

$$\text{subject to} \quad h(x_1, x_2, \ldots, x_n) \geq 0 \tag{4.2}$$

$$g(x_1, x_2, \ldots, x_n) \le 0 \qquad (4.3)$$

$$l(x_1, x_2, \ldots, x_n) = 0 \qquad (4.4)$$

$$(x_1, x_2, \ldots, x_n) \in F \subset \mathbb{R}^n, \qquad (4.5)$$

F is the feasible region.

**Definition 4.2:** In an OP, if both objective function $f$ and the constraints (4.2)–(4.5) are linear functions of the decision variables $x_1, x_2, \ldots, x_n$, then an OP is called a linear OP. Furthermore, if at least one of the constraints or the objective function is nonlinear, then the OP is called nonlinear OP.

**Definition 4.3:** When an OP is used for modeling a real-life problem which involves more than one objective function, the task to find the optimal solution is called a MOO problem.

The most general mathematical form of a MOO problem is depicted as follows:

$$\text{minimize or maximize} \quad f = f(f_1, f_2, \ldots, f_k)$$
$$\text{subject to the constraints (4.2)–(4.5)} \qquad (4.6)$$

where $f_1, f_2, \ldots, f_k$ are the objective functions containing the decision variables $x_1, x_2, \ldots, x_n$.

**Definition 4.4:** In a MOO problem, if all the objective functions $f_1, f_2, \ldots, f_k$ and the constraints (4.2)–(4.5) are linear functions in terms of decision variables $x_1, x_2, \ldots, x_n$, then the MOO problem is called a linear MOO problem. Furthermore, if at least one of the constraints or one of the objective functions becomes a nonlinear type, then the MOO problem is called a nonlinear MOO problem.

## 4.2.1 Concepts of solution

In a single-objective optimization problem, the "best" solution is defined in terms of "optimal solution" for which the value of the objective function is optimized satisfying the set of all feasible restrictions. In a MOO problem, the "best" solution is usually referred to as the "Pareto optimal solution." Here are some useful definitions related to the solution of a MOO problem.

**Definition 4.5 (Feasible Solution, FS):** A solution set $X = \{x : x \in \mathbb{R}^n\}$ is said to be a feasible solution to a MOO, if it satisfies all the constraints. A set $S$ consisting of all FSs is called a feasible solution set which lies in the space of action, where

$$S = \{x : x \in \mathbb{R}^n \quad \text{satisfying the constraints (4.2)–(4.5)}\}.$$

**Definition 4.6 (Optimal Solution):** An optimal solution of minimization problem MOO is a FS which gives the minimum value of each objective function simultaneously, i.e., if, $x^* \in S$ where $x^*$ is an optimal solution and $f_k(x^*) \le f_k(x)$, $k = 1, 2, \ldots, K$ for all $x^* \in S$.

In the space of objective function, the optimal solution is located within the boundary of the feasible space. Here, the optimal solution is also known as the inferior solution. Generally, there is no optimal solution to a MOO problem because the objective functions are conflicting in nature. In MOO with conflicting objective functions, the optimum solution corresponding to each objective function is obtained individually but the optimum solution of MOO problem reflecting optimum values of objective functions individually does not exist in general.

Optimal Compromise Solution: Compromise programming seeks the compromise solution among several objective functions of a MOO problem. The idea is based on the minimization of the distance between the ideal and the desired solutions.

**Definition 4.7:** The optimal compromise solution of a MOO is a solution $\underline{x} \in X$ which is preferred by the DM to all other solutions, taking into consideration all objectives contained in the several functions of the MOO problem.

It is generally accepted that an optimal compromise solution has to be an efficient solution according to the definition of an efficient solution. For a real-life practical problem, the complete solution (set of all efficient solutions) is not always necessary. We need only a procedure that finds an optimal compromise solution.

**Definition 4.8 (Pareto-optimal solution [Efficient solution]):** A feasible solution $x$ of a MOO problem is said to be a nondominated (noninferior) solution if there does not exist any other feasible solution $\underline{x}$ which dominates the solution obtained through $\underline{x}$. Therefore, for a nondominated solution, an increase in the value of any one objective function is not possible without decreases in the value of at least one other objective function. Mathematically, a solution $\underline{x} \in X$ is nondominated if there does not exist any $x \in X$, such that $f_k(x) \le f_k(\underline{x})$, $k = 1, 2, \ldots, K$; and $f_k(x) \ne f_k(\underline{x})$, at least one $k$.

**Fuzzy Programming (FP):** In real-life uncertain situations, the fuzzy set theory is an important and an effective tool to take into account for analyzing the MODM problem. Although the fuzzy set theory is rigorously used in the field of operations research as a tool for solving a MOO problem, here we intend to use a fuzzy set for accommodating real-life situations in a MOO problem through the fuzzy parameters.

**Definition 4.9 (Fuzzy set) [31]:** A fuzzy set $\tilde{A}$ is a pair $(A, \mu_{\tilde{A}})$ where $A$ is a crisp set that belongs to the universal set $X$ and $\mu_{\tilde{A}} : X \to [0,1]$ is a function, called a membership function.

**Fuzzy membership function:** Membership values are used in order to determine the degree of membership of the elements to the fuzzy set. The evaluation of a membership value is of critical importance in the application of fuzzy set theory in the field of engineering and science. The linear membership function is defined by two flexible points, such as upper and lower aspiration levels, or two bounds of tolerance intervals.

**Definition 4.10 (Membership function of triangular fuzzy number):** The membership function of a triangular fuzzy number $\tilde{A} = (a, b, c)$ is depicted as follows (Figure 4.2):

$$\mu_{\tilde{A}}(x) = \begin{cases} \dfrac{x-a}{b-a}, & \text{if } a \le x \le b \\[2mm] \dfrac{c-x}{c-b}, & \text{if } b \le x \le c \\[2mm] 0, & \text{if elsewhere} \end{cases}$$

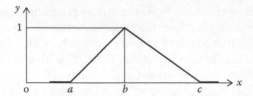

*Figure 4.2* Membership function of triangular fuzzy number.

In a usual sense, the goal of an objective function in an OP is a specified aspiration level determined by the DM. Historically, the main concept of GP was to minimize the deviation between the achievement goal and the achievement level. GP is also useful to optimize a MOO problem with goals to each of the objective functions.

**Definition 4.11:** The mathematical model of GP for solving a MOO problem can be considered in the following form:

**Model GP:**

$$\text{minimize} \quad \sum_{i=1}^{k} w_i \left| f_i(x) - g_i \right|$$

$$\text{subject to} \quad x \in F,$$

where $F$ is the feasible set and $w_i$ are the weights attached to the deviation of the achievement function. $f_i(x)$ is the $i$-th objective function of $i$-th goal, and $g_i$ is an aspiration level of the $i$-th goal. $\left| f_i(x) - g_i \right|$ represents the deviation of the $i$-th goal.

A modification of GP is provided and is denoted as weighted goal programming (WGP) which can be displayed in the following form:

**Model WGP:**

$$\text{minimize} \quad \sum_{i=1}^{k} w_i \left( d_i^+ + d_i^- \right)$$

$$\text{subject to} \quad f_i(x) - d_i^+ + d_i^- = g_i,$$

$$d_i^+ \geq 0, d_i^- \geq 0 \quad (i = 1, 2, \ldots, K),$$

$$x \in F,$$

where $d_i^+$ and $d_i^-$ are over and under achievements of the $i$-th goal, respectively. However, the conflicts of resources and the incompleteness of available information make it almost impossible for DMs to set the specific aspiration levels and choose a better decision. To overcome this situation, a revised multi-choice goal programming (RMCGP) approach was presented by Chang [30].

**Definition 4.12:** The mathematical model of RMCGP for solving a MOO problem can be defined as follows:

**Model RMCGP:**

$$\text{minimize} \quad \sum_{i=1}^{k} \left[ w_i \left( d_i^+ + d_i^- \right) + \alpha_i \left( e_i^+ + e_i^- \right) \right]$$

$$\text{subject to} \quad f_i(x) - d_i^+ + d_i^- = y_i \quad (i = 1, 2, \ldots, K),$$

$$y_i - e_i^+ + e_i^- = g_{i,\max} \text{ or } g_{i,\min} \quad (i = 1, 2, \ldots, K),$$

$$g_{i,\min} \le y_i \le g_{i,\max} \quad (i = 1, 2, \ldots, K),$$

$$d_i^+ \ge 0, d_i^- \ge 0, e_i^+ \ge 0, e_i^- \ge 0 \quad (i = 1, 2, \ldots, K),$$

$$x \in F.$$

Here $y_i$ is the continuous variable associated with the $i$-th goal which is restricted between the upper $(g_{i,\max})$ and lower $(g_{i,\min})$ bounds; $e_i^+$ and $e_i^-$ are positive and negative deviations attached to the $i$-th goal of $|y_i - g_{i,\max}|$; $\alpha_i$ is the weight attached to the sum of the deviations of $|y_i - g_{i,\max}|$; and other variables are defined as in WGP.

## 4.3   Mathematical model

The mathematical model of a transportation problem is as follows:

**Model 1**

$$\text{minimize} \quad Z = \sum_{i=1}^{m} \sum_{j=1}^{n} C_{ij} x_{ij}, \tag{4.7}$$

$$\text{subject to} \quad \sum_{j=1}^{n} x_{ij} \le a_i \quad (i = 1, 2, \ldots, m), \tag{4.8}$$

$$\sum_{i=1}^{m} x_{ij} \ge b_j \quad (j = 1, 2, \ldots, n), \tag{4.9}$$

$$x_{ij} \ge 0 \quad \forall \, i, j, \tag{4.10}$$

where $x_{ij}$ is the decision variable that represents the amount of goods delivered from $i$-th origin to $j$-th destination. $C_{ij}$ is the transportation cost per unit commodity. $a_i$ and $b_j$ are supply and demand at $i$-th origin and $j$-th destination, respectively, and $\sum_{i=1}^{m} a_i \ge \sum_{j=1}^{n} b_j$ is the feasibility condition.

A single-objective transportation problem does not serve to formulate all the real-life transportation problems. As, for example, if it is required to minimize cost of transportation, maximize profit, and minimize distance in a single TP, then it reduces to multi-objective ground. The transportation problem with multiple objective functions is considered as a MOTP. However, we deal with such kind of objective functions, which are conflicting and noncommensurable to each other involved in TP. A mathematical model of MOTP is as follows:

$$\text{minimize/maximize} \quad Z^t = \sum_{i=1}^{m} \sum_{j=1}^{n} C_{ij}^t x_{ij} \quad (t = 1, 2, \ldots, K)$$

subject to the constraints (4.8)–(4.10).

Here $C_{ij}^t = \left( C_{ij}^1, C_{ij}^2, \ldots, C_{ij}^K \right)$ is the cost per unit goods for each objective function when the total amount is transported from $i$-th origin to $j$-th destination.

## 4.4 Solution procedure

In this section, we introduce three well-known techniques for solving the MOTP: fuzzy programming, goal programming, and revised multi-choice programming. Thereafter, we propose a new algorithm using VAM to solve the MOTP.

### 4.4.1 Fuzzy programming

To solve the MOTP, we use the approach of fuzzy programming which reduces a multi-objective problem to a single-objective problem, then the single-objective problem is solved to find the compromise solution of the MOTP. The steps to be followed for converting a multi-objective problem to a single-objective problem are as follows:

**Step 1:** First, we solve each of the objective functions separately, under the demand and supply restrictions, and obtain optimal solutions of $K$ linear objective functions. Let $X_1^*, X_2^*, \ldots, X_K^*$ be the ideal solutions of the $K$ objectives $Z^t$, ($t = 1, 2, \ldots, K$).

**Step 2:** Each objective function is evaluated corresponding to the ideal solutions obtained in Step 1, and we formulate a pay-off matrix of order $K \times K$ as follows:
Table 4: Pay-off Matrix

$$\begin{bmatrix} Z^1(X_1^*) & Z^2(X_1^*) & \ldots & Z^K(X_1^*) \\ Z^1(X_2^*) & Z^2(X_2^*) & \ldots & Z^K(X_2^*) \\ Z^1(X_3^*) & Z^2(X_3^*) & \ldots & Z^K(X_3^*) \\ \vdots & \vdots & \ddots & \vdots \\ Z^1(X_K^*) & Z^2(X_K^*) & \ldots & Z^K(X_K^*) \end{bmatrix}$$

**Step 3:** When the $t$-th objective function is a minimizing type, obtain the lower bound $L_t$ (best solution) and upper bound $U_t$ (worst solution) corresponding to the $t$-th objective function. Then formulate the membership function using Zimmermann's [22] approach corresponding to each objective function $Z^t$ ($t = 1, 2, \ldots, K$) as follows:

$$\mu_{\tilde{A}}\left(Z^t(X)\right) = \begin{cases} 0, & \text{if } Z^t \geq U_t \\ \dfrac{U_t - Z^t(X)}{U_t - L_t}, & \text{if } L_t \leq Z^t(X) \leq U_t \\ 1, & \text{if } Z^t(X) \leq L_t \end{cases}$$

Again, if the $t$-th objective function is the maximizing type, obtain the lower bound $L_t$ (worst solution) and upper bound $U_t$ (best solution) corresponding to the $t$-th objective function. Then the membership function for the objective function $Z^t$ ($t = 1, 2, \ldots, K$) is formulated as follows:

$$\mu_{\tilde{A}}\left(Z^t(X)\right) = \begin{cases} 0, & \text{if } Z^t(X) \leq L_t \\ \dfrac{Z^t(X) - L_t}{U_t - L_t}, & \text{if } L_t \leq Z^t(X) \leq U_t \\ 1, & \text{if } Z^t \geq U_t \end{cases}$$

**Step 4**: Introduce an auxiliary variable $\lambda$ and formulate an equivalent fuzzy linear programming problem in the following form:

$$\text{maximize} \quad \lambda$$

$$\text{subject to} \quad \lambda \leq \mu_{\tilde{A}}\left(Z^t(X)\right) \quad (t = 1, 2, \ldots, K),$$

and the constraints (4.8)–(4.10).

Here, $\mu_{\tilde{A}}\left(Z^t(X)\right)$ is the membership function of the $t$-th objective function for ($t = 1, 2, \ldots, K$) as given in Step 3.

**Step 5**: Solve the crisp model obtained in Step 4, and derive the optimal compromise solution.

## 4.4.2   Goal programming

Goal programming is a useful approach for solving MOTP. Here we discuss the goal programming approach for solving MOTP (see Model 1A).

In this procedure, DM needs to decide goals corresponding to each of the objective functions. Consider $g_{i,\max}$ and $g_{i,\min}$ as the maximum and minimum aspiration values of the $t$-th objective function in MOTP. Consider $d_i^+$ and $d_i^-$ as positive and negative deviations corresponding to the $t$-th objective function. Then the mathematical model of GP is formulated as follows:

**Model 1A**

$$\text{minimize} \quad \sum_{t=1}^{k} w_t\left(d_t^+ + d_t^-\right) \tag{4.11}$$

$$\text{subject to} \quad f_t(x) - d_t^+ + d_t^- = y_t \quad (t = 1, 2, \ldots, K), \tag{4.12}$$

$$g_{t,\min} \le y_t \le g_{t,\max} \quad (t = 1,2,\ldots,K), \tag{4.13}$$

$$d_t^+ \ge 0, \ d_t^- \ge 0 \quad (t = 1,2,\ldots,K), \tag{4.14}$$

the constraints $(4.8)-(4.10)$.

### 4.4.3   Revised multi-choice programming

A modification of GP produces the RMCGP to solve the MOTP. Under the prediction of goals of objective functions, the RMCGP model is described in the following model:

**Model 1B**

$$\text{minimize} \quad \sum_{t=1}^{k} \left[ w_t \left( d_t^+ + d_t^- \right) + \alpha_t \left( e_t^+ + e_t^- \right) \right] \tag{4.15}$$

$$\text{subject to} \quad f_t(x) - d_t^+ + d_t^- = y_t \quad (t = 1,2,\ldots,K), \tag{4.16}$$

$$y_t - e_t^+ + e_t^- = g_{t,\min} \text{ or } g_{t,\max} \quad (t = 1,2,\ldots,K), \tag{4.17}$$

$$g_{t,\min} \le y_t \quad (t = 1,2,\ldots,K), \tag{4.18}$$

$$d_t^+ \ge 0, \ d_t^- \ge 0, e_t^+ \ge 0, e_t^- \ge 0 \quad (t = 1,2,\ldots,K), \tag{4.19}$$

the constraints $(4.8)-(4.10)$.

where $y_t$ is a continuous variable that lies between upper $(g_{t,\max})$ and lower $(g_{t,\min})$ bounds, and it is denoted as the aspiration level of the $t$-th objective function. Again $e_t^-$ and $e_t^+$ are negative and positive deviations attached to the $t$-th goal of $|y_t - g_{t,\max}|$, and $\alpha_t$ is taken as the weight attached with the sum of the deviations of $|y_t - g_{t,\max}|$.

### 4.4.4   Vogel approximation method

Here, we present a new approach to find a compromise solution of the MOTP. Especially, a new approach is defined to reduce the MOTP to a single-objective LPP. The cost parameters of different objective functions may be in different scales, and we propose an approach that reduces each of the cost parameters into a real value under the same unit scale. Following are the steps of the new algorithm:

- **Step 1:** Consider a MOTP with $K$ objective functions $Z^t$ $(t = 1,2,\ldots,K)$.
- **Step 2:** Assume $Z^p$ to be an objective function of maximization type. If $C_{i'j'}^p$ is the maximum value of the cost parameter, then take it by the value 1. After that each of the cost parameters is divided by $C_{i'j'}^p$ which gives the value $C_{ij}^p / C_{i'j'}^p = r_{ij}^p$ (say) $\forall i, j$ with $0 \le r_{ij}^p \le 1$. It means the allocation possibly made at the values of $r_{ij}^p$ where it is maximum. Then each of the cost parameters in the objective function of maximization type is reduced in the same scale.

  Again, choose $Z^q$ to be an objective function of minimization type. If $C_{i'j'}^q$ is the maximum value of cost parameter, then assume it by the value 1. After that each of the costs is assigned with the value $C_{ij}^q / C_{i'j'}^q = s_{ij}^q$ (say) $\forall i, j$ with $-1 \le s_{ij}^q \le 0$. As the objective function is a minimization type, so the allocations are made at the nodes where values of $s_{ij}^q$ are larger. Then each of the cost parameters in the objective functions of minimization type is reduced in the same scale.

- **Step 3**: Thereafter, consider the weights $w_t$ (normalized) for the objective functions $Z^t$ for all $t$.
- **Step 4**: Formulate the objective function, i.e., maximize $Z = \sum_{i=1}^{m} \sum_{j=1}^{n} \left( \sum_{t \in T'} w_t r_{ij}^t + \sum_{t \in T''} w_t s_{ij}^t \right) x_{ij}$ under the constraints (4.8)–(4.10) described in MOTP. Here $T'$ and $T''$ are the sets of objective functions with maximization and minimization types, respectively.
- **Step 5**: The formulated model in Step 4 is a LPP; solve the LPP by simplex algorithm.
- **Step 6**: Obtain the optimum solution of the LPP from Step 5 and notice the allocations made at the cells. Let $X_{ij}$ be the optimum solution. Then the compromise solution of the $t$-th objective function is $Z^t = \sum_{i=1}^{m} \sum_{j=1}^{m} C_{ij}^t X_{ij}$ $\forall t$.
- **Step 7**: Stop.

The presented algorithm is more effective for solving the MOTP based on the following reasons:

- In the proposed algorithm, the compromise solution of MOTP is introduced without using the solution procedure of MOTP like fuzzy programming, goal programming, revised multi-choice goal programming, etc.
- A bigger form of real-world MOO problem is made into a simple form with the single-objective optimization problem, which is easy to solve and to produce the solution with less computational burden.
- An important factor of the proposed algorithm is that there is not required any auxiliary variable, any additional constraint to reduce the problem into a TP, and it is easy to solve by VAM.
- The weight is an important factor for solving MOTP which is considered in the proposed algorithm. To increase the impact of any objective function, DM should change the weights, but the weights must be normalized with a better impact for the results.

## 4.4.5 Merits and demerits

Here, four approaches are considered to solve the MOTP for finding compromise solutions. Practically, it is not possible to choose a better solution or comparison among the optimal compromise solutions. But from the technical point of view, we are able to discuss the merits and demerits of the approaches and then classify the appropriateness for solving the MOTP among them. At first, let us see about the fuzzy programming. In fuzzy programming, it is required to solve several objective functions individually, and thereafter, the single-objective function is constructed to find the optimal compromise solution. Altogether, in this technique, we need to solve $(K + 1)$ objective functions. So, the technique is laborious and time consuming.

For GP, we need to consider the goals corresponding to the objective functions, and then a single objective function is formulated to obtain a compromise solution. But, the main drawback in GP is how to construct the goals corresponding to the objective function. Based on the priority, the goals are predicted by the DM regarding the appropriateness in the decision-making process.

Now for RMCGP, the drawbacks regarding the consideration of goals in RMCGP are similar to the GP. Again, the number of auxiliary variables and constraints in RMCGP is two times the number of those used in GP. But, it is proved that the RMCGP produces a better result than GP. The following theorem is presented in connection with GP and RMCGP.

**Theorem 4.1:** Solution of MOTP by RMCGP produces a better result than GP.

**Proof:** The mathematical model of GP for solving a MOTP is depicted as follows:

**GP**

$$\text{minimize} \quad \sum_{i=1}^{k} w_i \left| Z^i(x) - g_i \right| \tag{4.20}$$

$$\text{subject to } x \in F. \tag{4.21}$$

Let us consider $d_i^+ = Z^i(x) - g_i$, if $Z^i(x) \geq g_i = 0$, and $d_i^+ = 0$ otherwise ($i = 1, 2, \ldots, K$). Also, we take $d_i^- = g_i - Z^i(x)$, if $Z^i(x) \leq g_i = 0$, and $d_i^- = 0$ otherwise ($i = 1, 2, \ldots, K$). Therefore, $Z^i(x) - g_i = d_i^+ - d_i^-$, which produces the relation $Z^i(x) - d_i^+ + d_i^- = g_i$. Furthermore, $\left| Z^i(x) - g_i \right| = d_i^+ + d_i^-$.

Thus, the modification on the model GP reduces to the model WGP as follows:

**WGP**

$$\text{minimize} \quad \sum_{i=1}^{k} w_i \left( d_i^+ + d_i^- \right)$$

$$\text{subject to} \quad Z^i(x) - d_i^+ + d_i^- = g_i,$$

$$d_i^+ \geq 0, \quad d_i^- \geq 0 \quad (i = 1, 2, \ldots, K),$$

$$x \in F.$$

From the above discussion, we observe that the mathematical model GP is a function of the weights and goals along with the decision variables, whereas the objective function in the mathematical model of WGP contains goal deviations ($d_i^+$) and ($d_i^-$) as variable. It is observed that both models GP and WGP produce the same result, but the WGP model is easier to tackle than GP as its objective function includes a minimum number of variables compared to GP.

If the goal ($g_i$) for the $i$-th objective function is not a real value but it is taken as an interval $[g_{i,\min}, g_{i,\max}]$, then to obtain a better solution of a maximization-type objective function, $g_i$ should attain the maximum value of the range, and for objective function of a minimization type, it should take its minimum value of the specified range.

Thereafter, we introduce a new variable $y_i$ in the model WGP as $Z^i(x) - d_i^+ + d_i^- = y_i$, and two deviation variables such as $e_i^+$ and $e_i^-$ are similar to $d_i^+$ and $d_i^-$, respectively, along with the constraint $y_i - e_i^+ + e_i^- = g_{i,\min}$ or $g_{i,\max}$. Then, the objective function is converted into the form as minimize $\sum_{i=1}^{k} \left[ w_i \left( d_i^+ + d_i^- \right) + \alpha_i \left( e_i^+ + e_i^- \right) \right]$, where $\alpha_i$ are weights corresponding to the goal deviations. Using this objective function, we construct the model RMCGP. The RMCGP minimizes the deviations ($d_i^+ + d_i^-$) and ($e_i^+ + e_i^-$), whereas the GP minimizes the deviation of the value of objective function, i.e., $\sum_{i=1}^{k} w_i (d_i^+ + d_i^-)$. Thus, in fact of minimizing the objective function in the RMCGP model, the second part of objective function $\sum_{i=1}^{k} \alpha_i \left( e_i^+ + e_i^- \right)$ is also minimized. This implies that the value of $y_i$ tends to $g_{i,max}$ for a maximizing-type objective function, and $y_i$ tends to $g_{i,min}$ for a minimizing-type objective function. In WGP or GP, the goal deviations are only minimized, which does not consider the type of objective function. Subsequently, the additional variables $e_i^+$ and $e_i^-$ tackle the situation that minimizes the deviations according to the type of the objective function. Hence, we establish that RMCGP produces a better result compared to WGP or GP models.

Furthermore, in the mathematical model of RMCGP, if we consider the goal deviations $e_i^+$ and $e_i^-$ as 0, then, it reduces to the form of a WGP. Also, model WGP is a modification of the model GP. Therefore, it is cleared that the solution of RMCGP is better than the solution of WGP and GP. Hence, the arguments evince the proof of the theorem.

The efficiency of our proposed algorithm is presented followed by the algorithm, which establishes a better utility of our proposed algorithm in comparison to FP, GP, and RMCGP.

## 4.5    Numerical example

A rice merchant has three warehouses at three locations O1, O2, and O3. He delivers rice into three different markets D1, D2, and D3. In the warehouses, there are different capacities of rices with different prices in different locations. It is noticed by the merchant that the maximum supplying capacity of rice at the origins O1, O2, and O3 are 1100, 1250, and 1150 kg, respectively. Furthermore, the minimum demands of rice at the destinations D1, D2, and D3 are 1150, 1100, and 1225 kg, respectively.

The merchant wishes to deliver rice to the destinations keeping in mind that he has to minimize transportation cost but maximize the profit with the consideration that the transportation costs are paid by the customers. Basically every customer wishes to minimize transportation cost in time of purchasing rice, whereas the merchant wishes to maximize his profit. Conflicting situations occur, and the problem becomes a MOTP with two objective functions. We present the data regarding the transportation parameters in Tables 4.1 and 4.2.

Considering the aforementioned data, the following MOTP is formulated:

*Table 4.1* Transportation cost per unit kilogram rice (in $)

|     | D1  | D2  | D3  |
| --- | --- | --- | --- |
| O1  | 3.5 | 4.1 | 4.5 |
| O2  | 5.5 | 4.5 | 5.0 |
| O3  | 4.5 | 4.2 | 4.0 |

*Table 4.2* Profit per unit kilogram rice (in $)

|     | D1  | D2  | D3  |
| --- | --- | --- | --- |
| O1  | 2.5 | 1.5 | 2.0 |
| O2  | 2.1 | 1.8 | 1.5 |
| O3  | 1.5 | 2.2 | 1.9 |

**Model P**

$$\text{minimize} \quad Z^1 = 3.5x_{11} + 4.1x_{12} + 4.5x_{13} + 5.5x_{21} + 4.5x_{22}$$

$$+ 5.0x_{23} + 4.5x_{31} + 4.2x_{32} + 4.0x_{33} \tag{4.22}$$

$$\text{maximize} \quad Z^1 = 2.5x_{11} + 1.5x_{12} + 2.0x_{13} + 2.1x_{21} + 1.8x_{22}$$

$$+ 1.5x_{23} + 1.5x_{31} + 2.2x_{32} + 1.9x_{33} \tag{4.23}$$

$$\text{subject to} \quad x_{11} + x_{12} + x_{13} \le 1100 \tag{4.24}$$

$$x_{21} + x_{22} + x_{23} \le 1250 \tag{4.25}$$

$$x_{31} + x_{32} + x_{33} \le 1150 \tag{4.26}$$

$$x_{11} + x_{21} + x_{31} \ge 1150 \tag{4.27}$$

$$x_{12} + x_{22} + x_{32} \ge 1100 \tag{4.28}$$

$$x_{13} + x_{23} + x_{33} \ge 1225 \tag{4.29}$$

$$x_{ij} \ge 0 \quad \forall i \text{ and } j. \tag{4.30}$$

### 4.5.1   Fuzzy programming

The ideal solutions obtained by solving the objective functions $Z_1$ and $Z_2$ separately subject to the constraints (4.24)–(4.30) are given as $[X_1^*] = [1100,0,0,50,1100,75,0,0,1150]$, $[X_2^*] = [0,0,1100,1175,0,75,0,1100,50]$. Based on the ideal solutions, we formulate the pay-off matrix, which is shown in Table 4.3.

Using Table 4.3, we formulate the following membership functions corresponding to each objective function of the proposed problem as

$$\mu\left(Z^1(X)\right) = \begin{cases} 0, & \text{if } Z^1 \ge 16{,}607.5 \\ \dfrac{16{,}607.5 - Z^1(X)}{16{,}607.5 - 14{,}050}, & \text{if } 14{,}050 \le Z^1(X) \le 16{,}607.5 \\ 1, & \text{if } Z^1(X) \le 14{,}050, \end{cases}$$

$$\mu\left(Z^2(X)\right) = \begin{cases} 0, & \text{if } Z^2(X) \le 7132.5 \\ \dfrac{Z^2(X) - 7132.5}{7295 - 7132.5}, & \text{if } 7132.5 \le Z^2(X) \le 7295 \\ 1, & \text{if } Z^2 \ge 7295. \end{cases}$$

*Table 4.3* Pay-off matrix

|          | $Z^1$     | $Z^2$     |
|----------|-----------|-----------|
| $X_1^*$  | 14,050    | 16,607.5  |
| $X_2^*$  | 7,132.5   | 7,295     |

Using the procedure described in Section 4.4.1, finally we design the following model as:

**Model P1**

maximize $\lambda$

subject to $1957.5\lambda \le 16{,}607.5 - (3.5x_{11} + 4.1x_{12} + 4.5x_{13} + 5.5x_{21} + 4.5x_{22} + 5.0x_{23}$

$+ 4.5x_{31} + 4.2x_{32} + 4.0x_{33}), 162.5\lambda \le (2.5x_{11} + 1.5x_{12} + 2.0x_{13} + 2.1x_{21} + 1.8x_{22}$

$+ 1.5x_{23} + 1.5x_{31} + 2.2x_{32} + 1.9x_{33}) - 7132.5,$

and the constraints (4.24)–(4.30).

Model P1 is an LPP and using LINGO10 software, we obtain the compromise optimal solution as follows:

$$[X^*] = [559.53, 0, 540.47, 615.47, 634.53, 0, 0, 465.47, 684.53].$$

The minimum value of objective function $Z^1(X^*) = \$15{,}324.02$, and the maximum value of objective function $Z^2(X^*) = \$7239.05$. The value of the aspiration level is $\lambda = 0.66$.

## 4.5.2 Goal programming

According to the market situations, the DM has some knowledge of approximate profit connecting with the optimum transportation cost. In that situation, the DM wishes to solve the MOTP in such a way that the transportation cost belongs to the interval [15,000, 16,000] (lesser value is preferred by DM) and the profit belongs to [7100, 7500] (greater value is preferred by DM). So, it is required to schedule the amount of rice to be transported satisfying the predetermined goals assumed by the DM.

To achieve the goals in Model P, we formulate the model by goal programming in the following way: Assume the deviations of goal 1 and goal 2 are 1000 and 400, respectively. Consider the weights $w_1 = 1/1000$ and $w_2 = 1/400$ to Model P, then Model P reduces to Model P2 as follows:

**Model P2**

minimize $\dfrac{1}{1000}(d_1^+ + d_1^-) + \dfrac{1}{400}(d_2^+ + d_2^-)$

subject to $3.5x_{11} + 4.1x_{12} + 4.5x_{13} + 5.5x_{21} + 4.5x_{22} + 5.0x_{23} + 4.5x_{31} + 4.2x_{32}$

$+ 4.0x_{33} - d_1^+ + d_1^- = y_1, \ 15{,}000 \le y_1 \le 16{,}000,$

$2.5x_{11} + 1.5x_{12} + 2.0x_{13} + 2.1x_{21} + 1.8x_{22} + 1.5x_{23} + 1.5x_{31} + 2.2x_{32}$

$+ 1.9x_{33} - d_2^+ + d_2^- = y_2, \ 7100 \le y_2 \le 7500,$

$d_t^+ \ge 0, \ d_t^- \ge 0, \quad t = 1, 2,$

and the constraints (4.24)–(4.30).

The following compromise solution is obtained by solving Model P2:

[$X^*$] = [618.66, 0, 481.34, 450.70, 774.3, 0, 80.63, 325.7, 743.66]. The values of the objective functions are $Z^1(X^*) = \$15{,}000$ and $Z^2(X^*) = \$7100$, respectively.

### 4.5.3 Revised multi-choice goal programming

Assume the same deviations and weights as taken in Model P2 and construct the following Model P3 under RMCGP for solving MOTP.

**Model P3**

minimize    $\frac{1}{1000}\left(d_1^+ + d_1^-\right) + \frac{1}{400}\left(d_2^+ + d_2^-\right) + \frac{1}{1000}\left(e_1^+ + e_1^-\right) + \frac{1}{400}\left(e_2^+ + e_2^-\right)$

subject to   $3.5x_{11} + 4.1x_{12} + 4.5x_{13} + 5.5x_{21} + 4.5x_{22} + 5.0x_{23} + 4.5x_{31} + 4.2x_{32}$

$\qquad\qquad + 4.0x_{33} - d_1^+ + d_1^- = y_1,\ y_1 - e_1^+ + e_1^- = 15,000,$

$\qquad 15,000 \le y_1 \le 16,000,$

$\qquad 2.5x_{11} + 1.5x_{12} + 2.0x_{13} + 2.1x_{21} + 1.8x_{22} + 1.5x_{23} + 1.5x_{31} + 2.2x_{32} + 1.9x_{33} - d_2^+ + d_2^- = y_2,$

$\qquad y_2 - e_2^+ + e_2^- = 7500,$

$\qquad 7100 \le y_2 \le 7500,$

$\qquad d_t^+ \ge 0,\ d_t^- \ge 0,\quad e_t^+ \ge 0, e_t^- \ge 0,\quad t = 1,2,$

and the constraints (4.24)–(4.30).

The following compromise solution is obtained by solving Model P3:
    $[X^*] = [706.82, 0, 393.18, 468{:}18;\ 781.82, 0, 0, 318.18, 831.82]$. The values of the objective functions are $Z^1(X^*) = \$15,000$ and $Z^2(X^*) = \$7224.32$.

### 4.5.4 Vogel approximation method

According to the proposed algorithm, first we reduce the transportation parameters in normal form, which are shown in Tables 4.4 and 4.5. To do this, we divide each of the cost

Table 4.4 Normalized transportation cost per unit kilogram rice (in $)

|     | D1    | D2    | D3    |
|-----|-------|-------|-------|
| O1  | 0.636 | 0.745 | 0.816 |
| O2  | 1     | 0.818 | 0.909 |
| O3  | 0.818 | 0.764 | 0.727 |

Table 4.5 Normalized profit per unit kilogram rice (in $)

|     | D1   | D2   | D3   |
|-----|------|------|------|
| O1  | 1    | 0.6  | 0.8  |
| O2  | 0.84 | 0.72 | 0.6  |
| O3  | 0.6  | 0.88 | 0.76 |

*Table 4.6* Reduced value of transportation parameter

|  | D1 | D2 | D3 |
|---|---|---|---|
| O1 | −0.364 | 0.145 | 0.018 |
| O2 | 0.16 | 0.098 | 0.309 |
| O3 | 0.218 | −0.116 | −0.033 |

*Table 4.7* Optimal compromise solution

| Weight for $Z^1$ | Weight for $Z^2$ | Solution of $Z^1$ | Solution of $Z^2$ |
|---|---|---|---|
| 0.50 | 0.50 | 14,050 | 7,132.5 |
| 0.45 | 0.55 | 14,162.5 | 7,177.5 |
| 0.40 | 0.60 | 14,162.5 | 7,177.5 |
| 0.35 | 0.65 | 14,187.5 | 7,185 |
| 0.30 | 0.70 | 14,187.5 | 7,185 |
| 0.20 | 0.80 | 14,187.5 | 7,185 |
| 0.10 | 0.90 | 14,300 | 7,192.5 |
| 0.05 | 0.95 | 16,555.5 | 7,295 |
| 0.00 | 1.0 | 16,555.5 | 7,295 |

components of Tables 4.1 and 4.2 by the maximum cost component to each of them, and they are 5.5 and 2.5, respectively.

According to the proposed algorithm, if we take equal weights to the objective functions, then we find the transportation cost which is shown in Table 4.6 in normalized form corresponding to the equivalent single-objective function for solving the proposed problem.

From Table 4.6, we see that some transportation parameters take negative value, we add the magnitude of the most negative value to each of the cost parameters in Table 4.6 and solve it by VAM to get the optimal allocation in the transportation cell. Then the following compromise solution is obtained: $[X^*] = [1100, 0, 0, 50, 1100, 75, 0, 0, 1150]$. Finally, the values of the objective functions are $Z^1(X^*) = \$14,050$ and $Z^2(X^*) = \$7132.5$. This optimal compromise solution is the best solution preferred by buyer.

In a similar way, to choose different weights for the objective functions, we prepare Table 4.7 which contains optimal compromise solutions.

From Table 4.7, the DM can choose any one of the optimal compromise solutions, and we present the best solutions preferred by both merchant and buyer. According to the choice of the DM, he may take the solution $Z^1(X^*) = \$14,300$ and $Z^2(X^*) = \$7192.5$ corresponding to the weights 0.1 and 0.9, respectively, which is better for both merchant and buyer, not dominating any one by the other.

## 4.6    Comparison

According to the obtained solutions of formulated Model P by FP, GP, RMCGP, and VAM, it is clear that the algorithm using VAM produces a better solution than FP, GP, and RMCGP. Also, we have seen that there is no need for any auxiliary variable in VAM which is necessary in FP, GP, and RMCGP. In this regard, we may say that the proposed algorithm is more effective with less computation burden for solving MOTP. The mathematical model of GP is a special structure of RMCGP because the value of $\alpha_i = 0$ for all $i$ in RMCGP which produces GP. GP tries to optimize the goal values but not prefer the goals properly for maximization or minimization problems, whereas RMCGP treats these goals as the DM's choices. Also, in GP or RMCGP one of the most important drawbacks is how to select the goals. There may be a situation, if the goal is not selected in proper way, in which the solution is infeasible. If, for example, the DM selects the goals [12,000, 13,000] (lesser value is preferred by DM) and the profit belongs to [7400, 7600] (greater value is preferred by DM), then we cannot find any optimal compromise solution from both the GP and the RMCGP.

Again in FP, we have solved the objective functions separately and to form a pay-off matrix; and finally a single-objective function is derived and solved to find the optimal compromise solution. During the process, altogether, we have solved three objective functions to obtain an optimum solution. Also, we have used two additional constraints and one auxiliary variable to solve the MOTP. So, this approach is laborious to solve MOTP. In addition to that, it is seen from our proposed example that the solution of MOTP by FP does not depend on the expected solution by the DM, if there is to be any kind of expectation for optimum values of objective functions. That is why, in most of the real-life decision-making problems, the FP is less important to produce a more optimal compromise solution.

Finally, in the proposed approach through VAM, the obtained optimal compromise solution is better than the solutions of FP, GP, and RMCGP. Furthermore, there is no need to use any auxiliary variable or any additional constraints to solve the MOTP by the proposed algorithm. It is clear that the set of all normalized weights $w_i$ produces a set of optimal compromise solutions of the MOTP. The better optimal compromise solution is one of the compromise solution picked by the DM. Here, we derive a better solution compared to the solutions obtained by FP, GP, and RMCGP.

## 4.7 Conclusion and future study

This chapter has explored the study of MOTP under the approaches such as FP, GP, RMCGP, and VAM. FP, GP, and RMCGP have been well-known methods to formulate the mathematical model and solve multi-objective decision-making problems which are discussed in this chapter. Contrary to existing approaches, here we have presented a new algorithm to solve MOTP using VAM. Basically, the optimization models of MOTP through FP, GP, and RMCGP have been carried out by the help of software for compromise solution. But our presented algorithm has been good enough to solve MOTP without any help from mathematical software. Again, in this chapter, we presented a comparison among the solution approaches along with the merits and demerits of the proposed approaches.

In future study, multi-objective transportation planning should be integrated in the different areas of study, such as networks, stations, user information, and fare payment systems. Again, the proposed model of MOTP can be used in the selection of modes in a variety of transportation improvement policies, such as mobility management strategies, pricing reforms, and smart growth land use policies. In addition, the proposed study can be implemented in different uncertain environments to accommodate real-life situations for selecting optimum modes of transportation.

## Acknowledgment

The author Gurupada Maity acknowledges the University Grants Commission of India for supporting the financial grant to carry on this research work under JRF(UGC) scheme: Sanctioned letter number [F.17-130/1998(SA-I)] dated 26/06/2014.

# References

1. F.L. Hitchcock, The distribution of a product from several sources to numerous localities, *Journal of Mathematics and Physics* 20 (1941) 224–230.
2. T.C. Koopmans, Optimum utilization of the transportation system, *Econometrica* 17 (1949) 136–146.
3. A. Ebrahimnejad, An improved approach for solving fuzzy transportation problem with triangular fuzzy numbers, *Journal of Intelligent and Fuzzy Systems* 29(2) (2015) 963–974.
4. A. Kaur and A. Kumar, A new method for solving fuzzy transportation problems using ranking function, *Applied Mathematical Modelling* 35(12) (2011) 5652–5661.
5. D.R. Mahapatra, S.K. Roy and M.P. Biswal, Multi-choice stochastic transportation problem involving extreme value distribution, *Applied Mathematical Modelling* 37(4) (2013) 2230–2240.
6. G. Maity and S.K. Roy, Solving multi-objective transportation problem with interval goal using utility function approach, *International Journal of Operational Research* 27(4) (2016) 513–529.
7. G. Maity and S.K. Roy, Solving multi-choice multi-objective transportation problem: A utility function approach, *Journal of Uncertainty Analysis and Applications* (2014) DOI:10.1186/2195-5468-2-11.
8. G. Maity, S.K. Roy and J.L. Verdegay, Multi-objective transportation problem with cost reliability under uncertain environment, *International Journal of Computational Intelligence Systems* 9(5) (2016) 839–849.
9. S. Midya and S.K. Roy, Single-sink, fixed-charge, multi-objective, multi-index stochastic transportation problem, *American Journal of Mathematics and Management Sciences* 33 (2014) 300–314.
10. S. Midya and S.K. Roy, Analysis of interval programming in different environments and its application to fixed charge transportation problem, *Discrete Mathematics, Algorithms and Applications* 9(3) (2017) 1750040, 17 pages.
11. S.K. Roy, Multi-choice stochastic transportation problem involving Weibull distribution, *International Journal of Operational Research* 21(1) (2014) 38–58.
12. S.K. Roy and G. Maity, Minimizing cost and time through single objective function in multi-choice interval valued transportation problem, *Journal of Intelligent and Fuzzy Systems* 32(3) (2017) 1697–1709.
13. S.K. Roy, G. Maity and G.W. Weber, Multi-objective two-stage grey transportation problem using utility function with goals, *Central European Journal of Operations Research* 25(2) (2017) 417–439.
14. A. Charnes and W.W. Cooper, *Management Model and Industrial Application of Linear Programming*, 1, Wiley: New York (1961).
15. G. Maity and S.K. Roy, Solving a multi-objective transportation problem with nonlinear cost and multi-choice demand, *International Journal of Management Science and Engineering Management* 11(1) (2016) 62–70.
16. F.A.E.W. Waiel, A multi-objective transportation problem under fuzziness, *Fuzzy Sets and Systems* 117(1) (2001) 27–33.
17. S.K. Roy, G. Maity, G.W. Weber and S.Z. Alparslan Gök, Conic scalarization approach to solve multi-choice multi-objective transportation problem with interval goal, *Annals of Operations Research* 253(1) (2017) 599–620.
18. A. Kumar, S. Pant, M. Ram and S.B. Sing, On solving complex reliability optimization problem using multi-objective Particle Swarm optimization, *Mathematics Applied to Engineering*, Academic Press, (2017) 115–131.
19. A. Kumar, S. Pant and M. Ram, System reliability optimization using grey wolf optimizer algorithm, *Quality and Reliability Engineering International*, John Wiley & Sons 33 (2017) 1327–1335.
20. S. Pant, A. Kumar, S.B. Sing and M. Ram, A modified Particle Swarm optimization algorithm for nonlinear optimization, *Nonlinear Studies* 24(1) (2017) 127–138.
21. S. Pant, A. Kumar and M. Ram, Reliability optimization: A particle swarm approach, In: Ram M., and Davim J. (eds), *Advances in Reliability and System Engineering, Management and Industrial Engineering*, Springer, Cham,
22. H.J. Zimmermann, Fuzzy programming and linear programming with several objective functions, *Fuzzy Sets and Systems* 1 (1978) 45–55.

23. E.L. Hannan, An assessment of some of the criticisms of goal programming, *Computers & Operations Research* 12 (1985) 525–541.
24. J.P. Ignizio, *Goal Programming and Extensions*, Lexington Books: Lexington, MA (1976).
25. M. Tamiz, D.F. Jones and C. Romero, Goal programming for decision making: An overview of the current state-of-the-art, *European Journal of Operational Research* 111 (1998) 569–581.
26. C. Romero, A general structure of achievement function for a goal programming model, *European Journal of Operational Research* 153 (2004) 675–686.
27. C.N. Liao, Formulating the multi-segment goal programming, *Computers and Industrial Engineering* 56 (2009) 138–141.
28. B.B. Tabrizi, K. Shahanaghi and M.S. Jabalameli, Fuzzy multi-choice goal programming, *Applied Mathematical Modelling* 36 (2012) 1415–1420.
29. C.T. Chang, Multi-choice goal programming, *Omega* 35 (2007) 389–396.
30. C.T. Chang, Revised multi-choice goal programming, *Applied Mathematical Modelling* 32 (2008) 2587–2595.
31. L.A. Zadeh, Fuzzy sets, *Information and Control* 8 (1965) 338–353.

## chapter five

# An application of dual-response surface optimization methodology to improve the yield of pulp cooking process

**Boby John and K.K. Chowdhury**
*Indian Statistical Institute*

## Contents

## 5.1   Introduction

Many modern processes are reasonably complex and have multiple output characteristics. For example, a heat treatment process like induction hardening needs to be executed to meet the requirements on surface hardness and case depth. Even an agile software development process needs to be executed to simultaneously meet the goals set on performance characteristics like sprint productivity, spring velocity, defect density, etc. (John et al., 2017). In such a scenario, the process engineers need to execute the processes in such a way to meet the customer requirements on multiple characteristics. In other words, the engineers need to identify an optimum setting of process control factors, which would simultaneously optimize multiple output characteristics. This can be done using the application of simultaneous optimization of multiple characteristics methodology. A lot of research has been carried out in the past on simultaneous optimization of output characteristics, and many approaches have been proposed. The important among them are Derringer's desirability function approach, Taguchi's loss function approach, dual-response surface methodology and fuzzy logic–based approach. In this chapter, the authors present a case study on simultaneous optimization of multiple output characteristics of the pulp cooking process. The methodology used for simultaneous optimization is dual-response surface methodology.

This study is carried out at an organization manufacturing rayon grade pulp. The rayon grade pulp is the raw material for manufacturing viscous staple fiber. The viscous staple fiber is used for making clothes. The pulp cooking process is an important step in the manufacturing of rayon grade pulp. The pulp is the cellulose component of the wood. The cellulose is separated from other components and impurities of wood by cooking the wood chips in a highly pressurized chamber followed by multiple stages of washing and chemical treatments.

The company produces approximately 210 tons of pulp daily and sells at a price of Indian Rupees 28,000 per ton of pulp. Even a small increase in pulp yield can have huge economic benefits for the organization. The yield of the pulp cooking process is defined as

$$y = \frac{w_p}{w_c} \times 100 \qquad (5.1)$$

where $y$ is the pulp yield, $w_p$ is the weight of pulp produced, and $w_c$ is the weight of the wood chips loaded. Unfortunately, the pulp yield cannot be increased indefinitely as it will adversely affect the pulp viscosity. The pulp with viscosity beyond 52 centipoises (cp) is graded as low quality. One centipoise is equal to one millipascal-second. To quantify the current status of the pulp cooking process, the data on yield and viscosity of the past twenty batches are collected. The collected data are given in Table 5.1.

The descriptive summary of the pulp yield is given in Figure 5.1.

Figure 5.1 shows that the average pulp yield per batch is only 34.027. So there is a lot of scope for improvement. Figure 5.1 also revealed that the yield is normally distributed,

*Table 5.1* Yield and viscosity data of pulp cooking process

| Batch number | Yield | Viscosity |
| --- | --- | --- |
| 1 | 33.8 | 51.40 |
| 2 | 34.0 | 50.76 |
| 3 | 34.0 | 52.24 |
| 4 | 33.9 | 51.11 |
| 5 | 34.0 | 52.24 |
| 6 | 34.2 | 49.87 |
| 7 | 33.9 | 51.12 |
| 8 | 34.0 | 52.05 |
| 9 | 34.2 | 51.46 |
| 10 | 33.9 | 50.64 |
| 11 | 33.7 | 51.76 |
| 12 | 34.2 | 49.09 |
| 13 | 34.2 | 50.25 |
| 14 | 34.0 | 50.74 |
| 15 | 34.2 | 50.01 |
| 16 | 34.0 | 51.95 |
| 17 | 34.1 | 51.06 |
| 18 | 34.2 | 48.72 |
| 19 | 34.0 | 50.77 |
| 20 | 34.1 | 50.56 |

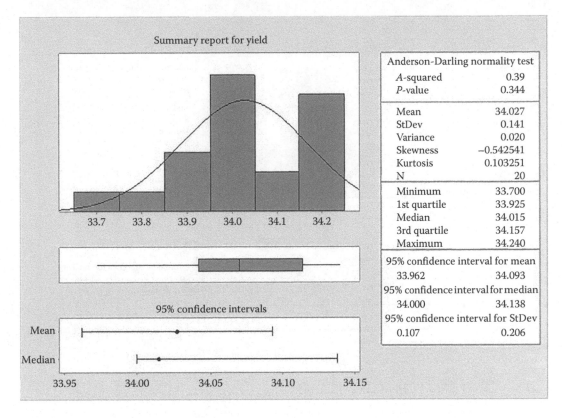

*Figure 5.1* Descriptive summary of yield of pulp cooking process.

as the *p*-value of Anderson–Darling normality test is greater than 0.05 (Mathews, 2005). Similarly, the descriptive summary of the viscosity is given in Figure 5.2.

Figure 5.2 shows that the average viscosity is 50.89 with a standard deviation of 0.973. The upper specification limit (USL) on pulp viscosity is 52 cp. Hence, it is very likely that the pulp cooking process is not capable of meeting the customer requirement of producing pulp with viscosity within 52 cp. Figure 5.2 also shows that the viscosity is normally distributed as Anderson–Darling normality test *p*-value > 0.05. Hence, the viscosity data are subjected to capability analysis. The process capability analysis result is given in Figure 5.3.

Figure 5.3 shows that the Ppk is only 0.38 which is less than 1.0 indicating that the pulp cooking process is not capable of meeting the customer requirements on viscosity. Hence, there is a need to make the pulp cooking process capable of meeting the customer requirement on viscosity as well as improving the yield of the process as far as possible.

The performance of the pulp cooking process can be unsatisfactory due to the presence of assignable causes. To check whether the pulp cooking process is in statistical control, control charts are constructed for the pulp yield and viscosity. Since both yield and viscosity are normally distributed, the individual *x* chart (Montgomery, 2002) is used. The individual *x* chart of yield is given in Figure 5.4 and that of viscosity is given in Figure 5.5.

Figures 5.4 and 5.5 show that none of the points plotted is beyond the upper or lower control limits (UCL or LCL). Moreover, none of the out-of-control run rules (Leavenworth and Grant, 2000) is violated in both the cases. This shows that the pulp cooking process is under control and free from the influence of assignable causes. In

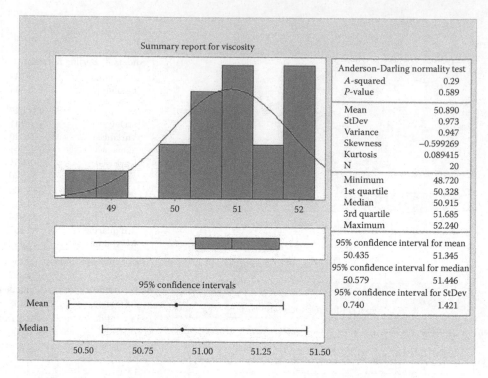

***Figure 5.2*** Descriptive summary of pulp viscosity.

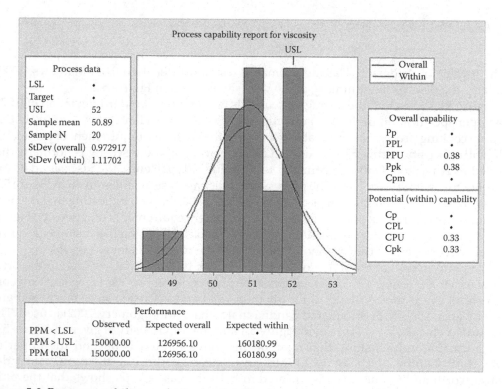

***Figure 5.3*** Process capability analysis of pulp viscosity.

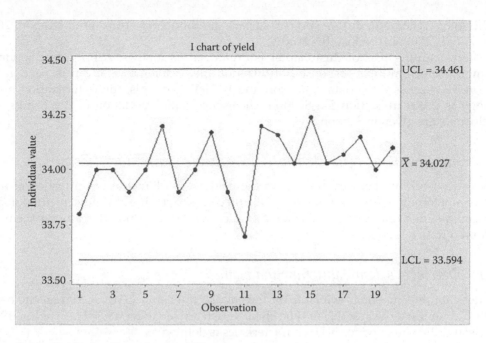

*Figure 5.4* Individual *x* chart of pulp yield.

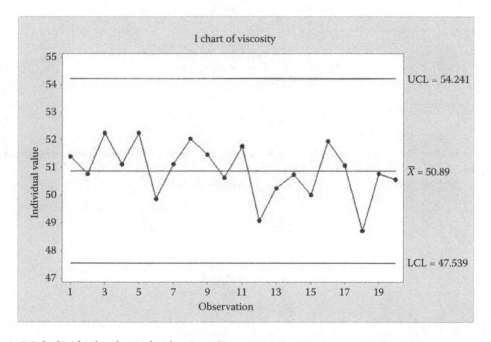

*Figure 5.5* Individual *x* chart of pulp viscosity.

other words, it is not just a process control problem but needs process optimization. Hence, it is decided to carry out the design of experiments to improve and optimize the pulp cooking process. Moreover, the pulp cooking process needs to be optimized to meet the requirements of two output characteristics, namely, pulp yield and viscosity. So a widely popular simultaneous optimization of multiple output characteristics methodology, namely, dual-response surface methodology (Box and Draper, 2007; Myers et al., 2009) is used in this study.

The remaining part of this chapter is organized as follows: A brief description of commonly used multiple response optimization methodologies is given in Section 5.2, Section 5.3 discusses the data collection and modeling details, the optimization methodology is given in Section 5.4, Section 5.5 discusses the validation of the results and conclusions are given in Section 5.6.

## 5.2   *Simultaneous optimization of multiple characteristics*

A lot of research has been carried out in the field of simultaneous optimization of multiple characteristics. Much research has resulted in suggesting different methodologies for simultaneous optimization of characteristics. The widely popular among them are discussed in this section.

### 5.2.1   *Derringer's desirability function method*

The desirability function is available for the smaller the better (STB), the larger the better (LTB) and nominal the best (NTB) type characteristics (Derringer, 1994; Harrington, 1965). The desirability function for NTB characteristics is defined as

$$d = \left| \frac{y - \text{LSL}}{T - \text{LSL}} \right|^{\alpha}, \quad \text{if } \text{LSL} < y \leq T \tag{5.2}$$

$$d = \left| \frac{y - \text{USL}}{T - \text{USL}} \right|^{\beta}, \quad \text{if } T \leq y < \text{USL} \tag{5.3}$$

$$d = 0, \quad \text{if } y \leq \text{LSL or } y \geq \text{USL} \tag{5.4}$$

where $y$ is the characteristic under study, $T$ is the target, LSL is the lower specification limit, and USL is the upper specification limit of $y$.

The desirability function for STB characteristics is defined as

$$d = \left| \frac{y - \text{USL}}{y_{\min} - \text{USL}} \right|^{\alpha}, \quad \text{if } y_{\min} < y < \text{USL} \tag{5.5}$$

$$d = 0, \quad \text{if } y \geq \text{USL} \tag{5.6}$$

$$d = 1, \quad \text{if } y \leq y_{\min} \tag{5.7}$$

where $y$ is the characteristic under study, USL is the upper specification limit of $y$ and $y_{\min}$ is the practically achievable most desirable minimum value or target of $y$.

Similarly, the desirability function for LTB characteristics is defined as

$$d = \left| \frac{y - \text{LSL}}{y_{\max} - \text{LSL}} \right|^{\alpha}, \quad \text{if } \text{LSL} < y < y_{\max} \tag{5.8}$$

$$d = 0, \quad \text{if } y \leq \text{LSL} \tag{5.9}$$

$$d = 1, \quad \text{if } y \geq y_{\text{max}} \tag{5.10}$$

where $y$ is the characteristic under study, LSL is the lower specification limit of $y$ and $y_{\text{max}}$ is the practically achievable most desirable maximum value or target of $y$. The weights $\alpha$ and $\beta$ in the desirability function need to be chosen based on the desirability of quality characteristic $y$ with respect to its target and specification limits.

Equations (5.2)–(5.10) show that the desirability value $d$ will be 1 when the characteristic $y$ is on the target. The desirability value decreases as $y$ moves away from the target. The desirability value will be 0 when the characteristic under study $y$ is on or beyond the specification limits. For simultaneous optimization of multiple characteristics, the desirability function value $d_i$, $i = 1, 2, ..., k$ is computed for each characteristic $y_i$, and the overall desirability $D$ is computed as the geometric mean of individual desirability values as shown in Equation (5.11):

$$D = (d_1 \times d_2 \times \cdots \times d_k)^k \tag{5.11}$$

Finally, the values of the factors that would simultaneously optimize multiple characteristics are found out by maximizing the overall desirability value $D$. Some of the important applications of the desirability function approach for simultaneous optimization of response variables are in CNC turning of AISI P-20 tool steel (Aggarwal et al., 2008), analytical methods development (Candioti et al., 2014), carbonitriding of bushes (John, 2013), etc.

### 5.2.2    Taguchi's loss function approach

Taguchi's loss function measures the weighted square of the deviation of the characteristic $y$ from its target (Taguchi et al., 2005). The loss function is available for NTB, STB, and LTB types of characteristics. Let $y_1, y_2, ..., y_n$ be $n$ observations of the response variable $y$, then the expected quality loss $l(y)$ is defined as

$$l(y) = \frac{1}{n} k \sum_{i=1}^{n} (y_i - T)^2, \text{ for NTB} \tag{5.12}$$

$$l(y) = \frac{1}{n} k \sum_{i=1}^{n} y_i^2, \quad \text{for STB} \tag{5.13}$$

$$l(y) = \frac{1}{n} k \sum_{i=1}^{n} \frac{1}{y_i^2}, \quad \text{for LTB} \tag{5.14}$$

where $T$ is the target and $k$ is a proportionality constant known as quality loss coefficient. For the STB-type response variable, the target is taken as zero and for the LTB-type response variable, the target is generally taken as infinity. From Equations (5.12) to (5.14), it is clear that the expected loss $l(y)$ will be zero when the responses are on target and the loss increases as the response variables move away from the respective targets. For simultaneous optimization of multiple responses, $k$ is often chosen in such a way that expected quality loss $l(y)$ will be equal to 1 when the response variables are either on upper or lower specification limits. For example, for NTB response variables, $k$ is chosen as

$$k = \left( \frac{2}{\text{USL}-\text{LSL}} \right)^2 \tag{5.15}$$

where USL is the upper specification limit and LSL is the lower specification limit. To use Taguchi's loss function approach for simultaneous optimization, the expected loss is computed for each response variable $y_j$, $j = 1, 2, \ldots, k$ using Equations (5.12)–(5.14), and the overall expected loss $L(y)$ is computed as the average of the individual expected losses as shown in Equation (5.16):

$$L(y) = \frac{1}{k} \sum_{j=1}^{k} l(y_j) \tag{5.16}$$

Finally, the values of the factors that would simultaneously optimize multiple responses are found out by minimizing the overall expected loss $L(y)$. Many applications of Taguchi's loss function approach for simultaneous optimization of multiple responses are available in the literature (Antony, 2001; John, 2012; Lin and Lin, 2002; Nian et al., 1999).

### 5.2.3 Fuzzy logic approach

The fuzzy logic approach for optimization is suggested by Kim and Lin (1998). In fuzzy logic approach, each characteristic $y$ is assigned a membership value using a membership function $m(z)$. Generally, an exponential membership function is used as given in Equations (5.17) and (5.18):

$$m(z) = \frac{e^d - e^{d|z|}}{e^d - 1}, \quad \text{if } d \neq 0 \tag{5.17}$$

$$m(z) = 1 - |z|, \quad \text{if } d = 0 \tag{5.18}$$

where $d$ is called the exponential constant, and $z$ measures the deviation of the characteristic $y$ from its target value. The $z$ is defined for all three types of characteristics, namely STB, LTB, and NTB are given in Equations (5.19)–(5.21):

$$z = \frac{y - T}{y_u - T} \quad \text{or} \quad \frac{y - T}{T - y_L}, \quad \text{for NTB} \tag{5.19}$$

$$z = \frac{y - y_{\min}}{y_u - y_{\min}}, \quad \text{for STB} \tag{5.20}$$

$$z = \frac{y_{\max} - y}{y_{\max} - y_L}, \quad \text{for LTB} \tag{5.21}$$

where $y$ is the characteristic under study, $T$ is the specified target on $y$, $y_u$, and $y_L$ are the upper and lower limits of $y$, $y_{\min}$ is the best possible minimum value $y$ can achieve in case of the STB and $y_{max}$ is the best possible maximum value $y$ can achieve in case of the LTB. The membership function $m(z)$ is assigned a value of 0 whenever $y$ is beyond the upper or lower limit. When $y = T$ in Equation (5.19) or $y = y_{min}$ in Equation (5.20) or $y = y_{max}$ in Equation (5.21), $z$ will be 0 and the membership function $m(z)$ achieves the maximum value of 1. In other words, $m(z)$ achieves the best value of 1 when $y$ is on target or at best possible value. Moreover, the rate of decrease of $m(z)$ will be high when $d < 0$, the rate of decrease will be low when $d > 0$ and the rate of decrease will be constant when $d = 0$. The user can choose the value of $d$ based on the desired rate of decrease of $m(z)$.

For simultaneous optimization of multiple characteristics using fuzzy logic, the membership function $m(z)$ is computed for all the characteristics, and the optimum values of the factors are identified by maximizing the minimum of $m(z)$ values (Lin et al., 2000).

## 5.2.4   Dual-response surface methodology

The response surface methodology aims to improve and optimize processes using statistical and mathematical techniques. In response surface methodology, the mean of the response variable is optimized (Ding et al., 2004). The emphasis of dual-response surface methodology is on simultaneously optimizing the mean and variance of the response variable (John, 2015). This is achieved by developing polynomial models for mean and standard deviation of the response variable as

$$\hat{y}_\mu = a_0 + \sum_{i=1}^{k} a_i x_i + \sum_{i-1}^{k} a_{ii} x_i^2 + \sum\sum_{i<j}^{k} a_{ij} x_i x_j \tag{5.22}$$

$$\hat{y}_\sigma = b_0 + \sum_{i=1}^{k} b_i x_i + \sum_{i=1}^{k} b_{ii} x_i^2 + \sum\sum_{i<j}^{k} b_{ij} x_i x_j \tag{5.23}$$

where is $\hat{y}_\mu$ is the estimated mean, $\hat{y}_\sigma$ is the estimated standard deviation of the response variable $y$ and $x_i$, $i = 1, 2, …, k$ are the explanatory variables. Then the optimum values of the explanatory variables that would simultaneously optimize the estimated mean and variance of the response variable (Vining and Myers, 1990) are obtained by formulating and solving the optimization problem of

$$\text{Minimize } \hat{y}_\sigma \tag{5.24}$$

$$\text{Subject to } \hat{y}_\mu = T \tag{5.25}$$

The aforementioned optimization problem can be solved using Microsoft Excel Solver (Del Castillo and Montgomery, 1993). The methodology can also be used for simultaneous optimization of multiple responses. For example, a STB-type response variable and a NTB-type response variable can be simultaneously optimized by solving

$$\text{Minimize } \hat{y}_1 \tag{5.26}$$

$$\text{Subject to } \hat{y}_2 = T \tag{5.27}$$

where $\hat{y}_1$ is the estimated value of the STB-type response variable, $\hat{y}_2$ is the estimated value of the NTB response variable, and $T$ is the specified target value of $\hat{y}_2$.

There are many other approaches also available for simultaneous optimization of multiple response variables, namely, grey relational analysis (Chiang and Chang, 2006), principal component analysis (Tong et al., 2005), artificial neural networks (Noorossana et al., 2009), genetic algorithm (Ortiz et al., 2004), etc. In this case study, the authors have used the dual-response surface methodology for simultaneous optimization of pulp yield and viscosity of the pulp cooking process.

## 5.3   Data collection and modeling

Through discussions with the technical professionals of the company, three factors, namely, percentage of sulfidity of cooking medium ($x_1$), percentage of black liquor in cooking medium ($x_2$) and cooking time ($x_3$) are selected for experimentation. The pulp yield ($y_1$) and pulp viscosity ($y_2$) are taken as the response variables. The engineers also suspected interaction between the factors; hence, an eight-run full factorial experiment is designed. The operational personnel also suggested that the relationship between the response variables and factors may not be linear, so four central points are also added to the design to verify the form of relationship between response variables and factors. The factors chosen for the experiments with its levels are given in Table 5.2.

The experiments are conducted as per the design, and the response variables are measured. The experimental layout with the response variables is given in Table 5.3.

The experimental data on response variable pulp yield ($y_1$) are subjected to analysis of variance. The analysis of variance (ANOVA) table is given in Table 5.4.

Table 5.4 shows that the $p$-values of regression and interaction are less than 0.05, indicating that one or more factors, as well as their interactions, are significant at the 5% level.

*Table 5.2* Factors and levels chosen for experiment

| Factor name | Factor code | Level 1 (−1) | Center point (0) | Level 2 (+1) |
| --- | --- | --- | --- | --- |
| Sulfidity (%) | $x_1$ | 15 | 18 | 21 |
| Black liquor (%) | $x_2$ | 0 | 5 | 10 |
| Cooking time (minutes) | $x_3$ | 55 | 60 | 65 |

*Table 5.3* Experimental layout with response values

| Experiment number | $x_1$ | $x_2$ | $x_3$ | $y_1$ | $y_2$ |
| --- | --- | --- | --- | --- | --- |
| 1 | −1 | −1 | −1 | 36 | 51.65 |
| 2 | −1 | −1 | +1 | 35.8 | 50.1 |
| 3 | −1 | +1 | −1 | 36.3 | 53.55 |
| 4 | −1 | +1 | +1 | 36.5 | 48.9 |
| 5 | +1 | −1 | −1 | 36.9 | 55.2 |
| 6 | +1 | −1 | +1 | 36.8 | 51.7 |
| 7 | +1 | +1 | −1 | 36.7 | 55.45 |
| 8 | +1 | +1 | +1 | 36.5 | 53.5 |
| 9 | 0 | 0 | 0 | 36.5 | 52.7 |
| 10 | 0 | 0 | 0 | 36.6 | 49.7 |
| 11 | 0 | 0 | 0 | 36.4 | 51.45 |
| 12 | 0 | 0 | 0 | 36.6 | 49.3 |

*Table 5.4* ANOVA table for pulp yield

| Source | df | SS | MS | F | p-value |
|---|---|---|---|---|---|
| Regression | 3 | 0.70375 | 0.23458 | 4.9010 | 0.03213 |
| Residual | 8 | 0.3829 | 0.04786 | | |
| Interaction | 3 | 0.30375 | 0.10125 | 11.045 | 0.03958 |
| Pure quadratic | 2 | 0.0204 | 0.01021 | 1.1136 | 0.43478 |
| Pure error | 3 | 0.0275 | 0.00917 | | |
| Total | 11 | 1.0867 | 0.09879 | | |

Table 5.4 also shows that the pure quadratic term is not significant at the 5% level as the corresponding $p$-value = 0.43478 > 0.05 (Montgomery, 2013). The ANOVA table is again constructed by dropping the insignificant quadratic term. The modified ANOVA table of pulp yield is given in Table 5.5.

Table 5.5 shows that the $p$-values for factor sulfidity ($x_1$) and interaction between sulfidity and black liquor ($x_1 x_2$) are significant ($p$-value < 0.05) at the 5% level. Hence, a model is developed for yield ($y_1$) using sulfidity ($x_1$) and interaction between sulfidity and black liquor ($x_1 x_2$) as explanatory variables (Draper and Smith, 2003). The coefficient table for the pulp yield model is given in Table 5.6.

From Table 5.6, the model for pulp yield ($y_1$) is identified as

$$\hat{y}_1 = 36.4375 + 0.2875 x_1 - 0.1875 x_1 x_2 \tag{5.28}$$

The model accuracy measures are given in Table 5.7.

*Table 5.5* Modified ANOVA table of pulp yield ($y_1$)

| Source | DF | SS | MS | F | p-Value |
|---|---|---|---|---|---|
| Model | 6 | 1.0075 | 0.16792 | 10.61 | 0.01 |
| Linear | 3 | 0.70375 | 0.23458 | 14.82 | 0.006 |
| $x_1$ | 1 | 0.66125 | 0.66125 | 41.76 | 0.001 |
| $x_2$ | 1 | 0.03125 | 0.03125 | 1.97 | 0.219 |
| $x_3$ | 1 | 0.01125 | 0.01125 | 0.71 | 0.438 |
| Two-way interaction | 3 | 0.30375 | 0.10125 | 6.39 | 0.037 |
| $x_1 x_2$ | 1 | 0.28125 | 0.28125 | 17.76 | 0.008 |
| $x_1 x_3$ | 1 | 0.01125 | 0.01125 | 0.71 | 0.438 |
| $x_2 x_3$ | 1 | 0.01125 | 0.01125 | 0.71 | 0.438 |
| Error | 5 | 0.07917 | 0.01583 | | |
| Lack-of-fit | 2 | 0.05167 | 0.02583 | 2.82 | 0.205 |
| Pure | | 3 | 0.0275 | 0.00917 | |
| Total | 11 | 1.08667 | | | |

*Table 5.6* Coefficient table of pulp yield

| Term | Coefficients | SE coefficient | t | p-Value |
|---|---|---|---|---|
| Constant | 36.4375 | 0.0491 | 742.81 | 0.00 |
| $x_1$ | 0.2875 | 0.0491 | 5.86 | 0.002 |
| $x_1 x_2$ | −0.1875 | 0.0491 | −3.82 | 0.012 |

*Table 5.7* Accuracy measures of pulp model

| Statistics | Value |
|---|---|
| $R^2$ | 90.73 |
| Adjusted $R^2$ | 87.03 |
| Standard error | 0.138744 |

Table 5.7 shows that the $R^2$ and adjusted $R^2$ are very high and standard error is reasonably close to zero. Hence, it is concluded that the model is accurate. The residual plots of pulp yield model are given in Figure 5.6.

Figure 5.6 shows that the residuals are more or less normally distributed, and there is no trend or pattern in the plot of residuals versus fitted values or observation order. So it is concluded that the model is adequate (Montgomery et al., 2003).

Similarly, the response variable viscosity is also subjected to ANOVA. The ANOVA table for viscosity is given in Table 5.8.

Table 5.8 shows that the regression is significant at the 5% significant level ($p$-value = 0.03104 < 0.05). The $p$-value for interaction is 0.89732 > 0.05 indicating that the interaction is not significant. But the $p$-value of the pure quadratic term is 0.0927 < 0.1, indicating the quadratic term is significant at the 10% level. So to develop a full polynomial model for viscosity, more experiments need to be carried out at factor axial points, which would make the study costlier. Hence, the possibility of developing a linear model for viscosity by transforming the response is explored. A linear model is found to be the best fit model for the logarithm of viscosity ($y_2'$). The experimental layout with the logarithm of viscosity ($y_2'$) is given in Table 5.9.

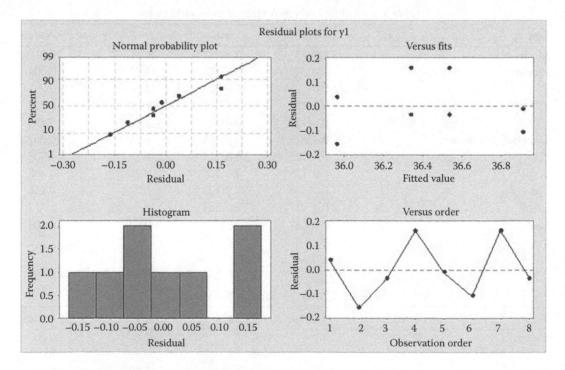

*Figure 5.6* Residual plots of pulp yield model.

*Table 5.8* ANOVA table of pulp viscosity

| Source | df | SS | MS | F | p-Value |
|---|---|---|---|---|---|
| Regression | 3 | 34.8759375 | 11.6253 | 4.9699 | 0.03104 |
| Residual | 8 | 18.7132 | 2.33915 | | |
| Interaction | 3 | 0.5984375 | 0.19948 | 0.190 | 0.89732 |
| Pure quadratic | 2 | 12.2551 | 6.12755 | 5.8231 | 0.09270 |
| Pure error | 3 | 3.156875 | 1.05229 | | |
| Total | 11 | 53.5892 | 4.87174 | | |

*Table 5.9* Experimental layout with logarithm of viscosity $(y_2')$

| Experimental number | $x_1$ | $x_2$ | $x_3$ | $y_2$ | $(y_2')$ |
|---|---|---|---|---|---|
| 1 | −1 | −1 | −1 | 51.65 | 3.944490 |
| 2 | −1 | −1 | +1 | 50.1 | 3.914021 |
| 3 | −1 | +1 | −1 | 53.55 | 3.980616 |
| 4 | −1 | +1 | +1 | 48.9 | 3.889777 |
| 5 | +1 | −1 | −1 | 55.2 | 4.010963 |
| 6 | +1 | −1 | +1 | 51.7 | 3.945458 |
| 7 | +1 | +1 | −1 | 55.45 | 4.015482 |
| 8 | +1 | +1 | +1 | 53.5 | 3.979682 |
| 9 | 0 | 0 | 0 | 52.7 | 3.931826 |
| 10 | 0 | 0 | 0 | 49.7 | 3.906005 |
| 11 | 0 | 0 | 0 | 51.45 | 3.940610 |
| 12 | 0 | 0 | 0 | 49.3 | 3.897924 |

The logarithm of viscosity is subjected to ANOVA. The ANOVA table for the transformed viscosity is given in Table 5.10.

Table 5.10 shows that neither the pure quadratic term nor the interaction is significant ($p$-value > 0.05). Hence, the ANOVA table is modified by dropping the insignificant terms. The modified ANOVA table is given in Table 5.11.

Table 5.11 shows that only the factors sulfidity ($x_1$) and cooking time ($x_3$) have a significant effect on the response variable. So the model is developed for the logarithm of viscosity ($y_2'$) using sulfidity ($x_1$) and cooking time ($x_3$) as explanatory variables. The coefficient table of the logarithm of viscosity ($y_2'$) model is given in Table 5.12.

From Table 5.12, the model for logarithm of pulp viscosity ($y_2'$) is identified as

*Table 5.10* ANOVA table for transformed viscosity $(y_2')$

| Source | df | SS | MS | F | p-Value |
|---|---|---|---|---|---|
| Regression | 3 | 0.012713196 | 0.00424 | 4.8483 | 0.03299 |
| Residual | 8 | 0.0070 | 0.00087 | | |
| Interaction | 3 | 0.000257734 | 8.6E-05 | 0.207 | 0.88578 |
| Pure quadratic | 2 | 0.0045 | 0.00224 | 5.3948 | 0.10147 |
| Pure error | 3 | 0.001244542 | 0.00041 | | |
| Total | 11 | 0.0197 | 0.00179 | | |

*Table 5.11* Modified ANOVA table of logarithm of viscosity ($y_2'$)

| Source | DF | SS | MS | F | p-Value |
|---|---|---|---|---|---|
| Model | 3 | 0.01271 | 0.00424 | 4.85 | 0.033 |
| Linear | 3 | 0.01271 | 0.00424 | 4.85 | 0.033 |
| $x_1$ | 1 | 0.0062 | 0.0062 | 7.09 | 0.029 |
| $x_2$ | 1 | 0.00032 | 0.00032 | 0.37 | 0.562 |
| $x_3$ | 1 | 0.0062 | 0.0062 | 7.09 | 0.029 |
| Error | 8 | 0.00699 | 0.00087 | | |
| Lack-of-fit | 5 | 0.00575 | 0.00115 | 2.77 | 0.215 |
| Pure | 3 | 0.00125 | 0.00042 | | |
| Total | 11 | 0.01971 | | | |

*Table 5.12* Effect and coefficient table of transformed viscosity

| Term | Coefficient | SE coefficient | t | p-Values |
|---|---|---|---|---|
| Constant | 3.96006 | 0.00631 | 627.66 | 0.00 |
| $x_1$ | 0.02783 | 0.00631 | 4.41 | 0.007 |
| $x_3$ | −0.02783 | 0.00631 | −4.41 | 0.007 |

$$\hat{y}_2' = 3.96006 + 0.02783x_1 - 0.02783x_3 \tag{5.29}$$

The accuracy statistics of the model are given in Table 5.13.

Table 5.13 shows that the $R^2$ and adjusted $R^2$ are high, and standard error is reasonably close to zero. Hence, it is concluded that the model is accurate. The residual plots of the model are given in Figure 5.7.

Figure 5.7 shows that the model residuals are more or less normally distributed, and there is no trend or pattern in residuals versus fitted value plot or residuals versus observation order plot indicating that the model is adequate.

## 5.4 Optimization

The optimum setting of the factors that would increase the pulp yield as much as possible without increasing the viscosity beyond 52 is identified by formulating the problem as a constraint optimization problem (Hillier and Lieberman, 2008; Taha, 2014). Since the model is developed for the logarithm of viscosity, the problem is formulated to maximize the pulp yield ($y_1$) subject to the constraint that the logarithm of pulp viscosity ($y_2'$) will not exceed the upper limit. The upper limit $k'$ is computed as

$$k' = k - 1.96s \tag{5.30}$$

*Table 5.13* Accuracy measures of pulp viscosity model

| Statistics | Value |
|---|---|
| $R^2$ | 88.61 |
| Adjusted $R^2$ | 84.06 |
| Standard error | 0.0178452 |

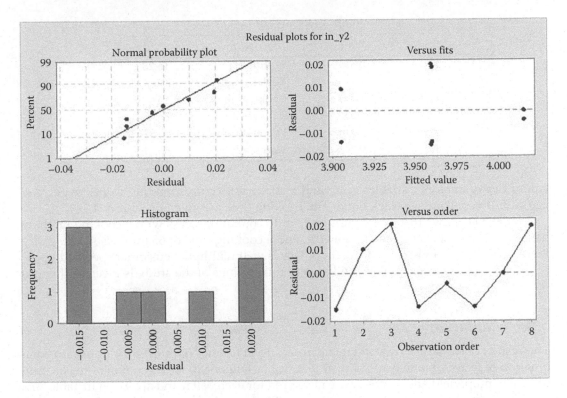

*Figure 5.7* The residual plots of viscosity model.

where $k$ is the logarithm of 52, the upper specification limit of viscosity, and $s$ is the standard error of viscosity model. The upper limit $k'$ is taken 1.96 standard deviations less than the logarithm of 52 cp. This is to ensure that even the individual values of viscosity are very unlikely to fall outside the specification limit. Substituting the values of $s$ and logarithm of 52 in Equation (5.30), $k'$ has become

$$k' = 3.95124 - 1.96 \times 0.01758 = 3.91627 \tag{5.31}$$

Using Equation (5.28)–(5.31), the optimization problem is formulated as follows:

$$\text{Maximize } 36.4375 + 0.2875x_1 - 0.1875x_1x_2 \tag{5.32}$$

$$\text{Subject to } 3.96006 + 0.02783x_1 - 0.02783x_3 \leq 3.91627 \tag{5.33}$$

$$-1 \leq x_1 \leq 1 \tag{5.34}$$

$$-1 \leq x_2 \leq 1 \tag{5.35}$$

$$-1 \leq x_3 \leq 1 \tag{5.36}$$

The optimization problem given in Equations (5.31)–(5.36) is solved using Microsoft Excel Solver utility (Fylstra et al., 1999). The solution obtained is given in Table 5.14. The optimum

*Table 5.14* Optimum solution

| Code | Value | Name | Value |
|------|-------|------|-------|
| $x_1$ | −0.5735 | % Sulfidity | 16.28 |
| $x_2$ | 1 | % Black liquor | 10 |
| $x_3$ | 1 | Cooking time | 65 |
| $y_1$ | 36.38 | Pulp yield | 36.38 |
| $y_2'$ | 3.91627 | Log viscosity | 3.91627 |
| $y_2$ | 50.21 | Viscosity | 50.21 |

values of explanatory variables $x_1$, $x_2$, and $x_3$ along with corresponding values of percentage sulfidity, percentage black liquor and cooking time are given in Table 5.14.

Table 5.14 shows that by executing the pulp cooking process with a cooking medium having sulfidity 16.28%, black liquor 10%, and a cooking time of 65 minutes would give a yield of 36.38 and a viscosity of 50.21 cp. This is well within the customer specified upper limit of 52 cp on viscosity. The validation of the findings of the study is given in the next section.

## 5.5 Validation

The results of the study are validated by cooking 14 batches of pulp at the optimum combination of factors, namely, sulfidity at 16.28%, black liquor 10%, and cooking time 65 minutes. The pulp yield and viscosity are measured for each batch and are given in Table 5.15.

The individual $x$ control chart comparing the pulp yield performance before and after the study is given in Figure 5.8 and that of pulp viscosity is given in Figure 5.9.

Figure 5.8 shows that executing the pulp cooking process with the optimum combination of factors suggested by the study would significantly improve the pulp yield. The validation data show that on an average the yield increased from 34% to 36.43%. Figure 5.9 shows that the optimum combination of factors reduced the mean as well as variation in

*Table 5.15* Validation results

| Batch | Yield | Viscosity |
|-------|-------|-----------|
| 1 | 36.5 | 49.91 |
| 2 | 36.4 | 49.76 |
| 3 | 36.3 | 51.36 |
| 4 | 36.4 | 50.36 |
| 5 | 36.4 | 49.55 |
| 6 | 36.5 | 49.94 |
| 7 | 35.9 | 50.25 |
| 8 | 36.4 | 50.16 |
| 9 | 36.3 | 49.83 |
| 10 | 36.3 | 50.6 |
| 11 | 36.7 | 50.48 |
| 12 | 36.3 | 49.73 |
| 13 | 36.4 | 49.56 |
| 14 | 36.2 | 50.48 |

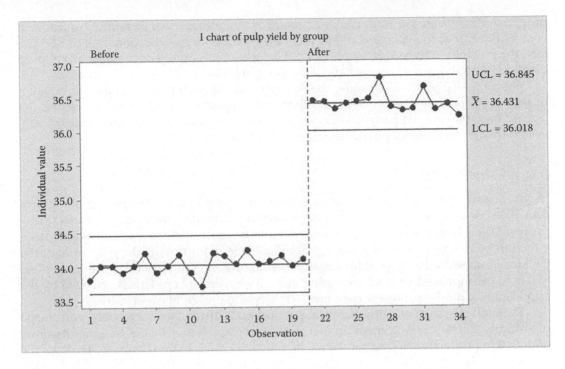

*Figure 5.8* Comparison of pulp yield before and after study.

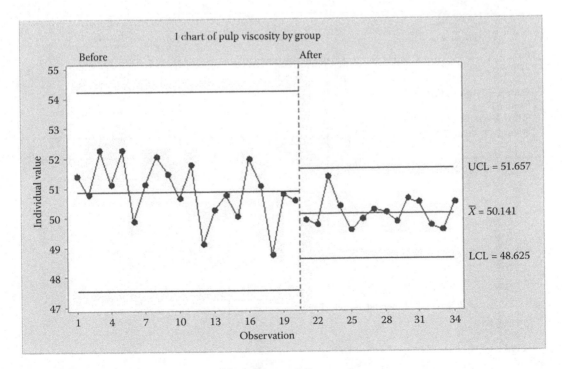

*Figure 5.9* Comparison of pulp viscosity before and after study.

pulp viscosity. When the pulp cooking process is operating under statistical control, it is very unlikely that the viscosity will be more than the upper specification limit of 52 cp, in fact, the viscosity will be less than 51.65 cp. The process capability analysis results of pulp viscosity with the validation data are given in Figure 5.10.

Figure 5.10 shows that running the cooking process with the optimum combination of factors would improve process capability with respect to viscosity to 1.25. Hence, the pulp manufacturing company has decided to use the optimum combination of factors suggested by the study for all future batches of pulp cooking process.

## 5.6   Conclusion

In this chapter, the authors presented a case study on optimizing the pulp cooking process. The cooking process is an important step in the rayon grade pulp manufacturing process. The rayon grade pulp is the raw material for manufacturing viscous staple fiber, which in turn is used for cloth making. The challenge in improving the pulp yield of the cooking process is that it would result in deteriorating the pulp viscosity. The pulp yield needs to be improved without increasing viscosity beyond 52 cp. This is achieved by the application of dual-response surface methodology and design of experiments.

Through discussions with technical professionals, three factors, namely, percentage of sulfidity in the cooking medium, percentage of black liquor in the cooking medium, and cooking time are selected for the study. The pulp yield and viscosity are taken as the

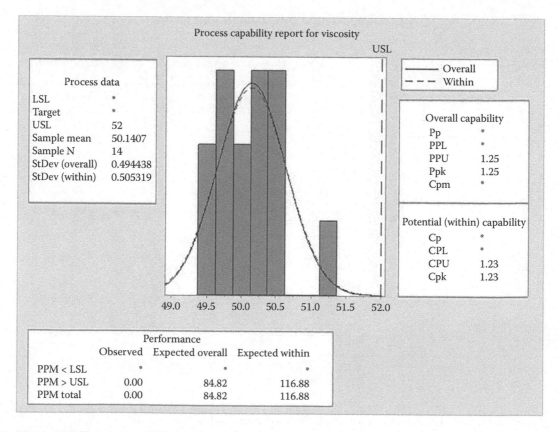

*Figure 5.10* Process capability analysis of viscosity.

response variables. Since the engineers suspected that interaction between the factors, a full factorial experiment is designed. To explore whether the relationship between the factors and response variables is nonlinear or not, four center points are also added to the design. The experiments are carried out as per the design and data on pulp yield and viscosity are collected. Based on the data, models are fitted for pulp yield and logarithm of pulp viscosity. Then the optimum combination of the factors that would maximize the pulp yield subject to the constraint on viscosity is obtained by formulating the problem as a constraint optimization problem and solving it using Microsoft Solver utility.

Validation of the results showed that the yield of the pulp cooking process has significantly improved from 34% to 36.43%. Moreover, the study also reduced pulp viscosity as well as variation in pulp viscosity. The process capability index Ppk of pulp viscosity improved to 1.25. Hence, the company decided to execute the pulp cooking process with the optimum combination of factors suggested by the study for the upcoming batches.

Many of the modern-day processes have multiple output characteristics. The process manager or engineer needs to find an optimum setting of process control factors that would result in simultaneously meeting the requirements on all the output characteristics. In this chapter, the authors have demonstrated the dual-response surface methodology for the simultaneous optimization of multiple output characteristics. Even though the case study is dealing with the optimization of the pulp cooking process for simultaneously meeting the requirements of the output characteristics, namely, pulp yield and viscosity, the methodology can be used for optimizing any process. Moreover, more than two output characteristics can also be simultaneously optimized using response surface methodology.

## References

Aggarwal, A., Singh, H., Kumar, P., and Singh, M. (2008). Optimization of multiple quality characteristics for CNC turning under cryogenic cutting environment using desirability function. *Journal of Materials Processing Technology, 205*(1), 42–50.

Antony, J. (2001). Simultaneous optimisation of multiple quality characteristics in manufacturing processes using Taguchi's quality loss function. *International Journal of Advanced Manufacturing Technology, 17*(2), 134–138.

Box, G. E. P., and Draper, N. R. (2007). *Response Surfaces, Mixtures and Ridge Analysis.* 2nd edition, New Jersey, NJ: John Wiley and Sons.

Candioti, L. V., De Zan, M. M., Cámara, M. S., and Goicoechea, H. C. (2014). Experimental design and multiple response optimization. Using the desirability function in analytical methods development. *Talanta, 124,* 123–138.

Chiang, K. T., and Chang, F. P. (2006). Optimization of the WEDM process of particle-reinforced material with multiple performance characteristics using grey relational analysis. *Journal of Materials Processing Technology, 180*(1), 96–101.

Del Castillo, E., and Montgomery, D. C. (1993). A nonlinear programming solution to the dual response problem. *Journal of Quality Technology, 25*(3), 199–204.

Derringer, G. (1994). A balancing act: Optimizing product's properties. *Quality Progress, 27*(6): 51–58.

Ding, R., Lin, D. K., and Wei, D. (2004). Dual-response surface optimization: A weighted MSE approach. *Quality Engineering, 16*(3), 377–385.

Draper, N. R., and Smith, H. (2003). *Applied Regression Analysis.* 3rd edition, Singapore: John Wiley and Sons (Asia) Pte Ltd.

Fylstra, D., Lasdon, L., Watson, J., and Waren, A. (1999). Design and use of the Microsoft Excel Solver. *Interfaces, 28*(5): 29–55.

Harrington, E. (1965). The desirability function. *Industrial Quality Control, 21*(10): 494–498.

Hillier, F. S., and Lieberman, G. J. (2008). *Operations Research – Concepts and Cases.* 8th edition. New Delhi: Tata McGraw-Hill Publishing Company Ltd.

John, B. (2012). Simultaneous optimization of multiple performance characteristics of carbonitrided pellets: A case study. *International Journal of Advanced Manufacturing Technology, 61*(5), 585–594.

John, B. (2013). Application of desirability function for optimizing the performance characteristics of carbonitrided bushes. *International Journal of Industrial Engineering Computations, 4*(3), 305–314.

John, B. (2015). A dual response surface optimization methodology for achieving uniform coating thickness in powder coating process. *International Journal of Industrial Engineering Computations, 6*(4), 469–480.

John, B., Kadadevaramath, R.S., and Edinbarough, I.A. (2017). Designing software development processes to optimise multiple output performance characteristics. *Software Quality Professional 19*(4): 16–24.

Kim, K.-J., and Lin, D. K. (1998). Dual response surface optimization: A fuzzy modeling approach. *Journal of Quality Technology, 30*(1), 1.

Leavenworth, R. S., and Grant, E. L. (2000). *Statistical Quality Control*. 7th edition, New Delhi: Tata McGraw-Hill Education.

Lin, J. L., and Lin, C. L. (2002). The use of the orthogonal array with grey relational analysis to optimize the electrical discharge machining process with multiple performance characteristics. *International Journal of Machine Tools and Manufacture, 42*(2), 237–244.

Lin, J. L., Wang, K. S., Yan, B. H., and Tarng, Y. S. (2000). Optimization of the electrical discharge machining process based on the Taguchi method with fuzzy logics. *Journal of Materials Processing Technology, 102*(1), 48–55.

Mathews, P. (2005). *Design of Experiments with Minitab*. New Delhi: Pearson Education (Singapore) Pte Ltd.

Montgomery, D. C. (2002). *Introduction to Statistical Quality Control*. 4th edition, New Delhi: Wiley India (P) Ltd.

Montgomery, D. C. (2013). *Design and Analysis of Experiments*. 8th edition, New Delhi: Wiley India (Pvt) Ltd.

Montgomery, D.C., Peck, E.A., and Vining, G.G. (2003). *Introduction to Linear Regression Analysis*. 3rd edition, Singapore: John Wiley and Sons (Asia) Pte Ltd.

Myers, R. H., Montgomery, D. C., and Anderson-Cook, C. M. (2009). *Response Surface Methodology: Process and Product Optimization Using Design of Experiments*. 3rd edition, New Jersey, NJ: John Wiley and Sons.

Nian, C. Y., Yang, W. H., and Tarng, Y. S. (1999). Optimization of turning operations with multiple performance characteristics. *Journal of Materials Processing Technology, 95*(1), 90–96.

Noorossana, R., Tajbakhsh, S. D., and Saghaei, A. (2009). An artificial neural network approach to multiple-response optimization. *International Journal of Advanced Manufacturing Technology, 40*(11–12), 1227–1238.

Ortiz Jr, F., Simpson, J. R., Pignatiello Jr, J. J., and Heredia-Langner, A. (2004). A genetic algorithm approach to multiple-response optimization. *Journal of Quality Technology, 36*(4), 432.

Taguchi, G., Chowdhury, S., and Wu, Y. (2005). *Taguchi's Quality Engineering Handbook*. Hoboken, NJ: John Wiley and Sons.

Taha, H. A. (2014). *Operations Research – An Introduction*. 9th edition, New Delhi: Pearson.

Tong, L. I., Wang, C. H., and Chen, H. C. (2005). Optimization of multiple responses using principal component analysis and technique for order preference by similarity to ideal solution. *International Journal of Advanced Manufacturing Technology, 27*(3), 407–414.

Vining, G., and Myers, R. (1990). Combining Taguchi and response surface philosophies – A dual response approach. *Journal of Quality Technology, 22*, 38–45.

*chapter six*

# Time-dependent conflicting bifuzzy set and its applications in reliability evaluation

**Shshank Chaube**
University of Petroleum and Energy Studies

**S.B. Singh**
G.B. Pant University of Agriculture and Technology

**Sangeeta Pant**
University of Petroleum and Energy Studies

**Anuj Kumar**
University of Petroleum and Energy Studies

## Contents

## 6.1  Introduction

The conventional reliability of a system is defined as the probability that the system will perform a predefined operation under some specified condition for a fixed time period. Traditionally, system reliability evaluation is dependent on the probabilistic approach. But this approach is not always valid, since in reality a lot of times data related to the system information do not represent the realistic situation correctly due to uncertainties present in it. Therefore, in many cases, reliability assessment of the system becomes a very difficult task. Hence, to evaluate reliability of a system when available information is uncertain, then we apply the fuzzy approach. Zadeh (1965) constituted the foundation for this approach by his works on fuzzy set theory with the assumption that the nonmembership degree is equal to one minus the membership degree. Here we can consider membership degree and nonmembership degree as positive and negative aspects of a situation. It implies if the membership is correct, then the nonmembership is wrong, which is a contrary relation. Over this theory Atanassov (1986) introduced the concept of intuitionistic fuzzy sets. He proposed the condition $0 \le \mu_A(x) + v_A(x) \le 1$, where $\mu_A(x)$ and $v_A(x)$ represent the degree of membership and the degree of nonmembership, respectively. Many researchers (Burillo & Bustinces, 1996; Li, Shan & Cheng, 2005; Supriya, Ranjit & Akhil, 2005; Gianpiero & David, 2006) have done work on intuitionistic fuzzy sets. Other theories like L-fuzzy sets (Goguen, 1967), Ying-Yang bipolar fuzzy logic (Zhang & Zhang, 2004), soft sets (Basu, Deb & Pattanaik, 1992), vague sets (Gau & Buehrer, 1993), and interval-valued intuitionistic fuzzy sets (Atanassov, 1999) also were introduced to handle the uncertainty. Then, Zamali, Lazim, and Osman (2008) introduced the concept of a conflicting bifuzzy set (CBFS), and proposed that the sum of membership degree and nonmembership degree can be more than one.

Several authors (Singer, 1990; Cai, Wen & Zhang, 1991; Chen, 1994, 1996; Roy et al., 2017) proposed and developed the fuzzy reliability theory. Extending these works, in this chapter, applications of conflicting bifuzzy sets are applied in fuzzy reliability theory.

In this chapter, some basic concepts of triangular CBFS are defined, and a procedure using triangular CBFS is introduced to estimate the fuzzy reliability of the system. Here, membership and nonmembership functions of fuzzy reliability of systems are constructed by considering the failure rate of each component as a time-dependent triangular CBFN.

## 6.2  Basic concept of time-dependent CBFS and some definitions

### 6.2.1  Time-dependent CBFS

Let $X$ denote the universe of discourse, and $T$ is a nonempty set whose elements are said to be time moments. A time-dependent CBFS in $X$ is defined as

$$A(t) = \left\{ \langle x, \mu_{A(t)}(x), v_{A(t)}(x) \rangle : x \in X, t \in T \right\}$$

where $\mu_{A(t)}(x) : X \to [0,1]$ and $v_{A(t)}(x) : X \to [0,1]$ are the degrees of membership and nonmembership, respectively, of the element $x \in X$ at time moment $t$ to $A \subseteq X$. For every $x \in X$ and $t \in T$,

$$0 \le \mu_{A(t)}(x) + v_{A(t)}(x) \le 2 \tag{6.1}$$

## 6.2.2   Normal CBFS

A CBFS $A$ in universe of discourse $X$ is normal if there exist at least two points $a, b \in X$ such that $\mu_A(a) = 1$ and $v_A(b) = 1$.

## 6.2.3   Convex CBFS

A CBFS $A$ in the universe of discourse $X$ is convex if and only if

i. Membership function $\mu_A(x)$ of $A$ is fuzzy convex, i.e.,

$$\mu_A\left(\lambda x_1 + (1-\lambda)x_2\right) \geq \min\left(\mu_A(x_1), \mu_A(x_2)\right) \quad \forall x_1,\, x_2 \in X,\, \lambda \in [0,1]$$

ii. Nonmembership function $v_A(x)$ of $A$ is fuzzy concave, i.e.,

$$v_A\left(\lambda x_1 + (1-\lambda)x_2\right) \leq \max\left(v_A(x_1), v_A(x_2)\right) \quad \forall x_1,\, x_2 \in X,\, \lambda \in [0,1]$$

## 6.2.4   Conflicting bifuzzy number

A conflicting bifuzzy subset $A = \left\{\langle x, \mu_A(x), v_A(x) : x \in R\rangle\right\}$ of the set of real numbers $R$, is said to be a conflicting bifuzzy number (CBFN) if

i. $A$ is normal and convex.
ii. $\mu_A$ is upper semicontinuous, and $v_A$ is lower semicontinuous.
iii. Support of $A = \left\{x \in X : \mu_A(x) > 0\right\}$ is bounded.

## 6.2.5   $(\alpha, \beta)$-Cut of a time-dependent CBFS

$(\alpha, \beta)$-Cut of a time-dependent CBFS $A(t)$ with time moment $t$ is defined as

$$A_{\alpha,\beta}(t) = \left\{x \in X : \mu_{A(t)}(x) \geq \alpha, v_{A(t)}(x) \leq \beta\right\}; 0 \leq \alpha, \beta \leq 1, \alpha + \beta \leq 1$$

## 6.2.6   Triangular time-dependent CBFS

A time-dependent triangular CBFS $A(t)$ is defined as

$$A(t) = \left(m(t) - l(t), m(t), m(t) + n(t),\ m(t) - l'(t), m(t), m(t) + n'(t)\right)$$

where $m(t) \in R$ is the center, $l(t) > 0$ and $n(t) > 0$ are the left and right spreads of the membership function of $A(t)$, and, $l'(t) > 0$ and $n'(t) > 0$ are the left and right spreads of the nonmembership function of $A(t)$, at time $t$.

## 6.3   Problem formulation

Let $X$ and $Y$ be two sets and failure rate function on $X$ is represented by a CBFS $F(t)$ as

$$F(t) = \left\{x, \mu_{F(t)}(x), v_{F(t)}(x) : x \in X, t \in T\right\}$$

Now for membership function, $\alpha$-cut of $F(t)$ is defined as

$$F_\alpha(t) = \left\{ x : \mu_{F(t)}(x) \geq \alpha, \quad \alpha \in [0,1] \right\}$$

for nonmembership $\beta$-cut of $F(t)$ is defined as

$$F_\beta(t) = \left\{ x : v_{F(t)}(x) \leq \beta, \quad \beta \in [0,1] \right\}$$

it is very obvious that both $F_\alpha(t)$ and $F_\beta(t)$ are crisp sets.

Assume that $F(t)$ is a CBFN then by the fuzzy-convexity property of the membership function of CBFN, we have

$$F_\alpha(t) = \left[ f_{1\alpha}(t), f_{2\alpha}(t) \right], \quad \forall \alpha \in [0,1]$$

and by the fuzzy-concavity property of the nonmembership function of CBFN, we have

$$F_\beta(t) = \left[ f_{1\beta}(t), f_{2\beta}(t) \right], \quad \forall \alpha \in [0,1]$$

where $f_{1\alpha}(t), f_{2\alpha}(t)$ are increasing functions of $\alpha$, and $f_{1\alpha}(t), f_{2\alpha}(t)$ are decreasing functions of $\beta$; $\alpha, \beta \in [0,1]$.

Define a bounded differential function $\psi$ from $X$ to $Y$ as

$$\psi : X \to Y \text{ such that } y = \psi(x) \quad \forall x \in X$$

Now to calculate reliability function $R(t)$ by applying $\psi$ to the set $F(t)$.

$$R(t) = \left\{ \langle y, \mu_{R(t)}(y), v_{R(t)}(y) \rangle : y \in Y \right\}$$

where membership and nonmembership functions for $R(t)$ are evaluated as

$$\mu_{R(t)}(y) = \sup \left\{ \mu_{F(t)}(x) : y = \psi(x), x \in X \right\}$$

$$v_{R(t)}(y) = \inf \left\{ v_{F(t)}(x) : y = \psi(x), x \in X \right\}$$

$$\text{and } \psi(x) = y \in \left[ a_{1\alpha}(t), a_{2\alpha}(t) \right], \text{where } x \in \left[ f_{1\alpha}(t), f_{2\alpha}(t) \right]$$

$$\psi(x) = y \in \left[ a_{1\beta}(t), a_{2\beta}(t) \right], \text{where } x \in \left[ f_{1\beta}(t), f_{2\beta}(t) \right]$$

where $a_{1\alpha}(t)$ and $a_{2\alpha}(t)$ are minimum and maximum of $\psi$ over $F(t)$, respectively, and $a_{1\beta}(t)$ and $a_{2\beta}(t)$ are minimum and maximum of $\psi$ over $F(t)$, respectively, i.e.,

$$a_{1\alpha}(t) = \min \psi(x); \quad f_{1\alpha}(t) \leq x \leq f_{2\alpha}(t) \tag{6.2}$$

$$a_{2\alpha}(t) = \max \psi(x); \quad f_{1\alpha}(t) \leq x \leq f_{2\alpha}(t) \tag{6.3}$$

$$a_{1\beta}(t) = \min \psi(x); \quad f_{1\beta}(t) \leq x \leq f_{2\beta}(t) \tag{6.4}$$

$$a_{2\beta}(t) = \max \psi(x); \quad f_{1\beta}(t) \leq x \leq f_{2\beta}(t) \tag{6.5}$$

If $a_{1\alpha}(t)$ and $a_{2\alpha}(t)$ are invertible, then left shape function $g_{R(t)}(y)$ and right shape function $h_{R(t)}(y)$ are obtained as

$$g_{R(t)}(y) = \left[a_{1\alpha}\right]^{-1} = \left[\min_{y_1 \leq y \leq y_2} u\right]^{-1} \tag{6.6}$$

$$h_{R(t)}(y) = \left[a_{2\alpha}\right]^{-1} = \left[\max_{y_2 \leq y \leq y_3} u\right]^{-1} \tag{6.7}$$

From Equations (6.6) and (6.7), the membership function can be constructed as

$$\mu_{R(t)}(y) = \begin{cases} g_{R(t)}(y), & y_1 \leq y \leq y_2 \\ h_{R(t)}(y), & y_2 \leq y \leq y_3 \\ 0, & \text{otherwise} \end{cases}$$

where $g_{R(t)}(y_1) = h_{R(t)}(y_3) = 0$ and $g_{R(t)}(y_2) = h_{R(t)}(y_2) = 1$.

Similarly, the nonmembership function can be constructed as

$$v_{R(t)}(y) = \begin{cases} g_{R(t)}(y), & y_1' \leq y \leq y_2 \\ h_{R(t)}(y), & y_2 \leq y \leq y_3' \\ 0, & \text{otherwise} \end{cases}$$

where $g_{R(t)}(y_1') = h_{R(t)}(y_3') = 1$ and $g_{R(t)}(y_2) = h_{R(t)}(y_2) = 0$.

## 6.4    Reliability evaluation with time-dependent CBFN

If $f(t)$ is the failure rate function, then the system reliability function can be obtained as

$$R(t) = \exp\left[-\int_0^t f(k)\,dk\right], \quad t > 0 \tag{6.8}$$

Let the failure rate function be represented by triangular CBFN as

$$F(t) = \left(m(t) - l(t), m(t), m(t) + n(t); m(t) - l'(t), m(t), m(t) + n'(t)\right)$$

For the membership function $\alpha$-cut of $F(t)$ is

$$F_\alpha(t) = \left(m(t) - l(t) + \alpha l(t), m(t) + n(t) - \alpha n(t)\right), \quad \forall \alpha \in [0,1]$$

Similarly, for the nonmembership function $\beta$-cut of $F(t)$ is

$$F_\beta(t) = \left(m(t) - \beta l'(t), m(t) + \beta n'(t)\right), \quad \forall \beta \in [0,1]$$

Now Equations (6.2)–(6.5) can, respectively, be written as

$$a_{1\alpha}(t) = \min\left[\exp\left(-\int_0^t x(k)\,dk\right)\right] \text{ s.t. } m(t) - l(t) + \alpha l(t) \le x(t) \le m(t) + n(t) - \alpha n(t) \qquad (6.9)$$

$$a_{2\alpha}(t) = \max\left[\exp\left(-\int_0^t x(k)\,dk\right)\right] \text{ s.t. } m(t) - l(t) + \alpha l(t) \le x(t) \le m(t) + n(t) - \alpha n(t) \qquad (6.10)$$

$$a_{1\beta}(t) = \min\left[\exp\left(-\int_0^t x(k)\,dk\right)\right] \text{ s.t. } m(t) - \beta l'(t) \le x(t) \le m(t) + \beta n'(t) \qquad (6.11)$$

$$a_{2\beta}(t) = \max\left[\exp\left(-\int_0^t x(k)\,dk\right)\right] \text{ s.t. } m(t) - \beta l'(t) \le x(t) \le m(t) + \beta n'(t) \qquad (6.12)$$

Here, $R(t)$ attains its extremes at the bounds. Therefore, we have

$$a_{1\alpha}(t) = \left[\exp\left(-\int_0^t \{m(k) + n(k) - \alpha n(k)\}dk\right)\right], \quad t > 0 \qquad (6.13)$$

$$a_{2\alpha}(t) = \left[\exp\left(-\int_0^t \{m(k) - l(k) + \alpha l(k)\}\,dk\right)\right], \quad t > 0 \qquad (6.14)$$

$$a_{1\beta}(t) = \left[\exp\left(-\int_0^t \{m(k) + \beta n'(k)\}\,dk\right)\right], \quad t > 0 \qquad (6.15)$$

$$a_{2\beta}(t) = \left[\exp\left(-\int_0^t \{m(k) - \beta l'(k)\}dk\right)\right], \quad t > 0 \qquad (6.16)$$

By taking the inverses of Equations (6.13)–(6.16), $\mu_{R(t)}$ and $\nu_{R(t)}$ can be obtained as

$$\mu_{R(t)} = \begin{cases} \dfrac{\ln(y) + \displaystyle\int_0^t (m(k) + n(k))\,dk}{\displaystyle\int_0^t n(k)\,dk}, & \exp\left[-\int_0^t \{m(k) + n(k)\}\,dk\right] \le y \le \exp\left[-\int_0^t m(k)\,dk\right] \\[4ex] -\dfrac{\ln(y) + \displaystyle\int_0^t (m(k) - l(k))\,dk}{\displaystyle\int_0^t l(k)\,dk}, & \exp\left[-\int_0^t m(k)\,dk\right] \le y \le \exp\left[-\int_0^t \{m(k) - l(k)\}\,dk\right] \end{cases}$$

$$(6.17)$$

$$
v_{R(t)} = \begin{cases}
-\dfrac{\ln(y) + \displaystyle\int_0^t m(k)\,dk}{\displaystyle\int_0^t n'(k)\,dk}, & \exp\left[-\displaystyle\int_0^t \{m(k)+n'(k)\}\,dk\right] \le y \le \exp\left[-\displaystyle\int_0^t m(k)\,dk\right] \\[2em]
\dfrac{\ln(y) + \displaystyle\int_0^t m(k)\,dk}{\displaystyle\int_0^t l'(k)\,dk}, & \exp\left[-\displaystyle\int_0^t \{m(k)\}\,dk\right] \le y \le \exp\left[-\displaystyle\int_0^t \{m(k)-l'(k)\}\,dk\right]
\end{cases}
\tag{6.18}
$$

It is very clear that $R(t)$ is CBFN. Now we can consider the following two models:

**Model 1.** When the failure rate function is fixed, i.e., $F(t) = F$, then $l(t) = l$, $m(t) = m$, $n(t) = n$, $l'(t) = l'$ and $n'(t) = n'$. Now we have

$$
F_\alpha(t) = [m - l + \alpha l, m + n - \alpha n], \quad \forall \alpha \in [0,1]
$$

and $F_\beta(t) = [m - \beta l', m + \beta n'], \quad \forall \beta \in [0,1]$

Since $R(0) = 1$ and $R(\infty) = 0$, from Equations (6.19) and (6.20) we obtain

$$
\mu_{R(t)} = \begin{cases}
\dfrac{\ln(y) + (m+n)t}{nt}, & \exp[-(m+n)t] \le y \le \exp[-mt], \ 0 < t < \infty \\[1.5em]
-\dfrac{\ln(y) + (m-l)t}{lt}, & \exp[-mt] \le y \le \exp[-(m-l)t], \ 0 < t < \infty
\end{cases}
\tag{6.19}
$$

$$
v_{R(t)} = \begin{cases}
-\dfrac{\ln(y) + mt}{n't}, & \exp[-(m+n')t] \le y \le \exp[-mt], \ 0 < t < \infty \\[1.5em]
-\dfrac{\ln(y) + (m-l)t}{l't}, & \exp[-mt] \le y \le \exp[-(m-l')t], \ 0 < t < \infty
\end{cases}
\tag{6.20}
$$

**Model 2.** When the failure rate function is not constant, i.e., $F(t)$ depends on $m(t)$, $l(t)$, $n(t)$, $l'(t)$, and $n'(t)$.

Let us assume that $l(t) = l = $ constant, $n(t) = n = $ constant, $l'(t) = l' = $ constant, $n'(t) = n = $ constant, and $m(t) = pe^{qt}$, where $p$ is a positive constant. Since $R(0) = 1$ and $R(\infty) = 0$, from Equations (6.19) and (6.20) we get

$$
\mu_{R(t)} = \begin{cases}
\dfrac{\ln(y) + \frac{p}{q}\left[\exp(qt)-1\right] + nt}{nt}, & \exp\left[-\frac{p}{q}\left[\exp(qt)-1\right]-nt\right] \le y \le \exp\left[-\frac{p}{q}\left[\exp(qt)-1\right]\right] \\[1.5em]
-\dfrac{\ln(y) + \frac{p}{q}\left[\exp(qt)-1\right] - lt}{lt}, & \exp\left[-\frac{p}{q}\left[\exp(qt)-1\right]\right] \le y \le \exp\left[-\frac{p}{q}\left[\exp(qt)-1\right]+lt\right]
\end{cases}
$$
for $0 < t < \infty$

$$\tag{6.21}$$

$$\mu_{R(t)} = \begin{cases} -\dfrac{\ln(y) + \dfrac{p}{q}\left[\exp(qt) - 1\right]}{n't}, & \exp\left[-\dfrac{p}{q}[\exp(qt)-1] - n't\right] \le y \le \exp\left[-\dfrac{p}{q}[\exp(qt)-1]\right] \\[3em] \dfrac{\ln(y) + \dfrac{p}{q}\left[\exp(qt)-1\right]}{l't}, & \exp\left[-\dfrac{p}{q}[\exp(qt)-1]\right] \le y \le \exp\left[-\dfrac{p}{q}[\exp(qt)-1] + l't\right] \end{cases} \quad \text{for } 0 < t < \infty$$

$$(6.22)$$

## 6.5 Reliability evaluation of series and parallel system having components following time-dependent conflicting bifuzzy failure rate

### 6.5.1 Series system

Consider a series system having "*j*" components. Let the failure rate of the *i*th component, $\gamma_i(t)$ be represented as

$$\gamma_i(t) = \left[ m_i(t) - l_i(t), m_i(t), m_i(t) + n_i(t); \, m_i(t) - l_i'(t), m_i(t), m_i(t) + n_i'(t) \right]$$

Let the reliability of the *i*th component at time *t* be $R_i(t)$ for $i = 1, 2, \ldots, j$.
   Therefore, at time *t*, reliability of the system is

$$R_S(t) = R_1(t)R_2(t)\ldots R_j(t) \tag{6.23}$$

From Equation (6.8), we have

$$R_i(t) = \exp\left[ -\int_0^t \gamma_i(k)\,dk \right]; \quad i = 1, 2, \ldots, j, \, t > 0$$

Hence, the system reliability becomes

$$R_S(t) = \exp\left[ -\int_0^t \left( \sum_{i=1}^j \gamma_i(k)\,dk \right) \right] = \exp\left[ -\int_0^t \gamma_S(k)\,dk \right] \tag{6.24}$$

where $\gamma_S(t) = \displaystyle\sum_{i=1}^j \gamma_i(t)$.

   $\alpha$-Cut of $\gamma_i(t)$, for the membership function, is

$$\gamma_i(t, \alpha) = \left[ m_i(t) - l_i(t) + \alpha l_i(t), m_i(t) + n_i(t) - \alpha n_i(t) \right], \quad \forall \alpha \in [0, 1]$$

$\beta$-Cut of $\gamma_i(t)$, for the nonmembership function, is

$$\gamma_i(t, \beta) = \left[ m_i(t) - \beta l_i'(t), m_i(t) + \beta n_i'(t) \right], \quad \forall \beta \in [0, 1]$$

Hence, for the membership function, $\alpha$-cut of $\gamma_S(t)$ is

$$\gamma_S(t,\alpha) = \left[\sum_{i=1}^{j} m_i(t) - l_i(t) + \alpha l_i(t), \sum_{i=1}^{j} m_i(t) + n_i(t) - \alpha n_i(t)\right], \quad \forall \alpha \in [0,1] \qquad (6.25)$$

Similarly,

$$\gamma_S(t,\beta) = \left[\sum_{i=1}^{j} m_i(t) - \beta l_i'(t), \sum_{i=1}^{j} m_i(t) + \beta n_i'(t)\right], \quad \forall \alpha \in [0,1] \qquad (6.26)$$

Now since $R_S(t)$ is also a CBFN, therefore, from Equations (6.11)–(6.14), we can have the $\alpha$-cut and $\beta$-cut of $R_S(t)$, respectively, for the membership function and the nonmembership function as

$$R_S(t,\alpha) = \left[\exp\left\{-\int_0^t \left(\sum_{i=1}^{j} m_j(k) + n_j(k) - \alpha n_j(k)\right)dk\right\}, \exp\left\{-\int_0^t \left(\sum_{i=1}^{j} m_j(k) - l_j(k) + \alpha n_j(k)\right)dk\right\}\right]$$

$$(6.27)$$

$$R_S(t,\beta) = \left[\exp\left\{-\int_0^t \left(\sum_{i=1}^{j} m_j(k) + \beta n_j'(k)\right)dk\right\}, \exp\left\{-\int_0^t \left(\sum_{i=1}^{j} m_j(k) - \beta l_j'(k)\right)dk\right\}\right] \qquad (6.28)$$

From Model 1, considering the failure rate as a constant, then the $\alpha$-cut of $R_S(t)$ for the membership function and $\beta$-cut of $R_S(t)$ for the nonmembership function are obtained as

$$R_S(t,\alpha) = \left[\exp\left\{-t\left(\sum_{i=1}^{j} m_i + n_j - \alpha n_j\right)\right\}, \exp\left\{-t\left(\sum_{i=1}^{j} m_j - l_j + \alpha l_j\right)\right\}\right] \qquad (6.29)$$

$$R_S(t,\beta) = \left[\exp\left\{-t\left(\sum_{i=1}^{j} m_i + \beta n_j'\right)\right\}, \exp\left\{-t\left(\sum_{i=1}^{j} m_j - \beta l_j'\right)\right\}\right] \qquad (6.30)$$

From the Model 2, when the failure rate is not fixed, then the $\alpha$-cut and $\beta$-cut of $R_S(t)$ for the membership function and the nonmembership function are calculated as

$$R_S(t,\alpha) = \left[\exp\left\{-\int_0^t \left(\sum_{i=1}^{j} p_i e^{q_i k} + n_i - \alpha n_i\right)dk\right\}, \exp\left\{-\int_0^t \left(\sum_{i=1}^{j} p_i e^{q_i k} + l_i + \alpha l_i\right)dk\right\}\right] \qquad (6.31)$$

$$R_S(t,\beta) = \left[\exp\left\{-\int_0^t \left(\sum_{i=1}^{j} p_i e^{q_i k} + \beta n_i'\right)dk\right\}, \exp\left\{-\int_0^t \left(\sum_{i=1}^{j} p_i e^{q_i k} - \beta l_i'\right)dk\right\}\right] \qquad (6.32)$$

## 6.5.2   Parallel system

Consider a parallel system with "*j*" components with failure rate function $\gamma_i(t)$ for the *i*th component, where

$$\gamma_i(t) = \left[ m_i(t) - l_i(t), m_i(t), m_i(t) + n_i(t); m_i(t) - l_i'(t), m_i(t), m_i(t) + n_i'(t) \right]$$

Let the reliability of the *i*th component at time *t* be $R_i(t)$ for $i = 1, 2, \ldots, j$.

It is well known that the reliability of system $R_P(t)$ at time *t* is

$$R_P(t) = 1 - \prod_{i=1}^{j} \left( 1 - R_i(t) \right)$$

$$= 1 - \prod_{i=1}^{j} \left( 1 - \exp\left( -\int_0^t \gamma_i(k)\,dk \right) \right) \tag{6.33}$$

Now since $R_P(t)$ is also a CBFN, therefore, from Equations (6.11)–(6.14), the $\alpha$-cut and $\beta$-cut of $R_P(t)$ for the membership function and the nonmembership function, respectively, are obtained as

$$R_P(t, \alpha) = \left[ 1 - \prod_{i=1}^{j} \left( 1 - \exp\left( -\int_0^t \left( m_i(k) + n_i(k) - \alpha n_i(k) \right) dk \right) \right), \right.$$

$$\left. 1 - \prod_{i=1}^{j} \left( 1 - \exp\left( -\int_0^t \left( m_i(k) - l_i(k) + \alpha l_i(k) \right) dk \right) \right) \right] \tag{6.34}$$

$$R_P(t, \beta) = \left[ 1 - \prod_{i=1}^{j} \left( 1 - \exp\left( -\int_0^t \left( m_i(k) + \beta n_i'(k) \right) dk \right) \right), 1 - \prod_{i=1}^{j} \left( 1 - \exp\left( -\int_0^t \left( m_i(k) - \beta l_i'(k) \right) dk \right) \right) \right]$$

$$\tag{6.35}$$

From Model 1, if the failure rate is constant, then we have

$$R_P(t, \alpha) = \left[ 1 - \prod_{i=1}^{j} \left( 1 - \exp\left\{ -t\left( m_i + n_j - \alpha n_j \right) \right\} \right), 1 - \prod_{i=1}^{j} \left( 1 - \exp\left\{ -t\left( m_i - l_j + \alpha l_j \right) \right\} \right) \right] \tag{6.36}$$

$$R_P(t, \beta) = \left[ 1 - \prod_{i=1}^{j} \left( 1 - \exp\left\{ -t\left( m_i + \beta n_j' \right) \right\} \right), 1 - \prod_{i=1}^{j} \left( 1 - \exp\left\{ -t\left( m_i - \beta l_j' \right) \right\} \right) \right] \tag{6.37}$$

Again, from Model 2, if the failure rate is not constant, then we have

$$R_P(t, \alpha) = \left[ 1 - \prod_{i=1}^{j} \left( 1 - \exp\left( -\int_0^t \left( p_i e^{q_i k} + n_i - \alpha n_i \right) dk \right) \right), \right.$$

$$\left. 1 - \prod_{i=1}^{j} \left( 1 - \exp\left( -\int_0^t \left( p_i e^{q_i k} - l_i + \alpha l_i \right) dk \right) \right) \right] \tag{6.38}$$

$$R_P(t,\beta) = \left[ 1 - \prod_{i=1}^{j}\left(1 - \exp\left(-\int_0^t \left(p_i e^{q_i k} + \beta n_i'\right) dk\right)\right), \; 1 - \prod_{i=1}^{j}\left(1 - \exp\left(-\int_0^t \left(p_i e^{q_i k} - \beta l_i'\right) dk\right)\right) \right] \quad (6.39)$$

### 6.5.3 Parallel-series system

Consider a parallel-series system with "$j$" branches which are in parallel configuration and there are "$i$" components in each branch connected in a series configuration as shown in Figure 6.1.

Let failure rate function of the $s$th component of the $r$th branch ($r = 1, 2, \ldots, i$ and $s = 1, 2, \ldots, j$) be represented by time-dependent CBFS $\gamma_{rs}(t)$ as

$$\gamma_{rs}(t) = \left( m_{rs}(t) - l_{rs}(t), m_{rs}(t), m_{rs}(t) + n_{rs}(t); \; m_{rs}(t) - l_{rs}'(t), m_{rs}(t), m_{rs}(t) + n_{rs}'(t) \right)$$

From Equation (6.27), reliability of the $s$th component of the $r$th branch is given by

$$R_{rs}(t) = \exp\left[ -\int_0^t \gamma_{rs}(k)\, dk \right]$$

It is well known that the reliability of a parallel-series system $R_{PS}(t)$ at time $t$ is

$$R_{PS}(t) = 1 - \prod_{s=1}^{j}\left(1 - \prod_{r=1}^{i} R_{rs}\right)$$

$$= 1 - \prod_{s=1}^{j}\left(1 - \prod_{r=1}^{i} \exp\left(-\int_0^t \gamma_{rs}(k)\, dk\right)\right) \quad (6.40)$$

Now since $R_{PS}(t)$ is also a CBFN, therefore, from Equations (6.31)–(6.34), the $\alpha$-cut and $\beta$-cut of $R_{PS}(t)$, respectively, for the membership and the nonmembership functions, are obtained as

$$R_{PS}(t,\alpha) = \left[ 1 - \prod_{s=1}^{j}\left(1 - \prod_{r=1}^{i} \exp\left(-\int_0^t \left(m_{rs}(k) + n_{rs}(k) - \alpha n_{rs}(k)\right) dk\right)\right), \right.$$

$$\left. 1 - \prod_{s=1}^{j}\left(1 - \prod_{r=1}^{i} \exp\left(-\int_0^t \left(m_{rs}(k) - l_{rs}(k) + \alpha l_{rs}(k)\right) dk\right)\right) \right] \quad (6.41)$$

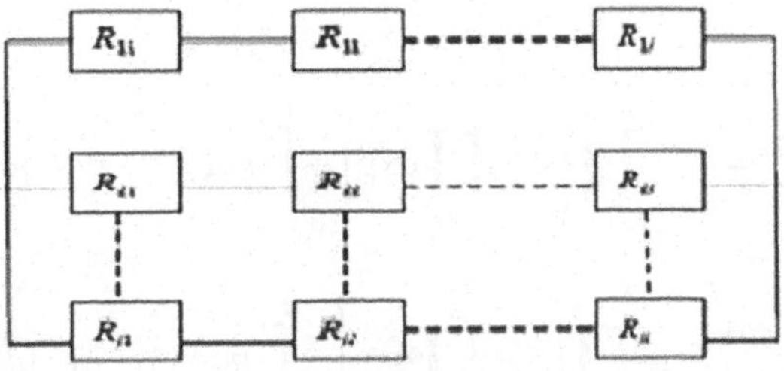

*Figure 6.1* Parallel-series system.

$$R_{PS}(t,\beta) = \left[ 1 - \prod_{s=1}^{j}\left( 1 - \prod_{r=1}^{i} \exp\left( -\int_{0}^{t}(m_{rs}(k) + \beta n'_{rs}(k))\,dk \right) \right),\right.$$

$$\left. 1 - \prod_{s=1}^{j}\left( 1 - \prod_{r=1}^{i} \exp\left( -\int_{0}^{t}(m_{rs}(k) - \beta l'_{rs}(k))\,dk \right) \right) \right] \qquad (6.42)$$

if the failure rate is constant, then we have

$$R_{PS}(t,\alpha) = \left[ 1 - \prod_{s=1}^{j}\left( 1 - \prod_{r=1}^{i}\left( \exp\{-t(m_{rs} + n_{rs} - \alpha n_{rs})\} \right) \right),\right.$$

$$\left. 1 - \prod_{s=1}^{j}\left( 1 - \prod_{r=1}^{i}\left( \exp\{-t(m_{rs} - l_{rs} + \alpha l_{rs})\} \right) \right) \right] \qquad (6.43)$$

$$R_{PS}(t,\beta) = \left[ 1 - \prod_{s=1}^{j}\left( 1 - \prod_{r=1}^{i} \exp\{-t(m_{rs} + \beta n'_{rs})\} \right),\right.$$

$$\left. 1 - \prod_{s=1}^{j}\left( 1 - \prod_{r=1}^{i} \exp\{-t(m_{rs} + \beta l'_{rs})\} \right) \right] \qquad (6.44)$$

Again, if the failure rate function is not constant, i.e., if the failure rate function $\gamma_{rs}(t)$ of the $s$th component of the $r$th branch is represented as

$$\gamma_{rs}(t) = \left( m_{rs}(t) - l_{rs}(t), m_{rs}(t), m_{rs}(t) + n_{rs}(t);\ m_{rs}(t) - l'_{rs}(t), m_{rs}(t), m_{rs}(t) + n'_{rs}(t) \right)$$

where $l_{rs}(t) = l_{rs} = $ constant, $n_{rs}(t) = n_{rs} = $ constant, $l'_{rs}(t) = l'_{rs} = $ constant, $n'_{rs}(t) = n'_{rs} = $ constant, and $m(t) = p_{rs}e^{q_{rs}t}$, here $p_{rs}$ is a positive constant, then we have

$$R_{PS}(t,\alpha) = \left[ 1 - \prod_{s=1}^{j}\left( 1 - \prod_{r=1}^{i} \exp\left( -\int_{0}^{t}(p_{rs}e^{q_{rs}k} + n_{rs} - \alpha n_{rs})\,dk \right) \right),\right.$$

$$\left. 1 - \prod_{s=1}^{j}\left( 1 - \prod_{r=1}^{i} \exp\left( -\int_{0}^{t}(p_{rs}e^{q_{rs}k} - l_{rs} + \alpha l_{rs})\,dk \right) \right) \right] \qquad (6.45)$$

$$R_{PS}(t,\beta) = \left[ 1 - \prod_{s=1}^{j}\left( 1 - \prod_{r=1}^{i} \exp\left( -\int_{0}^{t}(p_{rs}e^{q_{rs}k} + \beta n'_{rs})\,dk \right) \right),\right.$$

$$\left. 1 - \prod_{s=1}^{j}\left( 1 - \prod_{r=1}^{i} \exp\left( -\int_{0}^{t}(p_{rs}e^{q_{rs}k} - \beta l'_{rs})\,dk \right) \right) \right] \qquad (6.46)$$

## 6.5.4 Series-parallel system

Consider a system having "$i$" subsystems connected in series and each subsystem contains "$j$" components connected in parallel as in Figure 6.2.

Let the failure rate function of the $r$th component of the $s$th subsystem be represented by time-dependent CBFS $\gamma_{rs}(t)$ as

$$\gamma_{rs}(t) = \left( m_{rs}(t) - l_{rs}(t), m_{rs}(t), m_{rs}(t) + n_{rs}(t); m_{rs}(t) - l'_{rs}(t), m_{rs}(t), m_{rs}(t) + n'_{rs}(t) \right)$$

From Equation (6.10), reliability of the $r$th component of the $s$th subsystem is

$$R_{rs}(t) = \exp\left[ -\int_0^t \gamma_{rs}(k)\,dk \right]$$

It is well known that the reliability of system $R_{SP}(t)$ at time $t$ is

$$R_{SP} = \prod_{s=1}^{i}\left( 1 - \prod_{r=1}^{j}(1 - R_{rs}) \right)$$

$$= \prod_{s=1}^{i}\left( 1 - \prod_{r=1}^{j}\left( 1 - \exp\left( -\int_0^t \gamma_{rs}(k)\,dk \right) \right) \right) \tag{6.47}$$

Now since $R_{SP}(t)$ is also a CBFN, therefore, from Equations (6.11)–(6.14), the $\alpha$-cut and $\beta$-cut of $R_{SP}(t)$ for the membership function and the nonmembership function, respectively, are obtained as

$$R_{SP}(t,\alpha) = \left[ \prod_{s=1}^{i}\left( 1 - \prod_{r=1}^{j}\left( 1 - \exp\left( -\int_0^t (m_{rs}(k) + n_{rs}(k) - \alpha n_{rs}(k))\,dk \right) \right) \right), \right.$$

$$\left. \prod_{s=1}^{i}\left( 1 - \prod_{r=1}^{j}\left( 1 - \exp\left( -\int_0^t (m_{rs}(k) - l_{rs}(k) + \alpha l_{rs}(k))\,dk \right) \right) \right) \right] \tag{6.48}$$

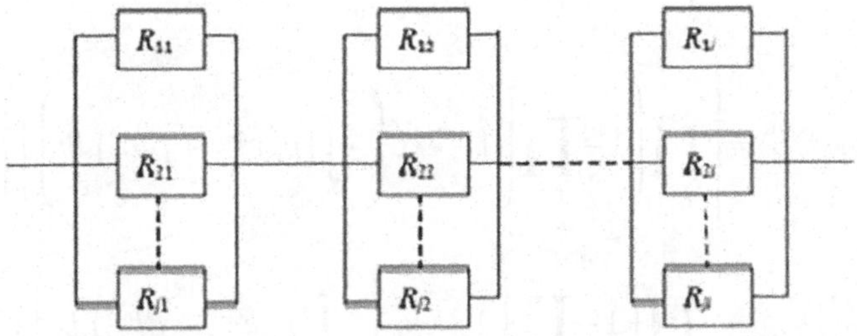

*Figure 6.2* Series-parallel system.

$$R_{SP}(t,\beta) = \left[ \prod_{s=1}^{i}\left( 1 - \prod_{r=1}^{j}\left( 1 - \exp\left( -\int_{0}^{t}\left( m_{rs}(k) + \beta n'_{rs}(k)\right)dk \right) \right) \right), \right.$$

$$\left. \prod_{s=1}^{i}\left( 1 - \prod_{r=1}^{j}\left( 1 - \exp\left( -\int_{0}^{t}\left( m_{rs}(k) - \beta l'_{rs}(k)\right)dk \right) \right) \right) \right] \tag{6.49}$$

Now, if the failure rate function is constant, then we have

$$R_{SP}(t,\alpha) = \left[ \prod_{s=1}^{i}\left( 1 - \prod_{r=1}^{j}\left( 1 - \exp\left(-t\left(m_{rs} + n_{rs} - \alpha n_{rs}\right)\right) \right) \right), \right.$$

$$\left. \prod_{s=1}^{i}\left( 1 - \prod_{r=1}^{j}\left( 1 - \exp\left(-t\left(m_{rs} - l_{rs} + \alpha l_{rs}\right)\right) \right) \right) \right] \tag{6.50}$$

$$R_{SP}(t,\beta) = \left[ \prod_{s=1}^{i}\left( 1 - \prod_{r=1}^{j}\left( 1 - \exp\left(-t\left(m_{rs} + \beta n'_{rs}\right)\right) \right) \right), \right.$$

$$\left. \prod_{s=1}^{i}\left( 1 - \prod_{r=1}^{j}\left( 1 - \exp\left(-t\left(m_{rs} - \beta l'_{rs}\right)\right) \right) \right) \right] \tag{6.51}$$

Again, if the failure rate function is not constant, i.e., if the failure rate function $\gamma_{rs}(t)$ is represented as

$$\gamma_{rs}(t) = \left( m_{rs}(t) - l_{rs}(t), m_{rs}(t), m_{rs}(t) + n_{rs}(t); m_{rs}(t) - l'_{rs}(t), m_{rs}(t), m_{rs}(t) + n'_{rs}(t) \right)$$

where $l_{rs}(t) = l_{rs} =$ constant, $n_{rs}(t) = n_{rs} =$ constant, $l'_{rs}(t) = l'_{rs} =$ constant, $n'_{rs}(t) = n'_{rs} =$ constant, and $m(t) = p_{rs}e^{q_{rs}t}$, where $p_{rs}$ is a positive constant, then we have

$$R_{SP}(t,\alpha) = \left[ \prod_{s=1}^{i}\left( 1 - \prod_{r=1}^{j}\left( 1 - \exp\left( -\int_{0}^{t}\left( p_{rs}e^{q_{rs}k} + n_{rs} - \alpha n_{rs}\right)dk \right) \right) \right), \right.$$

$$\left. \prod_{s=1}^{i}\left( 1 - \prod_{r=1}^{j}\left( 1 - \exp\left( -\int_{0}^{t}\left( p_{rs}e^{q_{rs}k} - l_{rs} + \alpha l_{rs}\right)dk \right) \right) \right) \right] \tag{6.52}$$

$$R_{SP}(t,\beta) = \left[ \prod_{s=1}^{i}\left( 1 - \prod_{r=1}^{j}\left( 1 - \exp\left( -\int_{0}^{t}\left( p_{rs}e^{q_{rs}k} + \beta n'_{rs}\right)dk \right) \right) \right), \right.$$

$$\left. \prod_{s=1}^{i}\left( 1 - \prod_{r=1}^{j}\left( 1 - \exp\left( -\int_{0}^{t}\left( p_{rs}e^{q_{rs}k} - \beta l'_{rs}\right)dk \right) \right) \right) \right] \tag{6.53}$$

## 6.6   Examples

Here in this section, some numerical examples are discussed to illustrate the new approach.

### 6.6.1   Series system

Consider an example of a hydro power plant. Let there be three turbines in a hydro power plant and all must work normally to generate a predefined amount of electricity. Let the failure rates of turbines be in the form of triangular CBFS as

$$\gamma_1 = (0.0015, 0.002, 0.0035; 0.001, 0.002, 0.0045)$$

$$\gamma_2 = (0.0017, 0.002, 0.003; 0.0015, 0.002, 0.004)$$

$$\gamma_3 = (0.0019, 0.002, 0.0037; 0.0017, 0.002, 0.005)$$

The system reliability of a power plant is calculated by using Equations (6.29) and (6.30). The reliabilities are obtained as triangular CBFN for a conflicting bifuzzy failure rate of turbines with different values of time $t$.

When time $t = 150$, then the system reliability is obtained as

$$R_S(150) = (0.2165, 0.4065, 0.4653; 0.1423, 0.4065, 0.5325)$$

When time $t = 175$, then the system reliability is obtained as

$$R_S(175) = (0.1677, 03499, 0.4096; 0.1027, 0.3499, 0.4795)$$

When time $t = 225$, then the system reliability is obtained as

$$R_S(225) = (0.1007, 0.2592, 0.3174; 0.0536, 0.2592, 0.3886)$$

### 6.6.2   Parallel system

Suppose there are three active and independent transmitters in a broadcast station. For a successful transmission at least one of these transmitters must work properly. Here, failure rates of transmitters are as

$$\gamma_1 = (0.05, 0.1, 0.13; 0.04, 0.1, 0.15)$$

$$\gamma_2 = (0.06, 0.1, 0.14; 0.05, 0.1, 0.16)$$

$$\gamma_3 = (0.07, 0.1, 0.15; 0.06, 0.1, 0.17)$$

The reliability of the system is evaluated using Equations (6.38) and (6.39). The system reliability is triangular CBFN for the conflicting bifuzzy failure rate of components with different values of time $t$.

At $t = 10$   $R_P(10) = (0.5742, 0.7474, 0.9106; 0.4932, 0.7474, 0.9414)$

At $t = 15$   $R_P(15) = (0.3266, 0.7964, 0.5311; 0.2489, 0.7964, 0.8802)$

At $t = 20$   $R_P(20) = (0.1738, 0.3535, 0.6672; 0.1189, 0.3535, 0.7567)$

### 6.6.3  Parallel-series system

Suppose there is a communication system that receives the input signal and transmits the output signal. For this there are two receivers and two transmitters in the system connected as is shown in Figure 6.3. For a successful communication, at least one receiver and one transmitter connected in series configuration must work properly.

The failure rates of receivers and transmitters are in the form of triangular CBFNs given as

$$\gamma_{11} = (0.05, 0.1, 0.15; 0.04, 0.1, 0.16)$$

$$\gamma_{12} = (0.06, 0.11, 0.16; 0.05, 0.11, 0.17)$$

$$\gamma_{21} = (0.07, 0.12, 0.17; 0.06, 0.12, 0.18)$$

$$\gamma_{22} = (0.08, 0.13, 0.18; 0.07, 0.13, 0.19)$$

The reliability of the system is evaluated using Equations (6.43) and (6.44).
At time $t = 10$ system reliability is

$$R_{PS}(10) = (0.1394, 0.6046, 0.9058; 0.0816, 0.6046, 0.9429)$$

At time $t = 20$ system reliability is

$$R_{PS}(20) = (0.4381, 0.9676, 0.9959; 0.2551, 0.9676, 0.9997)$$

### 6.6.4  Series-parallel system

Suppose there is a communication system, which receives the input signal and transmits the output signal. For this there are two receivers and two transmitters in the system connected as shown in Figure 6.4. For a successful communication, at least one receiver and one transmitter must work properly.

The failure rates of receivers and transmitters are in the form of triangular CBFNs given as

$$\gamma_{11} = (0.05, 0.1, 0.15; 0.04, 0.1, 0.16)$$
$$\gamma_{21} = (0.07, 0.12, 0.17; 0.06, 0.12, 0.18)$$
$$\gamma_{12} = (0.06, 0.11, 0.16; 0.05, 0.11, 0.17)$$
$$\gamma_{22} = (0.08, 0.13, 0.18; 0.07, 0.13, 0.19)$$

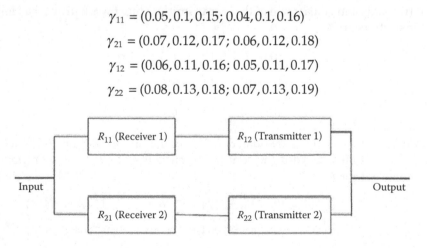

*Figure 6.3* Parallel-series system.

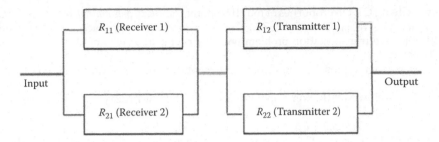

*Figure 6.4* Series-parallel system.

The reliability of the system is evaluated using Equations (6.51) and (6.52).
   At $t = 10$ system reliability is

$$R_{SP}(10) = (0.000001, 0.00021, 0.0092; 0.0000006, 0.00021, 0.0176)$$

At $t = 20$ system reliability is

$$R_{SP}(20) = (0.000004, 0.00040, 0.0513; 0.0000012, 0.00040, 0.0871)$$

## 6.7   Conclusion

In this study, a procedure is introduced to construct the membership and the nonmembership functions of the fuzzy reliability function, by considering the failure rates as time-dependent CBFN. With the introduced approach, reliability of different systems (series, parallel, parallel-series, and series-parallel systems) is evaluated in the form of a triangular CBFN. In all of these systems, the failure rate of each component is taken as time-dependent triangular CBFS. Since the fuzzy set and intuitionistic fuzzy set are the special cases of the CBFS, the proposed method is very nicely applicable in these types of sets also. Hence, we can conclude that this approach can easily be applied for the assessment of the reliability of the systems whenever there is some uncertainty in the information, available for the systems.

## References

Atanassov K. Intuitionistic fuzzy sets. *Fuzzy Sets and Systems* 1986; 20(1):87–96.

Atanassov K. *Intuitionistic Fuzzy Sets*. New York: Physica-Verlag; 1999.

Basu K, Deb R, Pattanaik PK. Soft sets: An ordinal formulation of vagueness with some application to the theory of choice. *Fuzzy Sets and Systems* 1992; 45:45–58.

Burillo P, Bustinces H. Construction theorem for intuitionistic fuzzy sets. *Fuzzy Sets and Systems* 1996; 84:271–281.

Cai KY, Wen CY, Zhang ML. Fuzzy variables as a basis for a theory of fuzzy reliability in the possibility context. *Fuzzy Sets and Systems* 1991; 42(2):145–172.

Chen SM. Fuzzy system reliability analysis using fuzzy number arithmetic operations. *Fuzzy Sets and Systems* 1994; 64(1):31–38.

Chen SM. New method for fuzzy system reliability analysis. *Cybernetics and Systems: An International Journal* 1996; 27:385–401.

Gau WL, Buehrer DJ. Vague sets. *IEEE Transactions on Systems, Man, and Cybernetics* 1993; 23:610–614.

Gianpiero C, David C. Basic intuitionistic principle in fuzzy set theories and its extension (A terminological debate on Atanassov IFS). *Fuzzy Sets and Systems* 2006; 157:3198–3219.

Goguen J. L-fuzzy sets. *Journal of Mathematical Analysis and Applications* 1967; 18:145–174.

Li DF, Shan F, Cheng CT. On properties of four IFS operators. *Fuzzy Sets and Systems* 2005; 154:151–155.

Roy SK, Maity G, Weber GW, Gok SZA. Conic scalarization approach to solve multi-choice multi-objective transportation problem with interval goal. *Annals of Operations Research* 2017; 253(1):599–620.

Singer D. A fuzzy set approach to fault tree and reliability analysis. *Fuzzy Sets and Systems* 1990; 34(2):145–155.

Supriya KD, Ranjit B, Akhil RR. Some operations on intuitionistic fuzzy sets. *Fuzzy Sets and Systems* 2005; 156:492–495.

Zadeh LA. Fuzzy sets. *Information and Control* 1965; 8(3):338–353.

Zamali T, Lazim MA, Osman MTA. An introduction to conflicting bifuzzy set theory. *International Journal of Mathematics and Statistics* 2008; 3(A08):86–95.

Zhang WR, Zhang L. Yin-Yang bipolar logic and bipolar fuzzy logic. *Information Sciences* 2004; 165:265–287.

*chapter seven*

# Recent progress on failure time data analysis of repairable system

**Yasuhiro Saito**
*Japan Coast Guard Academy*

**Tadashi Dohi**
*Hiroshima University*

## Contents

## 7.1   Introduction

In recent years, industrial systems have been more large scaled and more complex, and played a significant role to improve the quality of our daily life. For utilizing the ability of such systems, it is important to understand the behavior of failure phenomena of the industrial systems. People who operate the repairable systems may try to assess the system reliability and/or availability accurately for a long time. In addition, estimating the cost to maintain operation of the repairable systems is regarded as an important issue for practitioners. To describe the stochastic behavior of a cumulative number of failures occurring as the operating time progresses, we can apply the stochastic point processes as a powerful mathematical tool. In fact, there are many research results for life data analysis of the industrial systems including production machines, which are based on stochastic modeling of time-to-failure phenomena. By modeling a conditional intensity function that represents the rate of occurrence of failure at an arbitrary time, the various types of failure phenomena can be described by the stochastic point processes. These stochastic point processes can be characterized by the failure time (lifetime) distribution and/or the kind of repair operation (Ascher and Feingold 1984; Nakagawa 2005).

For non-repairable systems, the failed system or component is usually replaced by the new one or repaired appropriately, when a failure occurs. In the situation where the replacement time can be assumed to be negligible with respect to the failure time scale, the time evolution can be described by a renewal process (RP) with independent and identically distributed (i.i.d.) inter-failure time distributions (renewal distribution) (Cox 1972). We may also consider the failure occurrence phenomena in repairable systems by modeming with another representative stochastic point processes. It is well known that the nonhomogeneous Poisson process (NHPP) is the simplest but most useful tool for modeling such phenomena. In the repairable systems, we perform the repair action after failure occurs in order to return the failed component to the normal condition. Such an activity may restore only the damaged part of the failed component back to a working condition that is only as good as it was just before the failure in some cases. This repair action is called *minimal repair*. The minimal repair process with a negligible repair time sequence can be represented by a NHPP. Therefore, the analytical treatment of the minimal repair process is rather easier than that in the RP.

In this chapter we mainly focus on the failure time data analysis based on the NHPP and discuss several statistical estimation methods for a periodic replacement problem with minimal repair as the simplest application of life data analysis with NHPP. NHPP is characterized by the intensity function, which represents the rate at which events occur, or the corresponding mean value function, which is defined by the integral of the intensity function and means the expected cumulative number of events by an arbitrary time. Two types of statistical inference approaches for NHPP are considered according to the situation where we can know the information on intensity function (or equivalently mean value function) or cannot in advance. If the form of intensity function is known in advance, a parametric model having any parametric intensity function is usually applied. But, if the form of intensity function is unknown, nonparametric models may be applied to avoid the mis-specification of failure occurrence phenomena. In this chapter, we consider the well-known parametric model called the *power law process* as an example, and summarize two nonparametric methods called the *constrained nonparametric maximum likelihood estimation* (CNPMLE) and the *kernel-based estimation* for the NHPP.

As an application example of statistical inference of failure processes, a periodic replacement problem with minimal repair is one of the most fundamental, but most important maintenance solutions (Barlow and Proschan 1996). The original periodic replacement model with minimal repair has been extended from various points of view by several authors (Boland 1982; Colosimo et al. 2010; Nakagawa 1986; Park et al. 2000; Sheu 1990, 1991; Valdez-Flores and Feldman 1989) after the seminal contribution by Barlow and Hunter (1960). Recently, Okamura et al. (2014) developed a dynamic programming algorithm to compute effectively the optimal periodic replacement time in Nakagawa (1986).

We apply not only the parametric maximum likelihood estimation for the power law process but also CNPMLE and kernel-based estimation methods for estimating the cost-optimal periodic replacement problem with minimal repair, where single or multiple minimal repair data are assumed. The former means that a single time series of failure (minimal repair) time is observed; the latter implies that the multiple time series data are observed from multiple production machines, where the multiple data involve the single data case as a special case. In the numerical example, we conduct a simulation experiment of single minimal repair process and a real data analysis with the multiple field data sets of minimal repair of diesel engine in Nelson and Doganaksoy (1989).

## 7.2   Model description

Suppose that more than one failure may occur in each system component of a repairable system. Usually, two kinds of repair actions are performed to return the failed component state to the normal condition after each failure. One is called the *minimal repair* (Barlow and Proschan 1996). This repair activity restores only the damaged part of the failure component back to a working condition that is only as good as it was just before the failure. Another is called the *periodic replacement*. This is a preventive maintenance action which is planned in advance, where the used component is replaced by a new one at a prespecified time. For describing the failure occurrence phenomena under such repair activities, it is well known that an NHPP is useful. That is, NHPP can be used to model the stochastic behavior of a cumulative number of failures under the minimal repair.

More specifically, suppose that the failure time $T$ follows an absolutely continuous probability distribution function $\Pr(T \le t) = F(t)$ and a probability density function $dF(t)/dt = f(t)$. Define $\bar{F}(t) = 1 - F(t)$. If the minimal repair is made at the first failure, then the probability that the system does not fail beyond the time $t$ is given by $\bar{F}(t) + \int_0^t \left( \bar{F}(t)/\bar{F}(x) \right) dF(x)$. Continuing similar manipulations up to the $n$-th failure yields an NHPP, $\{N(t), t \ge 0\}$, with the mean value function (Baxter 1982):

$$E\left[ N(t) \right] = \Lambda(t) = -\log \bar{F}(t) \tag{7.1}$$

where $N(t)$ represents the cumulative number of minimal repairs by time $t$. It is well known that the NHPP $\{N(t), t \ge 0\}$ possesses the following properties:

- $N(0) = 0$
- $\{N(t), t \ge 0\}$ has independent increments
- $\Pr\{N(t + \Delta t) - N(t) \ge 2\} = o(\Delta t)$
- $\Pr\{N(t + \Delta t) - N(t) = 1\} = \lambda(t)\,\Delta t + o(\Delta t)$

where $o(\Delta t)$ is the higher term of $\Delta t$, and the function $\lambda(t)$ is called the *intensity function of* NHPP. The mean value function in Equation (7.1) is also defined as an integral of the intensity function:

$$\Lambda(t) = \int_0^t \lambda(x)\,dx \tag{7.2}$$

and means the expected cumulative number of failures occurred by time $t$. Then, the probability mass function (p.m.f.) of the NHPP is given by

$$\Pr\{N(t) = n\} = \frac{\{\Lambda(t)\}^n}{n!} \exp\{-\Lambda(t)\} \tag{7.3}$$

If the component fails before time $\tau\ (>0)$, then the failed component is restored to a working condition that is only as good as it was just before failure in the periodic replacement problem with minimal repair. Here, the minimal repair is made so that the failure rate remains undisturbed by repair after each failure. Also, after the operational time reaches $\tau$, we replace the used component by a new one preventively. Therefore, it is easier to plan the preventive replacement periodically than the age replacement since the past replacement history is not needed to record in the periodic replacement. However, an additional cost is necessary for the periodic preventive replacement since the used component is replaced by a new one at time $\tau$, where the repaired component before time $\tau$ is also used in operation. Define the time length from the beginning of the operation to the periodic replacement as one cycle. Then, the expected total cost for one cycle can be represented by $c_m \Lambda(\tau) + c_p$, where $c_m$ (>0) and $c_p$ (>0) represent the fixed cost of each minimal repair and a periodic replacement, respectively. Dividing the expected cost for one cycle by time $\tau$ leads to the long-run average cost per unit time in the periodic replacement with minimal repair:

$$C(\tau) = \frac{c_m \Lambda(\tau) + c_p}{\tau} = \frac{c_m \int_0^\tau \lambda(t)dt + c_p}{\tau} \tag{7.4}$$

where $\tau$ (>0) is a decision variable in our problem and denotes the periodic replacement time. Then, the purpose is to derive $\tau^*$ which minimizes Equation (7.4). Differentiating Equation (7.4) with respect to $\tau$ and setting it to zero implies the first-order condition of optimality:

$$\tau \lambda(\tau) - \int_0^\tau \lambda(t)dt = \gamma \tag{7.5}$$

where $\gamma$ is called the *cost ratio* and defined by $\gamma = c_p / c_m$. For solving the nonlinear Equation (7.5) with respect to $\tau$, the unknown intensity function has to be estimated via an either parametric or nonparametric way. This is usually done based on the information on the intensity function, which is available from the past minimal repair record or history. Under the strictly increasing intensity function, i.e., $d\lambda(t)/dt > 0$, if $\lim\limits_{\tau \to \infty} \left( \tau \lambda(\tau) - \int_0^\tau \lambda(t)dt \right) > \gamma$ holds, then a unique and finite optimal periodic replacement time $\tau^*$ $(0 < \tau^* < \infty)$ minimizing Equation (7.4) always exists.

## 7.3   *Parametric estimation method*

The maximum likelihood (ML) estimation is a commonly used technique to identify the failure time distribution or the minimal repair process. In this method, we assume that the form of intensity function is known in advance, and estimate the model parameters with the underlying minimal repair data. As the representative parametric models, the power-law process (Crow 1974) and Cox–Lewis process (Cox and Lewis 1966) are frequently assumed without justification for the failure occurrence phenomena under minimal repair. Once the intensity function is specified, the problem is reduced to estimate the model parameters included in the intensity function using the underlying minimal repair (failure time) data, and the point estimate of the optimal periodic replacement time (or the corresponding minimum long-run average cost per unit time) is derived as a plug-in estimate with model parameters estimated by the ML method.

## 7.3.1 Single failure-occurrence time data case

We assume at the moment that the form of intensity function is completely known from the past experience. Various failure occurrence phenomena in different situations can be considered by assuming the form of intensity function $\lambda(t)$ or the corresponding mean value function $\Lambda(t)$ appropriately. The *power law model* (Crow 1974) is also assumed for easy understanding:

$$\lambda(t;\eta,\beta) = \left(\frac{\beta}{\eta}\right)\left(\frac{t}{\eta}\right)^{\beta-1} \tag{7.6}$$

It is also known that the power law model represents two different situations where the repairable systems monotonically deteriorate or are monotonically improved over time $t$. The intensity function $\lambda(t)$ is decreasing with time under the model parameter $\beta < 1$, which means that the system is deteriorating in time. But, $\lambda(t)$ is increasing with time under the model parameter $\beta > 1$, which implies the system's improvement. Furthermore, $\lambda(t)$ is constant when $\beta = 1$. This corresponds to the special case where the underlying failure process reduces to a homogeneous Poisson process.

Once the model is selected, the next step is to estimate the model parameters $(\eta, \beta)$ with failure-occurrence time data. Suppose that $n$ failure-occurrence time data, which are the random variables, are given by $0 < T_1 \le T_2 \le \cdots \le T_n \le \tilde{T}$, where $\tilde{T}$ is the right censoring time of these observation data. That is, it is assumed that $n$ failures occur by time $\tilde{t}$, which is realization of $\tilde{T}$, and the realizations of $T_i$ $(i = 1, 2, \ldots, n)$, say, $t_i$, are observed, where $t_n \le \tilde{t}$. Then, the ML estimates $(\hat{\eta}, \hat{\beta})$ are defined as the parameters that maximize the following likelihood function:

$$\mathrm{LF}(\eta,\beta \mid t_i) = \prod_{i=1}^{n} \lambda(t_i;\eta,\beta) \exp\left(-\Lambda(\tilde{t};\eta,\beta)\right) \tag{7.7}$$

For easy calculation, it is usually considered to maximize the log-likelihood function, which is derived by taking the logarithm of both sides in Equation (7.7):

$$\mathrm{LLF}(\eta,\beta \mid t_i) = \sum_{i=1}^{n} \log \lambda(t_i;\eta,\beta) - \Lambda(\tilde{t};\eta,\beta) \tag{7.8}$$

From the first-order condition of optimality in Equation (7.8) for each parameter, the ML estimates $(\hat{\eta}, \hat{\beta})$ can be derived by solving the following simultaneous equations:

$$\hat{\eta} = \frac{\tilde{t}}{n^{1/\hat{\beta}}} \tag{7.9}$$

$$\hat{\beta} = \frac{n}{\displaystyle\sum_{i=1}^{n} \log \frac{\tilde{t}}{t_i}} \tag{7.10}$$

Following the above procedure, we obtain the ML-based plug-in point estimates, $\hat{\tau}^*$ and $C(\hat{\tau}^*)$, of the optimal periodic replacement time $\tau^*$ and its associated minimum long-run average cost per unit time by substituting the resulting intensity function with the estimates $\left(\hat{\eta}, \hat{\beta}\right)$ into Equation (7.5). From the property of intensity function of the power law process, if the parameter $\beta$ is greater than 1, then it is guaranteed to exist a unique and finite optimal periodic replacement time.

### 7.3.2  Multiple failure-occurrence time data case

In the failure time data analysis, it is often assumed that not only single but also multiple time-series data on minimal repair processes are observed. Suppose that $l$-independent sets of failure-occurrence time data for identical components are available. Let $t_{ji}$ $\left(j = 1, 2, ..., l; i = 1, 2, ..., n_j\right)$ represent $i$th failure-occurrence time data of $j$th data set, where $n_j$ means the number of failures occurred in the $j$th component. Also, each failure-occurrence time datum is assumed to be censored at time $\hat{t}_j$. The censoring time can be considered to be deterministic or regarded as an observation of random censoring time $\tilde{T}_j$. It is reasonable to assume that $\tilde{T}_1, \tilde{T}_2, ..., \tilde{T}_l$ follow an identically unknown probability distribution that is independent of the underlying Poisson process in the latter case. In other words, their probability distribution does not depend on the model parameters $(\eta, \beta)$ of the intensity function. Then the ML estimates of model parameters $\left(\hat{\eta}, \hat{\beta}\right)$, which are included in the intensity function or mean value function of NHPP, are defined by maximizing the following likelihood function:

$$\text{LLF}\left(\eta, \beta \mid t_{ji}\right) = \left(\prod_{j=1}^{l} \prod_{i=1}^{n_j} \lambda\left(t_{ji}\right)\right) \exp\left(\sum_{j=1}^{l} \int_0^{\hat{t}_j} \lambda(x) dx\right) f\left(\tilde{t}\right) \tag{7.11}$$

where $f\left(\tilde{t}\right)$ is the joint density function of $\tilde{T}_1, \tilde{T}_2, ..., \tilde{T}_l$. Since this density function is also independent of model parameters, the ML estimates $\left(\hat{\eta}, \hat{\beta}\right)$ can be derived by maximizing the following log-likelihood function similar to the single failure-occurrence time data case:

$$\text{LLF}\left(\eta, \beta \mid t_{ji}\right) = \sum_{j=1}^{l} \left\{ \sum_{i=1}^{nj} \log \lambda\left(t_{ji}; \eta, \beta\right) - \Lambda\left(\hat{t}_j; \eta, \beta\right) \right\} \tag{7.12}$$

From the first-order condition of optimality in Equation (7.12), ML estimates $\left(\hat{\eta}, \hat{\beta}\right)$ can be calculated with multiple failure-occurrence time data sets by solving the following simultaneous equations:

$$\hat{\eta} = \frac{1}{m} \left( \sum_{j=1}^{l} \frac{\hat{t}_j^{\hat{\beta}}}{n_j} \right)^{\frac{1}{\hat{\beta}}} \tag{7.13}$$

$$0 = \sum_{j=1}^{l} \left( \frac{n_j}{\hat{\beta}} + \sum_{i=1}^{nj} \log t_{ji} - \left( \frac{\hat{t}_j}{\hat{\eta}} \right)^{\hat{\beta}} \log \hat{t}_j \right) \tag{7.14}$$

Unfortunately, in this case, $\hat{\beta}$ cannot be derived in closed form. For the case of multiple minimal repair data sets, by assuming the form of intensity function $\lambda(t)$ or the corresponding mean value function $\Lambda(t)$, the model parameters that are included in these functions can be derived regardless of the form of intensity function. Once the estimates of intensity function are obtained, we can get the optimal periodic replacement time with Equation (7.5). Therefore, in a similar way with single minimal repair data, it is also easy to handle the multiple minimal repair data case.

Though a number of variations have been studied in the literature for the preventive maintenance problems including the age replacement problem and the periodic replacement problem, it is assumed in many cases that the failure time distribution or equivalently minimal repair process is completely known. Therefore, once the failure time distribution is specified, the problem is to estimate the model parameters from the underlying failure time data, and the point estimate of the optimal preventive maintenance time is derived as a plug-in estimate with the estimated model parameters. In other words, when the failure time distribution is unknown in advance, the analytical models in the literature cannot provide the optimal solution.

## 7.4   Nonparametric estimation methods

In related work on the preventive maintenance problem, Arunkumar (1972), Bergman (1979), and Ingram and Scheaffer (1976) were concerned with the nonparametric estimation for a simple age replacement model (Barlow and Proschan 1996). Under the assumption where the independent and identically distributed (i.i.d.) failure time data of the unknown probability distribution function are available, nonparametric point estimates of an optimal age replacement were derived in their works. In this section, we also consider nonparametric estimation methods for a periodic replacement problem with minimal repair, where the form of intensity function or the corresponding mean value function is completely unknown.

### 7.4.1   Constrained nonparametric ML estimator

In the case where nonparametric estimation methods are used to estimate the optimal periodic replacement time, Gilardoni et al. (2013) developed nonparametric estimation techniques for a periodic replacement problem with minimal repair. They assumed that the minimal repair process or its superposition is observed. The idea is based on the total time on test transform with multiple time series data (Bergman 1979) and a CNPMLE of NHPP by Boswell (1966). Of course, CNPMLE can be applied for the single minimal repair time data case as well.

#### 7.4.1.1   Single failure-occurrence time data case

Consider the case where the failure time distribution $F(t)$ and the intensity function $\lambda(t)$ are completely unknown. In this case, the common method of ML no longer works. In a fashion similar to the parametric case, suppose that the failure-occurrence time data are given by random variables $0 < T_1 \leq T_2 \leq \cdots \leq T_n \leq \tilde{T}$ with realizations $0 < t_1 \leq t_2 \leq \cdots \leq t_n \leq \tilde{t}$. At first, we introduce the most primitive but simplest method to estimate the intensity function of an NHPP. The basic idea is to use a piecewise-linear interpolation. The following piecewise-linear estimate with breakpoints $t_i$ can be defined with $n$ failure-occurrence time data for the mean value function of NHPP:

$$\hat{\Lambda}(t) = i + \frac{t - t_i}{t_{i+1} - t_i}, \quad t_i < t \leq t_{i+1}; \quad i = 0, 1, \ldots, n-1 \qquad (7.15)$$

where $t_0 = 0$. By plotting $n$ failure points and connecting them by line segments, we can obtain the resulting estimate of the mean value function in Equation (7.15). This is called the *naïve estimator* and has a property that the mean square error between the cumulative number of failures and the mean value function of NHPP is always zero. The corresponding naïve estimate of the intensity function can be defined as the slope of mean value function in each failure time interval by

$$\hat{\lambda}(t) = \frac{1}{t_i - t_{i-1}}, \quad t_{i-1} < t \leq t_i; \quad t_0 = 0 \tag{7.16}$$

Note that the naïve estimator in Equation (7.16) does not work well for the generalization ability. Although it can fit the past observation (training data) very well, the prediction result for the unknown (future) pattern data is rather poor. In addition, the naïve estimator in Equation (7.16) tends to fluctuate everywhere with big noise, and does not provide stable estimation results.

Contrary to this, Boswell (1966) introduced the idea on isotonic estimation and gave a CNPMLE. Suppose that the nondecreasing intensity function $\lambda(t)$ with respect to time $t$, i.e., the mean value function, is nondecreasing and convex in time. Boswell (1966) proved that the intensity function that maximizes the likelihood function of NHPP under the nondecreasing property is given by a step-function with breakpoints at any realization $t_i$. Therefore, we can define the likelihood function as a function of unknown intensity functions at respective time points:

$$\text{LF}\left(\lambda(t_i), i = 1, 2, \ldots, n\right) = \exp\left\{-\int_0^{\hat{i}} \lambda(t)dt\right\} \prod_{i=1}^n \lambda(t_i) \tag{7.17}$$

Then the nonparametric ML estimation method with Equation (7.17) is formulated as the following variational problem with respect to $\lambda(\cdot)$:

$$\text{arg max} \quad L\left(\lambda(t_i), i = 1, 2, \ldots, n\right)$$
$$\lambda(t_i), i = 1, 2, \ldots, n \tag{7.18}$$

Under the nondecreasing assumption, Boswell (1966) formulated it as a max–min problem and gave the solution:

$$\hat{\lambda}(t) = \begin{cases} 0, & 0 \leq t < t_1, \\ \max_{1 \leq p \leq i} \min_{i \leq q \leq n} \dfrac{q - p}{t_q - t_p}, & t_i \leq t < t_{i+1}; \quad i = 1, 2, \ldots, n-1 \end{cases} \tag{7.19}$$

It can be easily checked that Equation (7.19) leads to an upper bound of Equation (7.17) for an arbitrary nondecreasing intensity function by substituting Equation (7.19) into Equation (7.17). Although the resulting estimator in Equation (7.19) is still discontinuous, it is somewhat smoother than Equation (7.16). The computation cost is quite low for the CNPMLE compared with other representative nonparametric estimation methods. We introduce the following simple algorithm for the CNPMLE of an NHPP:

- Set $h = 1$ and $i_1 = 1$.
- Repeat until $i_{h+1} = n$:
  Set $i_{h+1}$ to be the index $i$, which minimizes the slopes between $(t_{i_h}, i_h - 1)$ and $(t_i, i - 1)$ $(i = i_h + 1, \ldots, n)$.
- The CNPMLE is then given by $\hat{\lambda}(t) = (i_{j+1} - i_j)/(t_{i_{j+1}} - t_{i_j})$ whenever $t_{i_j} \leq t < t_{i_{j+1}}$.

Since we assume at the moment that the intensity function is nondecreasing with respect to time, the resulting CNPMLE is regarded as a specific estimator that represents an increasing intensity trend. Therefore, if the data have such an increasing trend, then the nondecreasing CNPMLE is expected to be useful. However, when the assumption on an increasing intensity trend is violated, it may be less effective.

Contrary with the above discussion, we also consider a nonincreasing intensity function $\lambda(t)$. In this case, the mean value function is nondecreasing but concave in time. By solving the variational problem in Equation (7.18) under the condition that $\lambda(t)$ is nonincreasing, the corresponding CNPMLE can be derived as the following min–max solution:

$$\hat{\lambda}(t) = \begin{cases} \min_{1 \leq p \leq i} \max_{i \leq q \leq n} \dfrac{q - p}{t_q - t_p}, & t_{i-1} \leq t \leq t_i; \quad i = 1, 2, \ldots, n, \\ 0, & t \geq t_n \end{cases} \tag{7.20}$$

where $t_0 = 0$. By assuming the monotone trend (nondecreasing or nonincreasing) of intensity function in advance, CNPMLE can also represent the degradation or improvement of systems appropriately.

Since the left side of the first-order condition of optimality in Equation (7.5) with CNPMLE is a step function with breakpoints at any realization of failure-occurrence time, the resulting optimal periodic replacement time is also always equal to any realization of failure-occurrence time. This may be a drawback to apply the CNPMLE for the periodic replacement problem with minimal repair. If there does not exist the failure-occurrence time data that are closed to the true optimal periodic replacement time, then the resulting estimate of optimal periodic replacement time never provides the near value to the true one.

### 7.4.1.2 Multiple failure-occurrence time data case

Similar to the parametric ones, the idea of CNPMLE can be extended easily to the case with multiple failure-occurrence time data sets. Let $T_{ji}$ $(j = 1, 2, \ldots, l; i = 1, 2, \ldots, n_j)$ denote the random variable of $i$th failure-occurrence time data at $j$th Poisson process with realizations $t_{ji}$, and let $\tilde{T}^* = \max_{1 \leq j \leq l} \tilde{T}_j$ with realization $\tilde{t}^*$, where $\tilde{T}_j$ is the random censoring time of $j$th component with realizations $\tilde{t}_j$. Change the order of components' failures to a single-ordered sample $(0 =)t_0 < t_1 < \cdots < t_s < \tilde{t}^*$, where $t_i$ does not include the tie data and $s$ is the total number of time data sets after omitting the tie data. Following the idea by Zielinski et al. (1993), the CNPMLE under nondecreasing condition with multiple failure-occurrence time data sets is obtained by

$$\hat{\lambda}(t) = \begin{cases} 0, & 0 \leq t < t_1, \\ \max_{1 \leq p \leq i} \min_{i \leq q \leq f} \dfrac{\displaystyle\sum_{i=p}^{q} m_i}{\displaystyle\sum_{i=p}^{q} \Delta_i}, & t_i \leq t \leq t_{i+1}; \quad i = 1, 2, \ldots, s - 1 \end{cases} \tag{7.21}$$

where

$$\Delta_i = \sum_{i=1}^{l} \left\{ 0, \min\left\{ \left( t_{jn_j} - t_i \right), \left( t_{i+1} - t_i \right) \right\} \right\} \tag{7.22}$$

and $m_i$ means the number of failures occurred at time $t_i$. Similar to this, the CNPMLE under a nonincreasing condition with multiple failure-occurrence time data sets are also derived by

$$\hat{\lambda}(t) = \begin{cases} \min_{1 \le p \le i} \max_{i \le q \le f} \dfrac{\sum_{i=p}^{q} m_i}{\sum_{i=p}^{q} \Delta_i}, & t_{i-1} \le t \le t_i; \quad i = 1, 2, \dots, s, \\[4mm] 0, & t \ge t_s \end{cases} \tag{7.23}$$

where $t_0 = 0$, and $m_j$ and $\Delta_j$ are well defined.

Note that random censoring times $\tilde{T}_j$ are also independent of the underlying identical NHPPs and do not influence the resulting intensity functions in Equations (7.21) or (7.23). Furthermore, similar to the case of single minimal repair data, the candidates of estimated optimal periodic replacement time are obtained from single-ordered samples $t_i$ ($i = 1, 2, \dots, s$). Hence, the estimation results with CNPMLE may be far from the true optimal solution.

### 7.4.2 Kernel-based approach

We introduce an alternative nonparametric estimation technique, which is called the *kernel-based method*, for the periodic replacement problem with minimal repair. The kernel-based approach is well known to be quite useful since the convergence of nonparametric estimators can be improved (Rinsaka and Dohi 2005). Recently, Gilardoni and Colosimo (2011) applied a kernel-based estimation method to obtain the optimal periodic replacement time with multiple minimal repair data sets. Similar to the CNPMLE, they applied the so-called TTT method and transformed the multiple-case to the single one. In this section, we use the well-known Gaussian kernel function, and apply two bandwidth estimation methods with integrated least squares error (Diggle and Marron 1989) and log likelihood function (Guan 2007). Diggle and Marron (1989) proved the equivalence of smoothing parameter selection between the probability density function with i.i.d. samples and the intensity function estimation with the minimal repair data.

#### 7.4.2.1 Single failure-occurrence time data case

Of our interest here is the derivation of absolutely continuous nonparametric estimators. For this purpose, we apply the kernel-based approach to estimate the intensity function of an NHPP. Define

$$\hat{\lambda}(t) = \frac{1}{h} \sum_{i=1}^{n} K\left( \frac{t - t_i}{h} \right) \tag{7.24}$$

where $K(\cdot)$ denotes a kernel function and $h$ is a positive constant, called *smoothing parameter* or *bandwidth*. Roughly speaking, the kernel-based method approximates the intensity function with a superposition of kernel functions with location parameter at each failure-occurrence time. Since the choice of $h$ is more sensitive rather than the choice of kernel function to improve the accuracy of $\lambda(t)$, we deal with only a well-known Gaussian kernel function:

$$K(t) = \frac{1}{\sqrt{2\pi}}\left(-\frac{t^2}{2}\right) \tag{7.25}$$

It corresponds to the probability density function of the standard normal distribution. The main reason we prefer the absolutely continuous nonparametric estimator is that the naïve estimator in Equation (7.16) and the CNPMLE in Equations (7.19) or (7.20) are functions of $t_i$ $(i = 1, 2, \ldots, n)$. Therefore, the resulting estimate of the optimal periodic replacement time minimizing the long-run average cost per unit time has to be selected from the past failure-occurrence time. In other words, if a sufficiently large number of failure-occurrence time data is obtained, then both approaches may work better but will not in the small sample problems.

We consider two estimation methods for bandwidth $h$ for the kernel-based approach. Diggle and Marron (1989) determined a bandwidth with the least-squares cross-validation (LSCV) method by minimizing the relevant integrated least squares error between the kernel-based intensity function and unknown "real" intensity function of NHPP. The second approach is to apply the log-likelihood cross-validation (LLCV) method by Guan (2007). For preparation of both methods, we divide the underlying failure-occurrence time data into *training data* and *validation data*. By leaving out one of each $i$th $(i = 1, 2, \ldots, n)$ data from the $n$ original failure time data $t_i$ $(i = 1, 2, \ldots, n)$, we can make $n$ training data sets with $(n - 1)$ failure data. It is called the *leave-one-out cross-validation*. Although other cross-validation approaches can be considered even in our scheme, we concentrate the simplest case here. The integrated least squares error of the intensity function $\hat{\lambda}(t)$ in LSCV method is defined by

$$\text{ISE}(h) = \int_0^{\hat{t}}\left\{\hat{\lambda}(t) - \lambda(t)\right\}^2 dt$$
$$= \int_0^{\hat{t}} \hat{\lambda}(t)^2 dt - 2\int_0^{\hat{t}} \hat{\lambda}(t)\lambda(t)dt + \int_0^{\hat{t}} \lambda(t)^2 dt \tag{7.26}$$

where $\lambda(t)$ is the "true" but unknown intensity function. After omitting the last term, which is independent of $h$, and approximating the second term in Equation (7.26), it can be checked that the optimal bandwidth minimizing ISE $(h)$ is equal to $h$ minimizing the following function:

$$\text{CV}(h) = \int_0^{\hat{t}} \hat{\lambda}(t)^2 dt - 2\sum_{r=1}^{n} \hat{\lambda}_{h,r}(t_r) \tag{7.27}$$

where

$$\hat{\lambda}_{h,r}(t) = \frac{1}{h}\sum_{i=1,i\neq r}^{n} K\left(\frac{t_i - t}{h}\right) \tag{7.28}$$

In the LLCV method with the same $n$ training data sets, we consider obtaining the optimal bandwidth $h$ by maximizing the log-likelihood function with an unknown intensity function. The log-likelihood function based on the cross-validation approach is given by

$$\ln L(h) = \sum_{k=1}^{n} \sum_{r=1}^{n} \ln \hat{\lambda}_{h,r}(t_k) - \sum_{r=1}^{n} \hat{\Lambda}_{h,r}(\tilde{t}) \tag{7.29}$$

where

$$\hat{\Lambda}_{h,r}(t) = \int_0^t \hat{\lambda}_{h,r}(t) dt \tag{7.30}$$

and $\hat{\lambda}_{h,j}(t)$ is already defined in Equation (7.28).

It is noted that the above kernel estimates of intensity function are absolutely continuous, but may fluctuate everywhere similar to the naïve estimate. Nevertheless, the resulting estimates of optimal periodic replacement time and corresponding minimum long-run average cost per unit time can be expected to be much smoother than that. Of course, several kinds of kernel functions such as biweight kernel function and Epanechnikov kernel function (Rinsaka and Dohi 2005) can be applied for the periodic replacement problem with minimal repair as well. Furthermore, an adaptive choice of bandwidth is possible though we do not consider that here, i.e., $h = h(i)$ $(i = 1, 2, ..., n)$.

### 7.4.2.2  *Multiple failure-occurrence time data case*

Similar to the case of CNPMLE, suppose that $l$-independent sets of failure-occurrence time data $t_{ji}(j = 1, 2,..., l; i = 1, 2, ..., n_j)$ with censoring time $\tilde{t}_j(j = 1, 2, ..., l)$ are available. In the case of multiple failure-occurrence time data sets, several cross-validation techniques are also considered. In this section, we propose two estimation methods for intensity function of NHPP based on the kernel-based approach.

One is the simple extension of the single failure-occurrence time data case. For $j$th set of failure-occurrence time data, we define

$$\hat{\lambda}_j^1(t) = \frac{1}{h} \sum_{i=1}^{n_j} K\left(\frac{t - t_{ji}}{h}\right) \tag{7.31}$$

where $K(\cdot)$ is defined in Equation (7.25). Then the CV $(h)$ in Equation (7.27) with the LSCV method can be rewritten, by means of the leave-one-out cross-validation technique, as

$$\text{CV}(h) = \sum_{j=1}^{l} \int_0^{\tilde{t}_j} \left( \hat{\lambda}_j^1(t)^2 dt - 2 \sum_{r=1}^{n_j} \hat{\lambda}_{j,h,r}^1(t_{jr}) \right) \tag{7.32}$$

where

$$\hat{\lambda}_{j,h,r}^1(t) = \frac{1}{h} \sum_{i=1,i\neq r}^{n_j} K\left(\frac{t_{ji} - t}{h}\right) \tag{7.33}$$

Furthermore, the $\ln L(h)$ in Equation (7.29) can be rewritten as

$$\ln L(h) = \sum_{j=1}^{l} \left( \sum_{k=1}^{n_j} \sum_{r=1}^{n_j} \ln \hat{\lambda}_{j,h,r}^1 \left( t_{jk} \right) - \sum_{r=1}^{n_j} \hat{\Lambda}_{j,h,r}^1 \left( \hat{t}_j \right) \right) \tag{7.34}$$

where

$$\hat{\Lambda}_{j,h,r}^1(t) = \int_0^t \hat{\lambda}_{j,h,r}^1(t) dt \tag{7.35}$$

and $\hat{\lambda}_{j,h,r}^1(t)$ is already defined in Equation (7.33). Unfortunately, if only one failure occurred for $j$th component, then Equation (7.33) or Equation (7.35) based on leave-one-out cross-validation cannot work. Therefore, we remove such a data set in the analysis. By minimizing Equation (7.32) or maximizing Equation (7.34), we estimate the optimal bandwidth $h$. In this scheme, the intensity function of NHPP for component $j$ is based on the only $j$th set of failure-occurrence time data, since the intensity function is defined as Equation (7.31). Therefore, $l$-different intensity functions can be derived according to the behavior of each set of failure-occurrence time data. It may be useful to consider the arithmetic mean of all intensity functions by taking the average of the failure-occurrence phenomena. That is,

$$\hat{\lambda}(t) = \frac{1}{m} \sum_{j=1}^{m} \hat{\lambda}_j^1(t) \tag{7.36}$$

These bandwidth estimation methods are labeled as $LLCV_1$ and $LSCV_1$, respectively, in this chapter.

The second approach is based on the idea of superposition of NHPP (Arkin and Leemis 1998). Reorder all the failure-occurrence time data as a single-ordered sample $(0 =) t_0 \leq \cdots \leq t_{n^*} < \tilde{t}^*$, where $n^*$ is the total number of failure included the tie data, say, $n^* = \sum_{j=1}^{l} n_j$, and $\tilde{t}^*$ is the realization of the maximum of random censoring time, say, $\tilde{T}^* = \max_{1 \leq j \leq l} \tilde{T}_j$. For the single-ordered sample, the intensity function of superposition of NHPP is defined similar to the single failure-occurrence time data case:

$$\hat{\lambda}^2(t) = \frac{1}{h} \sum_{i=1}^{n^*} K\left( \frac{t - t_i}{h} \right) \tag{7.37}$$

where $K(\cdot)$ is defined in Equation (7.25). Then the CV $(h)$ in Equation (7.27) with the LSCV method can be rewritten with a superposed intensity function:

$$CV(h) = \int_0^{\tilde{t}^*} \left( \hat{\lambda}^2(t) \right)^2 dt - 2 \sum_{r=1}^{n^*} \hat{\lambda}_{h,r}^2(t_r) \tag{7.38}$$

where

$$\hat{\lambda}_{h,r}^2(t) = \frac{1}{h} \sum_{i=1, i \neq r}^{n^*} K\left( \frac{t_i - t}{h} \right) \tag{7.39}$$

Furthermore, the $\ln L(h)$ in Equation (7.29) can be rewritten with the superposed intensity function:

$$\ln L(h) = \sum_{k=1}^{n^*} \sum_{r=1}^{n^*} \ln \hat{\lambda}_{h,r}^2(t_k) - \sum_{r=1}^{n^*} \hat{\Lambda}_{h,r}^2(\hat{t}^*) \tag{7.40}$$

where

$$\hat{\Lambda}_{h,r}^2(t) = \int_0^t \hat{\lambda}_{h,r}^2(t) dt \tag{7.41}$$

and $\hat{\lambda}_{h,r}^2(t)$ is defined in Equation (7.39). By minimizing Equation (7.38) or maximizing Equation (7.40), we estimate the optimal bandwidth $h$. In this scheme, the function that is defined in Equation (7.37) is an intensity function of a superposition of NHPPs. Therefore, for each component, the intensity function can be defined by

$$\hat{\lambda}(t) = \frac{1}{l} \frac{1}{h} \sum_{i=1}^{n^*} K\left(\frac{t-t_i}{h}\right) \tag{7.42}$$

These bandwidth estimation methods are labeled $LLCV_2$ and $LSCV_2$, respectively. Actually, the estimation results of these four methods, $LLCV_1$, $LLCV_2$, $LSCV_1$, and $LSCV_2$, are slightly different from each other.

## 7.5   Numerical examples

### 7.5.1   Simulation experiments with single minimal repair data

We conduct the Monte Carlo simulation to investigate properties of parametric and non-parametric methods. Here, we focus on the single minimal repair data. The (true but unknown) minimal repair process is assumed to follow a power law NHPP model having the model parameters $(\beta, \eta) = (3.2, 0.23)$ of intensity function $\lambda(t) = (\beta/\eta)(t/\eta)^{\beta-1}$. By applying the thinning algorithm of NHPP (Lewis and Shedler 1979), the original failure (minimal repair) time data are generated as the pseudo random number. The "real" optimal periodic replacement time and its minimum long-run average cost per unit time are calculated numerically as $\tau^* = 0.44$ and $C(\tau^*) = 98.8$, under the fixed cost ratio $\gamma = 30$.

The optimal periodic replacement time and its minimum long-run average cost per unit time are estimated with both parametric and nonparametric methods for $n = 160$ minimal repair time data. Here, two parametric models are assumed: the power law (PL) NHPP model with $(t) = (\beta/\eta)(t/\eta)^{\beta-1}$, and the Cox–Lewis (CL) NHPP model with $\lambda(t) = \exp(\alpha + \beta t)$. Without any prior knowledge of the underlying system, it is very difficult to select the correct model (power law model) exactly. To investigate the effect of mis-specification of the underlying failure process model, we assume the CL model. Also, we apply the CNPMLE and the kernel-based approaches with LSCV and LLCV. In this example, we normalize the minimal repair data in order to reduce the computation cost. Therefore, all data $t_i$ $(i = 1, 2, \ldots, 160)$ are divided with the maximum value $t_{160}$.

For four cases with $n = 40, 80, 120,$ and $160$, the estimation results of the optimal periodic replacement time and the minimum long-run average cost per unit time are presented

*Table 7.1* Estimation results of the optimal periodic replacement time

| $n$ | PL | CL | CNPMLE | LSCV | LLCV |
|---|---|---|---|---|---|
| 40 | 0.42 | 0.44 | 0.43 | 0.44 | 0.43 |
| 80 | 0.44 | 0.48 | 0.43 | 0.44 | 0.43 |
| 120 | 0.44 | 0.50 | 0.43 | 0.43 | 0.43 |
| 160 | 0.44 | 0.50 | 0.43 | 0.44 | 0.43 |

*Table 7.2* Estimation results of the minimum long-run average cost per unit time

| $n$ | PL | CL | CNPMLE | LSCV | LLCV |
|---|---|---|---|---|---|
| 40 | 99.7 | 98.8 | 98.1 | 102.6 | 98.6 |
| 80 | 100.5 | 102.0 | 98.1 | 103.0 | 98.4 |
| 120 | 100.4 | 104.7 | 98.1 | 101.5 | 98.3 |
| 160 | 100.5 | 107.9 | 98.1 | 102.5 | 98.2 |

in Tables 7.1 and 7.2. From Table 7.1, it can be seen that three nonparametric estimation methods give very close results to the real optimal periodic replacement time $\tau^* = 44$. But, estimation results based on the mis-specification (CL model) are the worst among all cases. In this way, the influence by mis-specification of the parametric model is significant when the exact information of the underlying failure process is not available from the past experience. Focusing on Table 7.2, the kernel-based approach with LLCV can provide the best estimation results of the minimum long-run average cost per unit time. Also, CNPMLE shows the similar estimation results, because the original minimal repair data include a closed data 0.43 to the optimal periodic replacement time $\tau^* = 0.44$. Of course, this is a quite rare case. If there is no failure data near the optimal solution, CNPMLE cannot work in the small sample problem. Furthermore, it is known that two parametric models and a kernel-based approach with LSCV tend to give rather pessimistic estimates. This fact indicates that even if the real model (power law model) can be assumed, the estimation results may be biased. It may be caused that the minimum long-run average cost per unit time is very sensitive to the optimal periodic replacement time $\tau^*$. For the CL model, the difference from the real optimal solution tends to get larger as the number of minimal repair data increases.

Apart from the optimization, we estimate the long-run average cost per unit time at different periodic replacement time in order to examine the estimation accuracy of our methods. In addition to the naïve estimator, we compare the estimation results of long-run average cost per unit time with our six methods at arbitrary four time points, $t = 0.25$, 0.50, 0.75, and 1.00, where the number of minimal repairs is given by $n = 180$. In Table 7.3, "TRUE" represents the estimate calculated by the power law model with modeling of the parameters $(\beta, \eta) = (3.2, 0.23)$. It can be said that the model, which shows good results near to the theoretically optimal solution, provides good accuracy at different periodic replacement time points. In this table, we calculate the mean squared error between the estimation results and "true" values at 20 time points from $t = 0$ to $t = 1.00$ by 0.05. In this case, the power law model can provide the best accuracy performance among all models. But, four nonparametric models, especially for the kernel-based approach with LLCV and naïve estimator, show similar good accuracy performance. It is evident that the mis-specification of the parametric model also leads to the worst results as well.

**Table 7.3** Estimation accuracy of each method for arbitrary periodic replacement time

| $t$ | TRUE | PL | CL | CNPMLE | LSCV | LLCV | Naïve |
|-----|------|------|------|--------|------|------|-------|
| 0.25 | 128.8 | 130.0 | 148.5 | 120.7 | 126.0 | 128.0 | 128.1 |
| 0.5 | 100.5 | 102.5 | 109.8 | 101.5 | 106.2 | 104.0 | 104.3 |
| 0.75 | 138.8 | 138.8 | 132.4 | 134.8 | 140.0 | 140.0 | 140.7 |
| 1 | 216.1 | 210.0 | 210.0 | 209.0 | 200.9 | 209.5 | 210.0 |
| MSE | 0 | 3.53 | 140.8 | 17.8 | 19.4 | 13.4 | 14.1 |

### 7.5.2  Real example with multiple minimal repair data sets

The real failure data analysis is useful to give a reality to the mathematical modeling (Dohi et al. 2007). We also study a real example to demonstrate how to use the parametric methods and several nonparametric approaches for the periodic replacement problem with minimal repair.

We consider the multiple failure-occurrence time data sets of a diesel engine, which is shown in Nelson and Doganaksoy (1989) as the minimal repair data. Although there are repair records for 41 diesel engines in this data set, we select 24 diesel engines, in which more than one failure are observed. The data set is shown in Table 7.4.

**Table 7.4** Failure-occurrence time data of diesel engines

| Number | Censoring time | Replacement time | | | |
|--------|----------------|------------------|------|------|------|
| 1 | 389 | 166 | 206 | 348 | |
| 2 | 582 | 323 | 449 | | |
| 3 | 585 | 202 | 563 | 570 | |
| 4 | 589 | 573 | | | |
| 5 | 589 | 139 | 139 | | |
| 6 | 594 | 249 | | | |
| 7 | 595 | 265 | 586 | | |
| 8 | 601 | 410 | 581 | | |
| 9 | 603 | 367 | | | |
| 10 | 606 | 165 | 408 | 604 | |
| 11 | 613 | 344 | 497 | | |
| 12 | 614 | 120 | 479 | | |
| 13 | 641 | 635 | | | |
| 14 | 642 | 76 | 538 | | |
| 15 | 644 | 254 | 276 | 298 | 640 |
| 16 | 648 | 61 | 539 | | |
| 17 | 649 | 349 | 404 | 561 | |
| 18 | 650 | 258 | 328 | 377 | 621 |
| 19 | 653 | 646 | | | |
| 20 | 653 | 92 | | | |
| 21 | 663 | 87 | | | |
| 22 | 667 | 98 | | | |
| 23 | 667 | 326 | 653 | 653 | |
| 24 | 667 | 84 | | | |

The total number of failures is 48 in the entire data set. For this data set, we apply one parametric model (power law model) and two nonparametric models (CNPMLE and kernel-based approach). First, we estimate the intensity function and mean value function with each model in Figures 7.1 and 7.2. From Figure 7.1, it is seen that the intensity function with CNPMLE is represented by a nondecreasing step function. In other models, only the intensity function of the power law model increases as time goes by. $LLCV_1$ and $LSCV_1$ show the unimodal intensity functions. $LSCV_2$ gives the multimodal shape. $LLCV_2$ also indicates a similar trend to the naïve estimator and fluctuates everywhere. Looking at Figure 7.2, we can see that both the results of the power law model and CNPMLE have convex shapes. Although the mean value function of the power law model is a smoothed curve, CNPMLE constitutes the mean value function by several line segments. The mean value functions with $LLCV_2$ and $LSCV_2$ are not smooth compared with other models. Especially for $LLCV_2$, this property is remarkable.

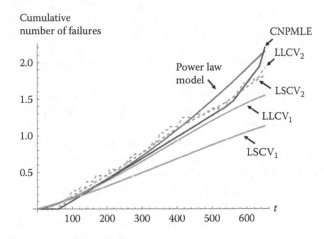

*Figure 7.1* Estimation results on intensity function.

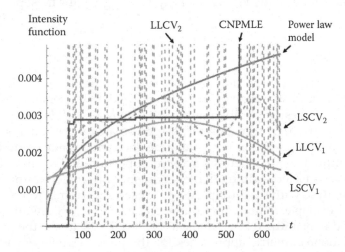

*Figure 7.2* Estimation results on mean value function.

***Table 7.5*** Estimation results on the optimal periodic replacement time

| $\gamma$ | PL | CNPMLE | $LLCV_1$ | $LLCV_2$ | $LSCV_1$ | $LSCV_2$ |
|---|---|---|---|---|---|---|
| 0.04 | 73 | 61 | 109 | 58 | 190 | 61 |
| 0.05 | 85 | 61 | 123 | 58 | 220 | 69 |
| 0.06 | 97 | 61 | 136 | 58 | 650 | 76 |
| 0.07 | 108 | 61 | 148 | 58 | 650 | 84 |
| 0.08 | 119 | 61 | 160 | 58 | 650 | 92 |

***Table 7.6*** Estimation results on the minimum long-run average cost per unit time

| $\gamma$ | PL | CNPMLE | $LLCV_1$ | $LLCV_2$ | $LSCV_1$ | $LSCV_2$ |
|---|---|---|---|---|---|---|
| 0.04 | 19.1 | 6.6 | 20.0 | 7.0 | 17.6 | 20.6 |
| 0.05 | 20.4 | 8.2 | 20.9 | 8.8 | 18.1 | 22.2 |
| 0.06 | 21.5 | 9.8 | 21.7 | 10.5 | 18.3 | 23.5 |
| 0.07 | 22.5 | 11.5 | 22.4 | 12.2 | 18.5 | 24.8 |
| 0.08 | 23.3 | 13.1 | 23.0 | 13.9 | 18.6 | 25.9 |

Suppose that the cost ratio $\gamma$ varies from 0.04 to 0.08 by 0.01. Tables 7.4 and 7.5 summarize the estimation results of the optimal periodic replacement time and the minimum long-run average cost per unit time.

Focusing on Table 7.5, we can know that the optimal periodic replacement time with CNPMLE and $LLCV_2$ is always constant and almost similar even though the cost ratio $\gamma$ increases. But, the results with other estimators are increasing as $\gamma$ increases. In this example, $LLCV_2$ gives the smallest optimal periodic replacement time and $LSCV_1$ gives the largest optimal periodic replacement time. The optimization results with the LSCV method are often influenced by the differences between two approaches ($LSCV_1$, $LSCV_2$), compared with the LLCV method. It is also observed that the minimum long-run average cost per unit time with the parametric power law model and $LLCV_1$ takes closed values from each other in Table 7.6. Furthermore, CNPMLE and $LLCV_2$ tend to show the relatively optimistic results among all models. Conversely, $LSCV_2$ gives the most pessimistic estimation results.

## 7.6   Conclusions

In this chapter we focused on the failure time data analysis based on the NHPP and discussed several statistical estimation methods for a periodic replacement problem with minimal repair as the simplest application of life data analysis with NHPP. We have applied not only the parametric ML estimation for the power law process but also CNPMLE and kernel-based estimation methods for estimating the cost-optimal periodic replacement problem with minimal repair, where single or multiple minimal repair data are assumed. Furthermore, we conducted a simulation experiment of a single minimal repair process and a real data analysis with the multiple field data sets of minimal repair of diesel engine in the numerical example. Throughout numerical examples, we investigated properties of parametric and nonparametric methods and showed how to use the parametric methods and several nonparametric approaches for the periodic replacement problem with minimal repair.

# References

Arkin, B. L. and Leemis, L. M. (1998). Nonparametric estimation of the cumulative intensity function for a nonhomogeneous Poisson process from overlapping realizations. *Management Science 46,* pp. 989–998.

Arunkumar, S. (1972). Nonparametric age replacement policy.*Indian Journal of Statistics Series A 34,* pp. 251–256.

Ascher, H. and Feingold, H. (1984). *Repairable Systems Reliability: Modeling, Inference, Misconceptions and Their Causes.* New York, Marcel Dekker.

Barlow, R. E. and Hunter, L. C. (1960). Optimum preventive maintenance policies. *Operations Research 8,* pp. 90–100.

Barlow, R. E. and Proschan, F. (1996). *Mathematical Theory of Reliability.* Philadelphia, SIAM.

Baxter, L. A. (1982). Reliability applications of the relevation transform. *Naval Research Logistics Quarterly 29,* pp. 323–330.

Bergman, B. (1979). On age replacement and the total time on test concept. *Scandinavian Journal of Statistics 6,* pp. 161–168.

Boland, P. J. (1982). Periodic replacement when minimal repair costs vary with time. *Naval Research Logistics Quarterly 29,* pp. 541–546.

Boswell, M. T. (1966). Estimating and testing trend in a stochastic process of Poisson type. *Annals of Mathematical Statistics 37,* pp. 1564–1573.

Colosimo, E. A., Gilardoni, G. L., Santos, W. B. and Motta, S. B. (2010). Optimal maintenance time for repairable systems under two types of failures. *Communications in Statistics—Theory and Methods 39,* pp. 1289–1298.

Cox, D. R. (1972). The statistical analysis of dependencies in point processes. *Stochastic Point Processes, Stochastic Point Processes: Statistical Analysis, Theory and Applications* (Lewis, P. A. W. ed.), pp. 55–66. New York, Wiley.

Cox, D. R. and Lewis, P. A. W. (1966). *The Statistical Analysis of Series of Events.* New York, Wiley.

Crow, L. H. (1974). Reliability analysis for complex repairable systems. *Reliability and Biometry: Statistical Analysis of Lifelength* (Proschan, F. and Serfling, R. J. eds.), pp. 379–410. Philadelphia, SIAM.

Diggle, P. and Marron, J. S. (1989). Equivalence of smoothing parameter selectors in density and intensity estimation. *Journal of the American Statistical Association 83,* pp. 793–800.

Dohi, T., Kaio, N. and Osaki, S. (2007). Optimal (T, S)-policies in a discrete-time opportunity-based age replacement: An empirical study. *International Journal of Industrial Engineering 14,* pp. 340–347.

Gilardoni, G. L. and Colosimo, E. A. (2011). On the superposition of overlapping Poisson processes and nonparametric estimation of their intensity function. *Journal of Statistical Planning and Inference 141,* pp. 3075–3083.

Gilardoni, G. L., De Oliveira, M. D. and Colosimo, E. A. (2013). Nonparametric estimation and bootstrap confidence intervals for the optimal maintenance time of a repairable system. *Computational Statistics & Data Analysis 63,* pp. 113–124.

Guan, Y. (2007). A composite likelihood cross-validation approach in selecting bandwidth for the estimation of the pair correlation function. *Scandinavian Journal of Statistics 34,* pp. 336–346.

Ingram, C. R. and Scheaffer, R. L. (1976). On consistent estimation of age replacement intervals. *Technometrics 18,* pp. 213–219.

Lewis, P. A. W. and Shedler, G. S. (1979). Simulation of nonhomogeneous Poisson processes by thinning. *Naval Research Logistics Quarterly 26,* pp. 403–413.

Nakagawa, T. (1986). Periodic and sequential preventive maintenance policies. *Journal of Applied Probability 23,* pp. 536–542.

Nakagawa, T. (2005). *Maintenance Theory of Reliability.* Berlin, Springer.

Nelson, W. and Doganaksoy, N. (1989). A computer program for an estimate and confidence limits for the mean cumulative function for cost or number of repairs of repairable products. *TIS report 89CRD239.* New York, General Electric Company Research and Development.

Okamura, H., Dohi, T. and Osaki, S. (2014). A dynamic programming approach for sequential preventive maintenance policies with two failure modes. *Reliability Modeling with Applications: Essays in Honor of Professor Toshio Nakagawa on His 70th Birthday* (Nakamura, S., Qian, C. H. and Chen, M. eds.), pp. 3–16. Singapore, World Scientific.

Park, D. H., Jung, G. M. and Yum, J. K. (2000). Cost minimization for periodic maintenance policy of a system subject to slow degradation. *Reliability Engineering & System Safety 68*, pp. 105–112.

Rinsaka, K. and Dohi, T. (2005). Estimating age replacement policies from small sample data. *Recent Advances in Stochastic Operations Research* (Dohi, T., Osaki, S. and Sawaki, K. eds.), pp. 145–158. Singapore, World Scientific.

Sheu, S. H. (1990). Periodic replacement when minimal repair costs depend on the age and the number of minimal repairs for a multi-unit system. *Microelectronics Reliability 30*, pp. 713–718.

Sheu, S. H. (1991). A generalized block replacement policy with minimal repair and general random repair costs for a multi-unit system. *Journal of the Operational Research Society 42*, pp. 331–341.

Valdez-Flores, C. and Feldman, R. M. (1989). A survey of preventive maintenance models for stochastically deteriorating single-unit systems. *Naval Research Logistics Quarterly 36*, pp. 419–446.

Zielinski, J. M., Wolfson, D. B., Nilakantan, L. and Confavreux, C. (1993). Isotonic estimation of the intensity of a nonhomogeneous Poisson process: The multiple realization setup. *Canadian Journal of Statistics 21*, pp. 257–268.

## chapter eight

# View-count based modeling for YouTube videos and weighted criteria–based ranking

**N. Aggrawal and A. Arora**
*Jaypee Institute of Information Technology*

**A. Anand and M.S. Irshad**
*University of Delhi*

### Contents

## 8.1   Introduction

Online social networks such as Facebook, Twitter, YouTube, Google+, etc., are part of our daily life now. Social networks are merged as a platform since they are able to bring people from varied backgrounds and common interests to the same place. Billions of people interact with each other on such platforms. It is also changing the way in which people create, share and consume information (Khan and Vong 2014). Information is not just shared in the form of text but also via images, gifs, videos, memes, etc. Social network sites are also instrumental in creating awareness on some topics, which directly or indirectly unites people and forces lawmakers and firms to make better decisions keeping the interests of the user in mind.

The social networking websites are emerging as an important factor in expanding a product's online market. There are several examples where special privileges are provided to the e-commerce customers over the traditional shop customers. Like the launching of certain products exclusively on e-commerce sites, namely, Flipkart, Amazon, Snapdeal, and so on. For the first few weeks they are exclusively sold through e-commerce. Every shopping cart allows users to share their experience on social media like Facebook, YouTube, Google+, etc., which works the word-of-mouth publicity in traditional media/markets.

Among various video sharing sites, YouTube has emerged as a leader in video sharing platforms. It started as a website to share short entertainment videos and has grown into a massive platform for people to connect freely. One can find videos of different genres like music, sports, comedy, recreational activities, religious and spiritual content, educational, etc. Since YouTube is acting more like a marketing forum, it is helping individuals as well as large production houses to increase their audience. Nowadays every news channel, production house, celebrity, and organization has their YouTube channels where they post their recent activities in the form of videos to stay connected to their fans, group members or general public. Independent content creators have built grassroots followings numbering in the thousands at very little cost or effort. YouTube's revenue-sharing "Partner Program" made it possible for people to earn a substantial living as a video producer alone – with each of its top 500 partners earning more than $100,000 annually and its 10 highest-earning channels grossing from $2.5 to $12 million (Berg 2015).

Since YouTube provides a glimpse of the product's success (in term of statistics), manufacturing companies and production houses prefer to launch previews of their products or trailers on YouTube first. The video's popularity would provide good business to not only YouTube but also the product manufacturer. YouTube's features, such as the "like," "dislike," "comments," "share," "subscribe," etc., go a long way in helping the customers/audiences to express themselves. A video's number of likes, dislikes, comments in favor and comments against them clearly shows a product's performance and the viewers' satisfaction. It provides a clear picture as to whether the product is going to be a hit in market or not. In the comments section one can also find the basis on which the customers compare two or more products. It clearly works in favor of the firm as firms get to know what to produce and what characteristics of the product lead a customer to buy that product. Manufacturers can easily estimate their potential buyers from available statistics and also know the "X-factor" of their product by deeply analyzing the comment or the review videos that are again posted by some YouTube user (uploader) on it. Therefore, it is quite clear that YouTube is not just advertising the product, but it is also giving a glimpse of its future. Features provided by YouTube help users as well as uploaders to interact better with each other.

According to an Internet survey conducted by Alexa in 2005, YouTube is the fastest growing website and was ranked second in traffic generation among all the websites surveyed (Cheng et al. 2008). Each minute 400 hours of content are uploaded and approximately 1 billion hours of content are viewed daily on YouTube (YouTube Press). Further, Alexa categorized YouTube's speed as "SLOW" as its average load time is 3.6 seconds and it is slower than 69% of their surveyed websites (Cheng et al. 2008). The results are almost similar even today. In October 2017, YouTube's average load time was 2.38 seconds, and it is slower than 68% of other websites.

The rest of the chapter is distributed as follows: Section 8.2 describes the twofold perspective of YouTube view count followed by a literature review in Section 8.3. In Section 8.4, a methodical approach to frame the view-count based models in a dynamic environment has been proposed. Section 8.5 contains the model validation carried out on YouTube video data sets. The conclusion and references are presented last.

## 8.2   *YouTube view count: A twofold perspective*

YouTube has gained much fame due to its user-friendly policy. It does not charge the users for uploading any content; in fact, it pays them if the content is popular. The popularity of the content is generally measured by the number of views on that video. The increased number of views on a video is both good as well as bad news for YouTube. On one hand it increases the monetary gains, while on the other it further slows down the platform due to the additional traffic generated. Therefore, understanding and predicting the view count is a twofold perspective: one that popular content generates more traffic, and hence understanding popularity has a direct impact on caching and replication strategy that the provider should adopt; and the other perspective that popularity has a direct economic impact (Richier et al. 2014).

A number of researchers have tried to understand and model the popularity and virality pattern of YouTube using various tools and techniques (Bauckhage et al. 2015; Richier et al. 2014; Vaish et al. 2012; Yu et al. 2015; Zhou et al. 2010). For predicting a video's popularity or the high view count, one must know how YouTube works and when a view count increases. The flowchart presented in Figure 8.1 explains legitimate view-count accountability.

It is important to predict a video's high demand and popularity and make better decisions of which videos should be cached on the limited data space of the proxy servers.

A view is a video playback that was requested by an actual user. A fake view includes misleading views, misleading titles and thumbnails that attract views. When a video has a large number of views that last for mere seconds after clicking, the views are not counted as legitimate. So if a video is viewed in its entirety by someone who clicked on it, it is counted as one view. But not all views are fully played. Google Ad Sense works only with videos that are over 30 seconds in length so that the click-through rates get registered. In fact, some videos are lucky enough to have just 10 seconds of play being considered as a view. Thereby, it can be understood that the amount of video played should be above a threshold percentage of the length of the video. The type and genre of the video also affect video length. YouTube considers views from the same IP address in a time interval of 6–8 hours. So one person viewing the same video repeatedly would only generate three to five views a day, even though he/she has viewed it over 300 times. A viewer being redirected to YouTube upon clicking an embedded video counts as one view. If there is an embedded video with auto play, it is *not* counted as a view. In December 2012, 2 billion views were removed from the view counts of Universal and Sony music videos on YouTube, provoking a claim by *The Daily Dot* that the views had been deleted because of infringement of the site's terms of service, which banned the use of automated processes

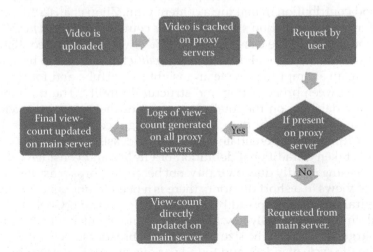

*Figure 8.1* View-count accountability.

to inflate view counts. In another incident on August 5, 2015, YouTube removed the feature that caused a video's view count to freeze at "301" (later "301+") until the time the actual count was verified to counteract view-count fraud. There might be many more restrictions and rules that go into categorizing a request as a view that might not have been looked into as for now.

A detailed number of attributes have been looked upon and carried on for analysis in terms of number of views. The next section provides a highlight of certain work in the related area.

## 8.3   Literature review

As far as a prediction of view count is concerned, there are several factors that can affect the number of views gathered by a video. There are several research proposals concerning the total number of view count (Bauckhage et al. 2015; Richier et al. 2014; Vaish et al. 2012; Yu et al. 2015) or predicting the growth rate of view count (Bauckhage et al. 2015; Richier et al. 2014; Yu et al. 2015). Then there is literature showing how recommendation in YouTube helps in increasing view count of a video (Yu et al. 2015). In some research, different caching techniques are given to reduce the caching time. Cheng et al. (2008) provided a glimpse of YouTube statistics. They found the active life span of the video in terms of caching, i.e., up to what time it is required to cache the video on the proxy server. They also showed how videos can be distributed on the basis of video category, the age of the video, video length, video file size, and video bit rate on YouTube. According to their results, 97.9% video lengths are within 600 seconds, and 99.1% are within 700 seconds in their entire data set. Since 22.9% of the videos fall in the music category and 17.8% of videos fall under the entertainment category, their segregation can be understood. If the rate of growth of view count is very high, then the content is considered viral, i.e., a large number of views in a very short period of time. Richier et al. (2014) fitted six bio-inspired models in their research to find the view count dynamics of YouTube videos for a fixed and growing population. Khan and Vong (2014) captured the effect of external influence on the virality of content. For this they used webometrics and network diagram and found a correlation between the various attributes of the video to the cause of virality. They also state that the external network (other social network sites except for YouTube) also have a good contribution in making a content viral. Zhou et al. (2010) highlighted the importance of a recommendation system on view count of a video. They found that the recommendation system is the cause for 30%–40% of views of a video. Rather researchers contradict with the traditional definition of *virality* and relate it to broadcasting. Goel et al. (2015) made an attempt to provide an insight to YouTube and found that there is a clear difference between broadcasting and structural virality. The traditional approach of virality largely depends on the total view count of the video, i.e., a video having a greater view count is more viral than a video having a comparatively less view count. Since it is very difficult to differentiate between broadcasting and virality, in this chapter we have also taken a traditional definition of virality and considered the number of views to be increasing rapidly due to virality, not because of broadcasting. As both yield a large number of views in a short duration, there is a need to understand that content can be said to be viral only if it is spread by the initial viewers, and Goel et al. (2015) found that structural virality is typically low, and remains independent of size, suggesting that popularity is largely driven by the size of the largest broadcast. Several researchers have also studied the life cycle of videos and found various phases that occur in the lifetime of a YouTube video. Yu et al. (2015) found phases as a description of the burst popularity of a video and found the multiple peaks of popularity in its life cycle. They also directly relate the phases to content type and evolution of popularity on the basis of the power law. Of late, Aggrawal et al. (2017) have proposed a novel approach of studying the life cycle of a YouTube video and categorization of viewers. Bauckhage et al. (2015) used a bio-inspired model in their research to determine "how viral were viral videos." They utilized convolution theory and took a joint effect of exponential infection and recovery rate in the Markov process to find the probability density function for the epidemic model (virality of videos). Ding et al. (2011) did their research for understanding the uploaders of the content where they demonstrate the positive reinforcement between online social behavior and uploading behavior. They also examined whether YouTube users are truly broadcasting themselves via characterizing and classified videos as user generated and user copied (2011). Their results claim that most of the content on YouTube is not user generated, and 63% of the most popular uploaders are just uploading the user-copied content. The UCC (user-copied content) uploaders upload more videos than the UGC (user-generated content). Further Vaish et al. (2012) used different factors like share count, number of views, number of likes and number of dislikes for calculating the virality index of the video and provide a conventional and hybrid asset valuation technique to demonstrate how virality can fit in to provide accurate results.

In this chapter, we have tried to capture the growth pattern of view count for certain videos on YouTube. The initial stage of the life cycle of a video starts with it being posted on YouTube (i.e., people are getting aware and then diffusing the information in the Internet market). As the video becomes more popular, it attracts a larger number of views, likes, dislikes, comments, shares, etc. This is considered as the growing phase of the video. After attaining maximum popularity or becoming viral or being viewed by most of its target viewers, the video is said to have matured, and its active life span is considered to be almost over. From this time onward, the video's growth will be very slow and steady as compared to the earlier phase (Cheng et al. 2008). A video is never deleted from YouTube, so there is no fixed life span, but the video's life is said to be over when the growth in a number of views is negligible over time. YouTube's revenue depends on advertisements, so it is very important to know the right time to introduce an advertisement on a video. The high rate of advertisement during the growth stage of the video is likely to have the maximum impact and yield high profits for the advertiser and YouTube. Therefore, our study of the view count growth shall prove to be a helpful contribution in this area.

## 8.4    Model development

In the proposed modeling framework, we have captured the view-count dynamics of YouTube viewers and have extended this framework for the dynamic Internet model. The popularity of YouTube can be judged by the number of likes, dislikes, comments, shares, views, etc. All the aforesaid attributes can be represented as a counting process. Out of these we have considered view counts as a counting process in the present research. As we know from the literature (Kapur et al. 1999), a counting process $(N(t), t > 0)$ is said to be a nonhomogenous Poison process with intensity function $\lambda(t)$ if it satisfies the following conditions:

i. $N(0) = 0$
ii. $\{N(t), t > 0\}$ has independent increments
iii. $P\{N(t + h) - N(h) \geq 2\} = o(h)$
iv. $P\{N(t + h) - N(h) = 1\} = \lambda(t)h + o(h)$

where, $o(h)$ denotes a quantity that tends to zero for small "$h$."

Let $v(t)$ represent the expected number of views by time $t$, i.e., $v(t) = \int_0^t \lambda(x)dx$, $t > 0$, then it can be shown that

$$\Pr[N(t) = k] = \frac{(v(t))^k e^{-v(t)}}{k!}, \quad k = 0, 1, 2, \ldots \tag{8.1}$$

In other words, $N(t)$ has a Poisson distribution with mean value function $v(t)$. Consider a case when the time scale of the content diffusion is very large as compared to the size of the potential population. Hence, we model the case where contents gain popularity through advertisement and other marketing tools: examples are when advertisement is done for a large pool of users of a social network and netizens access the content at random thereafter.

Hence, we assume that the expected number of views in $(t, t + \Delta t)$ is essentially proportional to the expected number of views left from total expectation at time $t$, i.e.,

$$v(t + \Delta t) - v(t) = b\{a - v(t)\}\Delta t + o(\Delta t) \tag{8.2}$$

where $o(\Delta t)/\Delta t \to 0$ as $\Delta t \to 0$, and $b$ is a constant of proportionality and can be explained in terms of viewing rate. Dividing Equation (8.2) by $\Delta t$ and letting $\Delta t \to 0$, a differential equation of the following form can be obtained:

$$\frac{dv(t)}{dt} = b(a - v(t)) \tag{8.3}$$

Solving Equation (8.3) under initial condition $v(0) = v_0$, we have

$$v(t) = v(0) + (a - v(0))(1 - e^{-bt}) \tag{8.4}$$

In Equation (8.4), $v(0)$ is nothing but the number of view counts when the process has started. Here, it is assumed that there is some number of views when the actual noting starts. This model is similar to that proposed by Richier et al. (2014). In Equation (8.4), we assume $v(0) = 0$, i.e., when the system starts there are no views. So we get an expression given by Equation (8.5):

$$v(t) = a\{1 - e^{-bt}\} \tag{8.5}$$

where $a$ is the expected number of total views that can be observed in a video.

The model explains how the rate of view count is directly linked with the leftover views of a video. This model can act as a very strong forecasting tool to predict the level a video can reach.

It is to further note that in today's market when everything is dynamic, one cannot actually talk about fixed market size. Especially in the Internet market where the market varies significantly because of various reasons, for example, with an increase in the popularity of a video, the number of viewers increases and so does the number of views. To inculcate this dynamic behavior, Equation (8.3) can be redesigned as

$$\frac{dv(t)}{dt} = b(a(t) - v(t)) \tag{8.6}$$

where $a(t)$ represents dynamic Internet market size. In the present chapter we have explicitly taken three forms of varying market size: linear growth, exponential growth, and growth because of repeat viewership. The various forms that $a(t)$ can take are systematically discussed in the following sections.

### 8.4.1 Model I: Linear growth

In YouTube, viewers generally grow in spurts that are dependent on both environment and people's influence. However, one can often observe a constant rate of growth. These periods of constant growth are often referred to as the linear portions of the growth curve. In this case, we can construct a linear equation to model the linear phase of growth for viewers or we can say view count.

Suppose the number of views of a particular video grows linearly during the last phase, from $t = 0$ to $t = 48$ hours. Let us take 50K (50,000) views on a video on Monday morning (at $t = 0$) and then 80K (80,000) views on Tuesday morning (at $t = 24$) and then 110K (110,000) views on Wednesday morning (at $t = 48$), and so on. So it is growing by 30K (30,000) views a day, a fixed amount.

So we can define $a(t)$ as

$$a(t) = a(1 + \alpha t) \tag{8.7}$$

And thereby Equation (8.3) can be restructured as follows:

$$\frac{dv(t)}{dt} = b\big(a(1 + \alpha t) - v(t)\big) \tag{8.8}$$

### 8.4.2 Model II: Exponential growth

Consider an example of exponential growth which is seen in view count. View count is increased by the viewers. One viewer influences the other viewer to watch the video by sharing it and by word-of-mouth. This influence takes time depending on each person. If 1000 viewers are placed in a large hall with an unlimited supply of people who have not watched that video (i.e., they can influence as many people as they wish), after an hour there will be the first round of influence (with each influencing one), resulting in 2000 viewers. In another hour, each of the 2000 viewers will influence double, resulting in 4000 viewers; after the third hour, there should be 8000 viewers in the hall; and so on. The important concept of exponential growth is that the population growth rate, the number of viewers increased after every hour, is accelerating; that is, it is increasing at a greater and greater rate. After 1 day and 24 of these cycles, the viewers would have increased from 1000 to more than 16 billion. When the population size, $N$, is plotted over time, a J-shaped growth curve is produced.

So, $a(t)$ can be defined as

$$a(t) = ae^{\alpha t} \tag{8.9}$$

And Equation (8.3) can be designed as

$$\frac{dv(t)}{dt} = b\big(ae^{\alpha t} - v(t)\big) \tag{8.10}$$

### 8.4.3   Model III: Repeat viewing

An important phenomenon observed during the view-count process is the possibility of repeat viewing. The existing viewers may re-view a number of videos for the second or for more number of times. The increase in a number of view counts of a video can be due to both initial viewing and repeat viewing. Several firms are interested in estimating the increase in the number of viewers due to repeat viewing, since the advertisement displays on the video have reduced the impact factor, because the viewer has already watched that advertisement of the video but firms have to pay for the advertisement. For online videos, repeat viewing can be observed during the later stages of a video's life cycle. The first two models (as discussed) might provide an ambiguous fit to the data in that scenario. The present case emphasizes the repeat viewing phenomenon which is a realistic scenario in today's market. In the model we have considered that at a given time $t$, the proportion of viewers, say $\alpha (0 \leq \alpha \leq 1)$, is susceptible to repeat viewing, and repeat viewing is influenced by all factors (both internal and external) affecting final views. So we can define $a(t)$ as

$$a(t) = a + \alpha v(t) \tag{8.11}$$

And likewise Equation (8.3) can be described as follows:

$$\frac{dv(t)}{dt} = b\big(a + \alpha v(t) - v(t)\big) \tag{8.12}$$

Making use of the initial condition that in starting there are no views, the functional form of all the aforesaid models can be represented in Table 8.1.

## 8.5   Data analysis and model validation

We collected data from YouTube manually at 72 different time points for 7 videos, namely, DS I, DS II, DS III, DS IV, DS V, DS VI, and DS VII, where DS I, II, III, and VI correspond to the music category; DS IV and V correspond to the entertainment category; and DS VII corresponds to the technology category. The interval between the two time points is approximately the same. All the view counts are in the thousands, i.e., 233 represents 233,000 views. We have used the software package SPSS (least square method and maximum likelihood method) to estimate the parameters $a$, $b$, and $\alpha$ that are shown in Table 8.2.

The aforesaid predicted and actual view counts are evaluated and shown in Tables 8.3 through 8.9 for the knowing performance of the three models based on the following comparison criteria: bias, variance, RMPSE, M.S.E., $R^2$, and S.S.E.

**Table 8.1**  Models for different forms of $a(t)$

| Model | $a(t)$ form | Final model |
|---|---|---|
| Model I (linear growth) | $a(t) = a(1 + \alpha t)$ | $v(t) = \left[ \left\{ a\left(1 - e^{-bt}\right)\left(1 - \dfrac{\alpha}{b}\right) \right\} + \alpha at \right]$ |
| Model II (exponential growth) | $a(t) = ae^{\alpha t}$ | $v(t) = \dfrac{a}{\alpha + b}\left[ e^{\alpha t} - e^{-bt} \right]$ |
| Model III (repeat viewing) | $a(t) = a + \alpha v(t)$ | $v(t) = \dfrac{a}{1 - \alpha}\left[ 1 - e^{-b(1-\alpha)t} \right]$ |

*Table 8.2* The estimated parameters of the data sets on all three models

| Parameter | DS I | DS II | DS III | DS IV | DS V | DS VI | DS VII |
|---|---|---|---|---|---|---|---|
| | | | | Dataset | | | |
| | **Parameter estimation for different dataset using Model I** | | | | | | |
| $a$ | 5,499.146 | 4,038.262 | 1,402.6 | 17,684.89 | 18,936.66 | 2,032.839 | 4,609.456 |
| $b$ | 0.279 | 0.549 | 0.284 | 0.189 | 0.36 | 0.216 | 0.144 |
| $\alpha$ | 0.013 | 0.003 | 0.007 | 0.004 | 0.02 | 0.01 | 0.004 |
| | **Parameter estimation for different dataset using Model II** | | | | | | |
| $a$ | 5,894.747 | 4,058.969 | 1,454.733 | 17,852.3 | 18,978.54 | 2,151.82 | 4,660.231 |
| $b$ | 0.241 | 0.539 | 0.255 | 0.186 | 0.358 | 0.193 | 0.142 |
| $\alpha$ | 0.008 | 0.002 | 0.005 | 0.003 | 0.002 | 0.007 | 0.003 |
| | **Parameter estimation for different dataset using Model III** | | | | | | |
| $a$ | 8,493.813 | 2,782.836 | 11,82.168 | 20,123.59 | 1,845.306 | 2,251.78 | 4,777.419 |
| $b$ | 0.08 | 0.602 | 0.187 | 0.135 | 0.322 | 0.109 | 0.115 |
| $\alpha$ | 0.062 | 0.376 | 0.362 | 0.014 | 0.909 | 0.259 | 0.12 |

*Table 8.3* Values of comparison parameters on the proposed models for Data Set I

| Comparison parameters | Models | | |
|---|---|---|---|
| | Model I | Model II | Model III |
| Bias | 4.229 | 1.447 | −88.783 |
| Variance | 250.337 | 249.531 | 749.213 |
| RMPSE | 250.372 | 249.535 | 754.455 |
| M.S.E | 62,686.291 | 62,267.613 | 569,202.180 |
| $R^2$ | 0.983 | 0.983 | 0.845 |
| S.S.E | 4,513,412.925 | 4,483,268.110 | 40,982,556.960 |

*Table 8.4* Values of comparison parameters on the proposed models for Data Set II

| Comparison parameters | Models | | |
|---|---|---|---|
| | Model I | Model II | Model III |
| Bias | −3.286 | −3.397 | −9.024 |
| Variance | 103.677 | 106.259 | 205.414 |
| RMPSE | 103.729 | 106.313 | 205.612 |
| M.S.E | 10,759.629 | 11,302.558 | 4,2276.422 |
| $R^2$ | 0.944 | 0.941 | 0.780 |
| S.S.E | 774,693.266 | 813,784.143 | 3,043,902.371 |

*Table 8.5* Values of comparison parameters on the proposed models for Data Set III

| Comparison parameters | Models | | |
|---|---|---|---|
| | Model I | Model II | Model III |
| Bias | −5.525 | −6.231 | −17.574 |
| Variance | 76.304 | 79.675 | 130.630 |
| RMPSE | 76.504 | 79.919 | 131.806 |
| M.S.E | 5,852.879 | 6,386.973 | 17,372.916 |
| $R^2$ | 0.938 | 0.933 | 0.817 |
| S.S.E | 421,407.265 | 459,862.083 | 1,250,849.931 |

*Table 8.6* Values of comparison parameters on the proposed models for Data Set IV

| Comparison parameters | Models | | |
|---|---|---|---|
| | Model I | Model II | Model III |
| Bias | −33.935 | −35.183 | −82.531 |
| Variance | 363.805 | 371.675 | 763.370 |
| RMPSE | 365.384 | 373.336 | 767.818 |
| M.S.E | 133,505.451 | 139,379.865 | 589,544.964 |
| $R^2$ | 0.990 | 0.989 | 0.955 |
| S.S.E | 9,211,876.145 | 9,617,210.648 | 40,678,602.510 |

*Table 8.7* Values of comparison parameters on the proposed models for Data Set V

| Comparison parameters | Models | | |
|---|---|---|---|
| | Model I | Model II | Model III |
| Bias | 20.157 | 19.929 | 21.208 |
| Variance | 133.237 | 132.567 | 145.133 |
| RMPSE | 134.753 | 134.056 | 146.674 |
| M.S.E | 18,158.358 | 17,971.066 | 21,513.276 |
| $R^2$ | 0.957 | 0.957 | 0.949 |
| S.S.E | 1,307,401.808 | 1,293,916.786 | 1,548,955.894 |

*Table 8.8* Values of comparison parameters on the proposed models for Data Set VI

| Comparison parameters | Models | | |
|---|---|---|---|
| | Model I | Model II | Model III |
| Bias | −4.873 | −5.109 | −20.301 |
| Variance | 347.482 | 350.105 | 642.360 |
| RMPSE | 347.516 | 350.142 | 642.680 |
| M.S.E | 120,767.653 | 122,599.623 | 413,038.037 |
| $R^2$ | 0.980 | 0.980 | 0.932 |
| S.S.E | 8,695,271.028 | 8,827,172.856 | 29,738,738.660 |

*Table 8.9* Values of comparison parameters on the proposed models for Data Set VII

| Comparison parameters | Models | | |
|---|---|---|---|
| | Model I | Model II | Model III |
| Bias | −2.323 | −3.116 | −26.352 |
| Variance | 56.387 | 66.752 | 191.522 |
| RMPSE | 56.435 | 66.825 | 193.326 |
| M.S.E | 3,184.932 | 4,465.605 | 37,374.959 |
| $R^2$ | 0.992 | 0.989 | 0.904 |
| S.S.E | 226,130.190 | 317,057.955 | 2,653,622.061 |

The seven data sets have been graphically analyzed as shown in Figures 8.2–8.8. The plots show the actual data and the estimated values from the three models (Model I, Model II, and Model III).

All three models give equally fine results. Moreover, looking at the graphs, it is very difficult to determine which model is performing the best. To sort out the query, we used the weighted criteria approach given by Anand et al. (2014). The weighted criteria approach

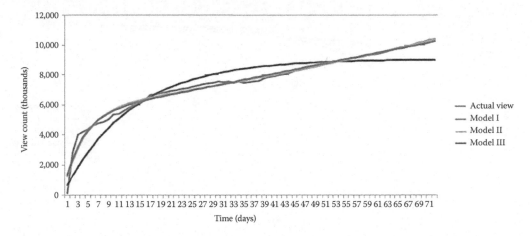

*Figure 8.2* Graphical analysis of Data Set I.

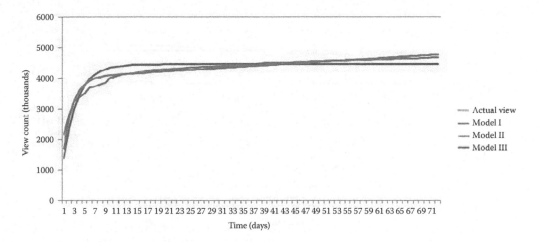

*Figure 8.3* Graphical analysis of Data Set II.

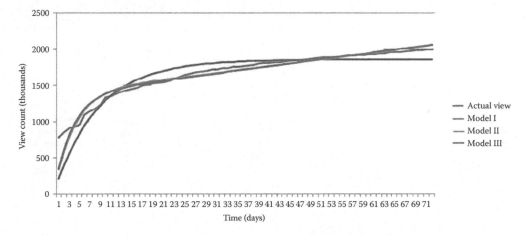

*Figure 8.4* Graphical analysis of Data Set III.

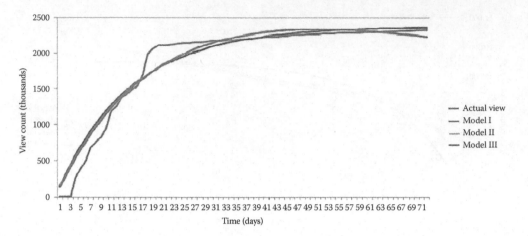

*Figure 8.5* Graphical analysis of Data Set IV.

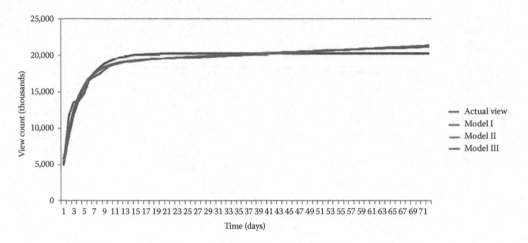

*Figure 8.6* Graphical analysis of Data Set V.

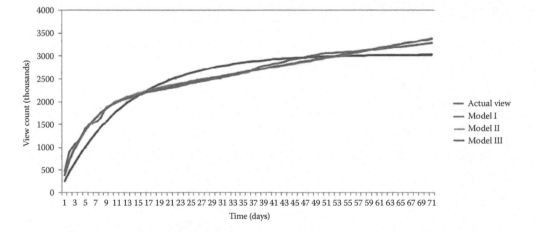

*Figure 8.7* Graphical analysis of Data Set VI.

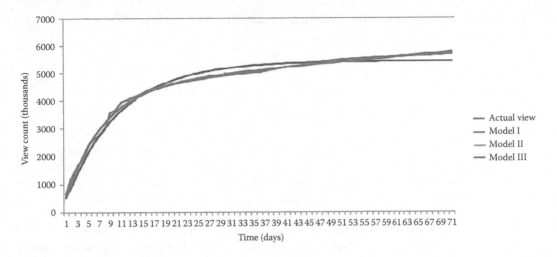

*Figure 8.8* Graphical analysis of Data Set VII.

is a ranking tool that helps us to determine the best fit among various models on the basis of the comparison parameters for each data set.

The algorithm of the approach is as follows:

- In the criteria value matrix, each element $a_{ij}$ shows the value of the $j$th criteria of the $i$th model.
- For each model compute the attribute value, i.e., the maximum value and a minimum value of each criterion.
- The criteria ratings are determined as under:

Case 1. When the smaller value of the criterion represents appropriate fitting to the actual data, i.e., best value, then

$$X_{ij} = \frac{\text{Max. value in dataset} - \text{Criteria value}}{\text{Max. value in dataset} - \text{Min. value in dataset}} \qquad (8.13)$$

Case 2. When the bigger value of the criterion represents appropriate fitting to the actual data, i.e., best value, then

$$X_{ij} = \frac{\text{Criteria value} - \text{Min. value in dataset}}{\text{Max. value in dataset} - \text{Min. value in dataset}} \qquad (8.14)$$

- The weight matrix can be represented as follows:

$$W_{ij} = 1 - X_{ij} \qquad (8.15)$$

- The weighted criteria value matrix is computed by the product of weight of each criterion with the criteria value, i.e.,

$$A_{ij} = W_{ij} * a_{ij} \qquad (8.16)$$

where $a_{ij}$ = value of the $j$th criteria of the $i$th model.

- The permanent value is the weighted mean value of all criteria:

$$Z_i = \frac{\sum_{i=1}^{m} A_{ij}}{\sum_{i=1}^{m} W_{ij}} \tag{8.17}$$

The ranking of models is done on the basis of the expression obtained in Equation (8.17) (i.e., based on permanent value). The smaller permanent value of the model represents good rank as compared to the bigger permanent value of the model. So all permanent values are compared, and ranks for each model are provided. The analysis of DS-I is shown in Tables 8.10–8.13, and the algorithm can be carried for the rest of the data sets.

In Data Set I, Model II is performing best. We found that the best performing model differs from data set to data set. The ranking of each model for different data sets is shown in Table 8.14.

Table 8.14 shows the models and their ranks corresponding to various data sets under consideration. We found that Model I (linear increment) is performing best in five data sets (DS II, DS III, DS V, DS VI, and DS VII), and Model II (exponential increment) is performing best on two data sets (DS I and DS IV).The relevance of studying these scenarios lies in the fact that the first two models, Model I and Model II, can be understood in terms of virality of the video and with what rate the videos are becoming viral. The third model (i.e., Model III) helps to model a certain real-life scenario where a video might be getting some repeat

*Table 8.10* Comparison parameter matrix (DS-I)

| Comparison parameters | Models | | |
|---|---|---|---|
| | Model I | Model II | Model III |
| Bias | 4.229 | 1.447 | −88.783 |
| Variance | 250.337 | 249.531 | 749.213 |
| RMPSE | 250.372 | 249.535 | 754.455 |
| M.S.E | 62,686.291 | 62,267.613 | 569,202.180 |
| $R^2$ | 0.983 | 0.983 | 0.845 |
| S.S.E | 4,513,412.925 | 4,483,268.110 | 40,982,556.960 |

*Table 8.11* Criteria matrix (DS-I)

| Comparison parameters | Models | | |
|---|---|---|---|
| | Model I | Model II | Model III |
| Bias | 0.000 | 0.030 | 1.000 |
| Variance | 0.998 | 1.000 | 0.000 |
| RMPSE | 0.998 | 1.000 | 0.000 |
| M.S.E | 0.999 | 1.000 | 0.000 |
| $R^2$ | 1.000 | 1.000 | 0.000 |
| S.S.E | 0.999 | 1.000 | 0.000 |

*Table 8.12* Weighted value criteria matrix (DS-I)

| Weighted matrix | Model I | Model II | Model III |
|---|---|---|---|
| Bias | 1.0000 | 0.9701 | 0.0000 |
| Variance | 0.0016 | 0.0000 | 1.0000 |
| RMPSE | 0.0017 | 0.0000 | 1.0000 |
| M.S.E | 0.0008 | 0.0000 | 1.0000 |
| $R^2$ | 0.0000 | 0.0000 | 1.0000 |
| S.S.E | 0.0008 | 0.0000 | 1.0000 |
| Sum of weights | 1.0049 | 0.9701 | 5.0000 |

*Table 8.13* Ranking (DS-I)

| Models | $Z_i$ | Rank |
|---|---|---|
| Model I | 3,765.913 | 2 |
| Model II | 1.447 | 1 |
| Model III | 8,310,653.000 | 3 |

*Table 8.14* Ranking of models for different data sets

| Datasets | Models | | |
|---|---|---|---|
| | Model I | Model II | Model III |
| DS I | 3 | 1 | 2 |
| DS II | 1 | 2 | 3 |
| DS III | 1 | 2 | 3 |
| DS IV | 2 | 1 | 3 |
| DS V | 1 | 2 | 3 |
| DS VI | 1 | 2 | 3 |
| DS VII | 1 | 2 | 3 |
| | Best in 5 data sets | Best in 2 data sets | Best in 0 data set |

counts and thereby increasing the view count to the permissible amount. Practically all of these scenarios can exist in the market, and the analysis is clearly able to represent that.

## 8.6   Conclusion

The fact that the population is fixed, is often a reasonable approximation when the evolution of the popularity of content increases quickly and then dies out within a short span of time. In this chapter, we considered the case in which the Internet market growth and the dynamic of view counts of content are in literacy linked. To compute such a dependence, different growth scenarios have been considered. Further, the models have been ranked based on a weighted criteria technique to know which type of video undergoes which type of growth.

In the future, it would be interesting to know how the concept works in the environment of irregular fluctuations in diffusion rate and about its connectivity based on attributes apart from view count.

# References

Anand, Adarsh, Parmod Kumar Kapur, Mohini Agarwal, and Deepti Aggrawal. "Generalized innovation diffusion modeling & weighted criteria based ranking." In *Reliability, Infocom Technologies and Optimization (ICRITO) (Trends and Future Directions)*, 2014 3rd International Conference on, pp. 1–6. IEEE, 2014.

Aggrawal, Niyati, Anuja Arora, and Adarsh Anand. "Modelling and characterizing viewers of YouTube videos." *International Journal of System Assurance Engineering and Management*, 2018. doi:10.1007/s13198-018-0700-6.

Bauckhage, Christian, Fabian Hadiji, and Kristian Kersting. "How viral are viral videos?" In *ICWSM*, pp. 22–30. 2015.

Berg, Madeline. "The World's Top-Earning YouTube Stars 2015". *Forbes*. (November 2015)

Cheng, Xu, Cameron Dale, and Jiangchuan Liu. "Statistics and social network of YouTube videos." In *Quality of Service, 2008. IWQoS 2008. 16th International Workshop on*, pp. 229–238. IEEE, 2008.

Ding, Yuan, Yuan Du, Yingkai Hu, Zhengye Liu, Luqin Wang, Keith Ross, and Anindya Ghose. "Broadcast yourself: Understanding YouTube uploaders." In *Proceedings of the 2011 ACM SIGCOMM Internet Measurement Conference*, pp. 361–370. ACM, 2011.

Khan, Gohar Feroz, and Sokha Vong. "Virality over YouTube: An empirical analysis." *Internet Research* 24, no. 5 (2014): 629–647.

Goel, Sharad, Ashton Anderson, Jake Hofman, and Duncan J. Watts. "The structural virality of online diffusion." *Management Science* 62, no. 1 (2015): 180–196.

Kapur, Parmod Kumar., R. B. Garg, and Santosh Kumar. *Contributions to Hardware and Software Reliability*. Vol. 3. World Scientific, Singapore, 1999.

Richier, Cédric, Eitan Altman, Rachid Elazouzi, Tania Altman, Georges Linares, and Yonathan Portilla. "Modelling view-count dynamics in YouTube." arXiv preprint arXiv:1404.2570 (2014).

Vaish, Abhishek, Rajiv Krishna, Akshay Saxena, Mahalingam Dharmaprakash, and Utkarsh Goel. "Quantifying virality of information in online social networks." *International Journal of Virtual Communities and Social Networking* 4, no. 1 (2012): 32–45.

YouTube Press [https://www.youtube.com/yt/about/press/]

Yu, Honglin, Lexing Xie, and Scott Sanner. "The lifecycle of a YouTube video: Phases, content and popularity." In *ICWSM*, pp. 533–542. 2015.

Zhou, Renjie, Samamon Khemmarat, and Lixin Gao. "The impact of YouTube recommendation system on video views." In *Proceedings of the 10th ACM SIGCOMM Conference on Internet Measurement*, pp. 404–410. ACM, 2010.

# chapter nine

# Market segmentation-based modeling: An approach to understand multiple modes in diffusion curves

*A. Anand, R. Aggarwal, and O. Singh*
*University of Delhi*

## Contents

## 9.1 Introduction

The diffusion of innovation is an essential topic of research in the field of marketing management. Since the 1960s, plenty of innovation diffusion models have been introduced to study the diffusion process of a product. A plethora of diffusion models based on the highly pertinent work of Bass (1969) are available in literature. Bass (1969) has contributed greatly to the understanding of a variety of diffusion models. The simple structure of the Bass model has led to its higher number of applications over the last few decades. The Bass model perceives that an innovation spreads throughout the market by two main channels: mass media (external influence) and word of mouth (internal influence). His model assumes the nature of consumers to be homogenous with respect to their response behavior (Agliari et al. 2009). The inexorable expansion of the market forces the researchers to explore alternative diffusion models with high explanatory power. The variation in customers' buying behavior requires a renewed focus toward the segmented market structure, which directly affects the expected profit of the firm (Wedel and Kamakura 2012).

In today's era of competition to build long-lasting relations and gain trust with consumers, it becomes mandatory for management to take into account different characteristics and adoption behaviors of customers in various segments of markets. Hence, it becomes vital for marketers to understand the concept of multisegmented marketing (Singh et al. 2015). However, the launch of a new product would raise awareness about its usage and may trigger demand among their potential customers (Aggrawal et al. 2014). Furthermore, the availability of the product/service centers of the technological products would also impact the number of adopters of the product. Due to the different adoption behavior, the studies suggest the presence of a dual market: an "early" market corresponding to the high needs and less price sensitivity and a "main" market corresponding to the relatively less needs and high price sensitivity.

Main market adopters are different from early market adopters. Recent literature suggests that, at least regarding high-tech products, main market adopters are not opinion generators; moreover, they do not influence the potential customers of the product (Moore 1991, 1995). In addition, industry studies ascertain different motives for adoption of innovative products among early and main market consumers. Early adopters are mainly technophiles attracted to a product for its competitive edge over similar products in the segment; main market consumers are primarily more interested in the product's enduring functions (Goldenberg et al. 2002).

Existing dual-market models tend to overlook price sensitivity when it comes to considering adoption behaviors of early and main market adopters. Early market adopters are higher risk takers as they endorse a product in spite of unpredictability and possible misgivings/imperfections at the initial stage of introduction of the product (Kim et al. 2014). In comparison, main market adopters are more calculative and are rationalists who weigh out the benefits offered by the product in the given price bracket before they make the final purchasing decision (Rogers 1995). This reasoning accounts for entrance of main market adopters late into the market. However, the existing diffusion models presume their entry at the earliest stages of the market. Hence, there is essentially some time of consideration after a main market adopter comes to know about the innovation and before he adopts it.

Another limitation of the above-mentioned models is that they take into account a single market for a single product. However, potential adopters are not strictly observant of various factors of the adoption system and may respond differently over time (Anand et al. 2016b). In addition, as the demographic distribution over population and potential adopters might be spread across vast regions, this may introduce some time lag. Hence, the introduction of product to a new customer through a mass-mediated process or personalized interaction is bound to take time. Thus, inclusion of the feature of time lag between early and main market for diffusion of product is necessary for a comprehensive understanding of the diffusion model.

The understanding of the dual-market structure in the initial stages of product life cycle has invoked marketing practitioners to introduce the concept of time lag between early and main markets of the product. High-tech executives have increasingly come to use terms like "early market/main market" or "visionaries/pragmatists" to comprehend the diffusion process. According to Moore (1991), this difference necessitates change in marketing strategy including product launch. The main market varies from the early market with respect to its magnitude, population distribution, nature, customer expectations, price sensitivity, and major benefits derived from the product (Gatingon and Robertson 1985). For effective product management, marketing agents should be sensitive to the time difference of the new and main markets, to the extent of being able to predict the time at which the mainstream consumers take over early adopters. Accordingly, marketing strategies can be orchestrated to suit the initiation of early and main market buyers. The interim period is also significant as it is closely related to other early product life cycle modifications.

Product life cycle (PLC) is considered to be the trajectory of sales of a product from its genesis to its final stages (Chandrasekaran and Tellis 2007). Considering it from a macro perspective, other researchers describe it as the fluctuations in the market during the product's lifetime (Helfat and Peteraf 2003). Hence, PLC can help to determine product-related strategy decisions for the company (Wong and Ellis 2007). Hofer (1975) studied and reemphasized the importance of PLC on business planning. Forrester was a pioneer in studying PLC and its applicability as a tool for management analysis and managerial modeling. He assumes the industry and products to be homogenous in terms of their characteristics and customer viewpoint to analyze the PLC stages. Hence, it is quintessential in mapping the development of innovation and its market opportunities. Rogers regards the diffusion curve based on potential adopters into five market segments: innovators, early adopters, early majority, late majority, and laggards. Subsequently, Moore worked on Rogers' normal diffusion curve and adopters' categories to describe expansion of the new products in the market. Moore detects a break in the process as the later consumer or mainstream market does not necessarily depend on the earlier adopters for product information. However, it can be perceived that Moore's purported "break" is not as sharp as he would make us believe. After the initial life span of the innovation, there may be a slump in the market, yet the other market comes up simultaneously before the previous market has died down. Similarly, there may be entry of the other market during the decline phase of the foregoing market. Therefore, assuming time lag might be misleading as at some point of time two or more markets exist side by side. Introduction of a multimodal product life cycle curve for the simultaneous multimarkets phenomena is more imperative as it is more realistic. But in this study, we consider the existence of two simultaneous markets to study the bimodal structure as a particular case of multimodal curves. The improved curve is going to have new long-bearing repercussions on marketing strategies for both the dying early market and the mainstream market.

Although a new trend in marketing literature differentiates between early and main markets for new products requiring separate treatment by marketers (Mahajan and Muller 1998), the existence of discontinuity in the diffusion process has not been sufficiently explored. This discontinuity may be due to insufficient transmission of product information between early market adopters and mainstream consumers (Moore 1991). Any significant difference in the adoption rates of the two markets will inevitably affect the overall sales. It may result in a temporary decline in the sales of the product at the intermediate stage (Goldenberg et al. 2002). Inclination or reluctance of consumers of the two segments of the market may be markedly different (Rogers 1995). This differentiation calls for bimodal curves for corresponding dual markets. This chapter proposes a new dual-market innovation diffusion model framework that considers the division of consumers as early adopters and mainstream consumers. The main adopters are assumed to enter the market after a certain period of time. Considering different influences of the product over potential buyers, we study the different adoption behaviors using distribution functions for main market adopters. We use the new dual-market model to study the pattern of product life cycles of innovations.

The remainder of this chapter is as follows: in Section 9.2, we present the details and mathematical framework of our model, followed by empirical analysis and validation of our proposal in Section 9.3. At last, we discuss the implications of our findings and conclude this chapter in Section 9.4.

## 9.2 Mathematical modeling

The proposed methodology is based on the following set of assumptions:

- The diffusion process is subject to adoption due to the remaining number of adopters in the market.

- The adoption process between the early and main market are disconnected.
- Both the markets have their own potential buyers based on their buying behavior.
- One market adoption is not influenced by the other, i.e., there is no cross-market influence.
- Market size (potential adopters) is fixed during the diffusion process.
- There is a time lag between both the markets.

In this section, we present the dual-market model guided under the above-mentioned assumption. As available in literature, according to Bass (1969) the adoption process occurs because of two adopter groups: innovators (external influentials) and imitators (internal influentials). The mathematical representation given by Bass is

$$n(t) = \frac{dN(t)}{dt} = \left[ p + q\frac{N(t)}{M} \right][M - N(t)] \tag{9.1}$$

where $p$ and $q$ are the coefficients of external and internal influence, respectively. The cumulative number of adopters at time $t$, $N(t)$ can be obtained over the remaining adopters of potential market size, $M$.

Building on the Bass model, Kapur et al. (2004) proposed an alternative formulation of the Bass model by replacing $p + q\dfrac{N(t)}{M}$ as $b(t)$ in Equation (9.1) to avoid the distinction between innovators and imitators, as innovator for one product may be imitator for the other. Equation (9.1) can be thus rewritten as

$$\frac{dN(t)}{dt} = b(t)[M - N(t)] \tag{9.2}$$

where $b(t)$ defines the rate of adoption of an innovation at time $t$.

The uniqueness of the adoption behavior of both early and main markets is worthy of elaboration. Hence, the dual-market innovation diffusion model assumes that adopters are highly affected by the information that transfers from their own peer group rather than the same information disseminated throughout the entire population (Goldenberg et al. 2002). Here we use the index $i$ to define the notations of early market and the index $m$ defines the main market segment.

The adoption process in the early market and in the main market segment progresses as follows:

*For early market:*

$$\frac{dI(t)}{dt} = b_i(t)[N_i - I(t)] \tag{9.3}$$

*For main market:*

$$\frac{dM(t)}{dt} = b_m(t)[N_m - M(t)] \tag{9.4}$$

Here, $b_i(t)$ denotes the hazard rate function that an early market consumer will adopt the product as a result of external and internal forces of marketing, and $b_m(t)$ is the rate at which the main market consumer will adopt the product as a result of external and internal forces of marketing. $N_i$ describes the market potential of the early market, and $N_m$ defines the market potential of the main market. $I(t)$ stands for the cumulative number of adopters at time $t$ for the early market population, and $M(t)$ is the cumulative number of adopters of the main market population at time $t$.

The early market adoption process is similar to the differential equation as defined in Equation (9.3). But for main market adoption, we employ a new parameter $\tau$ for delayed entry of the main market adopters. It is widely accepted in the literature that the early and main market adopters differ with respect to their adoption behavior and also have different levels of price sensitivity. Hence, the entry of main market adopters after a certain time $\tau$ can be represented in the following way:

$$\frac{dM(t-\tau)}{dt} = b_m(t-\tau)[N_m - M(t-\tau)] \tag{9.5}$$

If $\tau$ equals 0, Equation (9.5) is equivalent to Equation (9.4). Let $t' = t - \tau$, and Equation (9.5) can be rewritten as

$$\frac{dM(t')}{dt} = b_m(t')[N_m - M(t')] \tag{9.6}$$

## 9.2.1 Early market adoption model

As early market adopters are more open to choices, we assume the early market behaves the same as that of Bass and follows the S-shaped adoption curve. Also the information about the new product spread with time (Anand et al. 2016a). Hence, we assume that initially only the innovators of the early market will adopt the product, but later on imitators will enter into the market. Also by definition of an S-shaped diffusion pattern, it is clear that the diffusion initially expands at a slow rate and later on, the number of adopters increases with time. Therefore, with this mindset it is justifiable to consider the early market adoption pattern to be logistic viz. S-shape. The S-curve is a long-standing methodology used to predict sales of products in the logistic model. Hence, adoption in this segment can be best described by considering the logistic distribution adoption function, i.e., the adoption function can be expressed as $b_i(t) = \dfrac{b}{1 + \beta e^{-bt}}$.

By substituting the value of $b_i(t)$ in Equation (9.3), the cumulative number of early market adopters can be given as

$$N_i(t) = N_i \left( \frac{1 - e^{-b_i t}}{1 + \beta e^{-b_i t}} \right) \tag{9.7}$$

One should note that when there is only one market, i.e., $M(t) = 0$, our model converges to a special case of logistic growth model obtained by Kapur et al. (2004).

## 9.2.2 Main market adoption model

Main market adopters are genetically utilitarian and much more interested in the applicability of the innovation (Goldenberg et al. 2002). They appraise the benefits of adopting a product and also wait until the utility of the product overrides its price before entering into the market (Rogers 1995). In other words, we can say that the main market buyers are highly practical and price sensitive. This is the time point at which the cost of the product declines and becomes affordable or less expensive relative to its utility at that point that main market adopters enter the market and the sales curve increases dramatically (Golder and Tellis 1998). We assume that the main market is fully developed at the time of its introduction. Based on this assumption we can say that adoption here can take more or less time vis-a-vis early market depending on the product's availability in market and its utility. To address the heterogeneity of the main market, we have mentioned different types of S-shaped distribution functions. We have also considered the scenario when the main market adopters enter into market with the fastest pace, due to major change in the marketing policy of the firm. For that, we have considered the exponential growth function for main market adopters. In addition to this, the different adoption distribution functions of the main market have yielded the following expressions:

*Case 1:* Here we are taking into consideration of $b_m(t)$ as the exponential distribution function:

$$b_m(t) = b_m$$

Using this in Equation (9.7), the total number of main market adopters at time $t'$ is found as

$$M(t') = N_m\left(1 - e^{-b_m t'}\right) \tag{9.8}$$

*Case 2:* Using two-stage Erlang function as the rate of adoption, i.e.,

$$b_m(t') = \frac{b_m^2 t'}{1 + b_m t'}$$

Then the solution of Equation (9.7), corresponding to the above defined $b_m(t')$ is

$$M(t') = N_m\left(1 - (1 + b_m t')e^{-b_m t'}\right) \tag{9.9}$$

*Case 3:* Considering the logistic rate of adoption for main market adopters, i.e.,

$$b_m(t') = \frac{b_m}{1 + \beta_m e^{-b_m t'}}$$

Substituting it in Equation (9.7), then the corresponding total number of main market adopters is given as

$$M(t') = N_m\left(\frac{1 - e^{-b_m t'}}{1 + \beta_m e^{-b_m t'}}\right) \tag{9.10}$$

*Case 4:* Assuming the rate as two-stage Erlang logistic function in Equation (9.7), i.e.,

$$b_m(t') = \frac{b_m(b_m t' + (1 - e^{-b_m t'})\beta)}{(1 + \beta_m + b_m t')(1 + \beta_m e^{-b_m t'})}$$

then the corresponding total number of main market adopters is calculated as

$$M(t') = N_m\left(\frac{1 - (1 + b_m t')e^{-b_m t'}}{1 + \beta_m e^{-b_m t'}}\right) \tag{9.11}$$

We now define a function $L(t)$ as the cumulative number of adopters of the main market at time $t$, which starts from the initial time point 0, as follows:

$$L(t) = \begin{cases} M(t-\tau) & \text{for } t \geq \tau, \\ 0 & \text{for } t < \tau. \end{cases} \tag{9.12}$$

The different values of function $M(t)$ have been taken from Equations (9.8) to (9.11).

### 9.2.3   Total adoption modeling

Using the unified modeling approach, the dual-market innovation diffusion model (DMIDM) has been formulated. By adding the cumulative number of adopters of early and main market by time $t$, we can find the cumulative number of adopters at any given time as

$$N(t) = I(t) + L(t) \tag{9.13}$$

Here it is noted that the market potentials of early market $N_i$ and main market $N_m$ have been obtained from the market potential of total market, $M$. Assume $\theta$ defines the proportion of the early market in the population of the total market; as such,

$$N_i = \theta M \text{ and } N_m = (1-\theta)M \quad \text{where } (0 \leq \theta \leq 1) \tag{9.14}$$

Then by substituting the expressions of $N_i$ and $N_m$ in the above expression proposed in the last section, and then putting the values of $I(t)$ and $L(t)$, we summarize all the dual-market innovation diffusion models to find total sales $N(t)$ in Table 9.1, corresponding to various early and main market adoption functions.

## 9.3   Parameter estimation

Parameter estimation for all above-defined dual models has been adjudged in this section by using the techniques of the nonlinear least squares method. For the practical affirmation,

*Table 9.1* Dual market innovation diffusion models

| Models | $I(t)$ | $L(t)$ | $N(t)$ |
|---|---|---|---|
| DMIDM-I | Logistic | Exponential | $M\left[\theta\left(\dfrac{1-e^{-b_it}}{1+\beta_ie^{-b_it}}\right)+(1-\theta)\left(1-e^{-b_m(t-\tau)}\right)\right]$ |
| DMIDM-II | Logistic | Two-stage Erlang | $M\left[\theta\left(\dfrac{1-e^{-b_it}}{1+\beta_ie^{-b_it}}\right)+(1-\theta)\left(1-(1+b_m(t-\tau))e^{-b_m(t-\tau)}\right)\right]$ |
| DMIDM-III | Logistic | Logistic | $M\left[\theta\left(\dfrac{1-e^{-b_it}}{1+\beta_ie^{-b_it}}\right)+(1-\theta)\left(\dfrac{1-e^{-b_m(t-\tau)}}{1+\beta_me^{-b_m(t-\tau)}}\right)\right]$ |
| DMIDM-IV | Logistic | Two-stage Erlang logistic | $M\left[\theta\left(\dfrac{1-e^{-b_it}}{1+\beta_ie^{-b_it}}\right)+(1-\theta)\left(\dfrac{1-(1+b_m(t-\tau))e^{-b_m(t-\tau)}}{1+\beta_me^{-b_m(t-\tau)}}\right)\right]$ |

it is approachable to show the fit of the proposed models on the real-life data sets in terms of cumulative distribution functions. The empirical analysis of proposed models has been done over real-life Data Set I (DS I), which refers to cable TV, and Data Set II (DS II), which refers to cloth dryer sales data taken from Van den Bulte and Lilien (1997). In this study, statistical software package SPSS (Statistical Package for Social Sciences) nonlinear regression models have been used to estimate the parameters and their standard errors for the above-defined four models. The SPSS is an interactive and user-friendly software to apply more sophisticated models to the data. Also, the statistical software R has been used to draw the box plots of relative errors for all defined models. For the above-mentioned models in Table 9.1, the estimates of parameters are summarized in Table 9.2. It is assumed that the delay entry time parameter $\tau$ is a fixed number for all the models. In the case of DS I, the value of the parameter $\tau$ is taken as 5, and it is fixed as 8 for DS II.

The value of weight parameter $\theta$ is given in the eighth column of Table 9.2 and signifies that for DS I, the main market is more significant than the early market. But in the case of DS II, it implies that the early market dominates the main market. Hence, it is not justifiable to declare the importance of one segment of the market over the other without using a mathematical model. The performance analysis of the proposed models is measured by using the most common goodness-of-fit criteria as MSE (mean square error), $R^2$ (coefficient of determination), bias, and variation. The values of these comparison criteria are shown in Table 9.3, confirming the robustness of the approach.

For practical purposes, it is mandatory to find the better-fitted model to the given data sets. Hence, we have shown the cumulative sales data and predicted sales of the defined data sets in Figures 9.1 and 9.2. It can be observed that all the models are indistinguishable and equally fit to the actual sales data set as all graphs of predicted sales are overlapping

*Table 9.2* Estimates of the parameters for the dual models

| Data sets | Models | $M$ | $b_i$ | $b_m$ | $\beta_i$ | $\beta_m$ | $\theta$ |
|---|---|---|---|---|---|---|---|
| DS I | DMIDM-I | 121.1794 | 0.85 | 0.08238 | 7.906133 | | 0.298833 |
| | DMIDM-II | 87.78899 | 0.832 | 0.368664 | 9.252265 | | 0.449032 |
| | DMIDM-III | 92.7719 | 0.75 | 0.26261 | 6.460756 | 1.781948 | 0.414223 |
| | DMIDM-IV | 85.9224 | 0.705 | 0.417982 | 7.142053 | 1.1 | 0.497557 |
| DS II | DMIDM-I | 30.11925 | 0.384933 | 0.1 | 11.0378 | | 0.65 |
| | DMIDM-II | 26.85485 | 0.379993 | 0.436111 | 7.021542 | | 0.65 |
| | DMIDM-III | 25.97044 | 0.434153 | 0.736788 | 16.0431 | 29.73104 | 0.77473 |
| | DMIDM-IV | 31.54691 | 0.379977 | 0.376455 | 14.60838 | 18.10933 | 0.75 |

*Table 9.3* Goodness-of-fit measures

| Data sets | DS I | | | | DS II | | | |
|---|---|---|---|---|---|---|---|---|
| Models | DMIDM-I | DMIDM-II | DMIDM-III | DMIDM-IV | DMIDM-I | DMIDM-II | DMIDM-III | DMIDM-IV |
| MSE | 4.415 | 3.563 | 5.02 | 5.14 | 0.229 | 0.544 | 0.055 | 0.068 |
| Bias | 0.065 | 0.04 | 0.078 | 0.055 | 0.018 | 0.261 | −0.014 | 0.007 |
| Variance | 9.915 | 7.983 | 10.164 | 10.375 | 0.516 | 1.9 | 0.113 | 0.138 |
| $R^2$ | 0.995 | 0.996 | 0.995 | 0.995 | 0.998 | 0.994 | 0.999 | 0.999 |

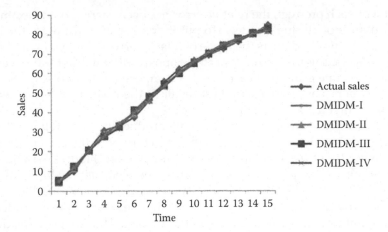

***Figure 9.1*** Actual versus predicted sales for DS I.

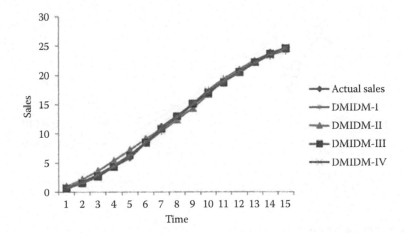

***Figure 9.2*** Actual versus predicted sales for DS II.

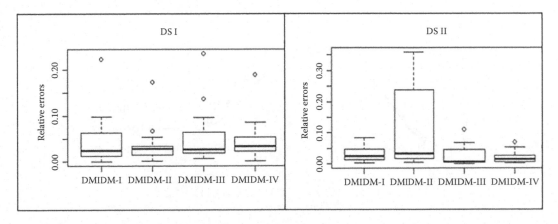

***Figure 9.3*** Box plot of the relative errors for DS I and DS II.

to each other. For each product, the relative error has been carried out to examine the predictive performance of all dual models. To put it another way, we draw the box plots as shown in Figure 9.3, which depicts the range of relative errors of all estimated dual models for both data sets. This figure shows that the proposed dual-market innovation diffusion model with logistic rate growth of early market composite with two-stage Erlang growth function in the main market gives the best result in the case of DS I, whereas the same model gives the worst result for DS II.

In innovation diffusion literature, most of the product life cycle follows the bell-shape structure. But in our study, we have shown that the concept of the dual market brings the multimodal structure of the diffusion curve of innovation. Figures 9.4 and 9.5 plot the noncumulative sales of the proposed models. It can be seen that the bimodal structure of innovation is well captured in these figures. As the early segment market is introduced, the sales of the product initially increase and reach a peak and afterward decrease with time, until the main market is introduced in the market. This shape helps us to explain why it is essential for a firm to choose the introduction of the main market after a certain

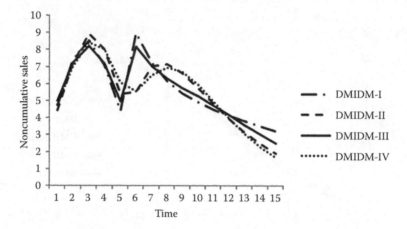

**Figure 9.4** Noncumulative sales curve for DS I.

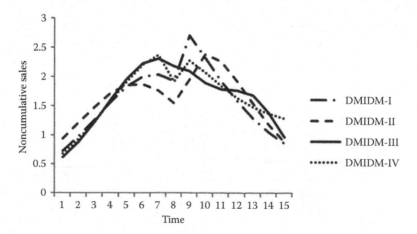

**Figure 9.5** Noncumulative sales curve for DS II.

period of time. This sales growth curve also drives the firm to allocate the promotional efforts and sales strategies on time because the market growth can also be captured by the introduction of the main market of the product and that helps them to increase their product revenue.

## 9.4   Discussion and summary

In this study, by considering the dual-market phenomenon, we have examined how a new product is diffused among potential adopters. We have shown that the dual-market phenomenon is one of the key factors in setting diffusion strategies for some durable products. Another key driver of the dual-market innovation diffusion model is the shape of the product life cycle. We specifically find that the multimodal shape of the product life cycle results due to multimarket structure models. By considering different adoption distribution functions, we represent the different multimodal diffusion curve of the product life cycle.

In this study we propose a new dual-market model that considers the heterogeneous rate of main market adopters by keeping the rate of early market adoption to be logistic. Moreover, we have used the S-shaped logistic adoption function for early market adopters, as early market adopters are fixed and slow at the initial stage, and later on imitators of the early market enter into the market. Main market adoption can be a slow or a fast rate of adoption as compared to early market. With this mindset, we have considered various distribution functions for main market adopters. A unified approach has been applied for modeling the adoption of these two different market segments. Based on our study, it can be seen that the proposed model based on the dual-market assumption can be used effectively to predict the product growth curve into the market.

Some of the applications of our study are in determining that the main market adopters are different from the early market adopters. We find that when the diffusion of the early market slows down, it is beneficial for the firm to target the main market by keeping in mind various factors such as price sensitivity, reliability, utility, etc., of product. We also find that when diffusion slows down at some time point, it is beneficial for the firm to introduce marketing strategies about the product to capture the high margin from the main market. During the phase of the early market, the firm needs to ready the platform for the main market. The firm can generate more revenue with a decision to make a delayed purchase.

The proposed diffusion model may not alter decisions such as introduction of product in the market, but certainly would affect the decision-making process for optimal time and minimum cost of promotional strategies. Another possible implication of our findings is that a firm may achieve a significant strategic advantage by driving the market at an appropriate phase in the product life cycle. It is also of interest to marketing managers that they study the behavior of their customers as to whether they are interested in immediate purchase or wait until the price decreases. We might expect that decisions of delayed purchase would slow the diffusion process, but after some time it again increases at a fast pace. Taking into account this structure helps in impacting the firm's financial performance.

Our model has some possible extensions. Our model does not consider the cross-market influence. However, to extend this model we can consider the interaction between both markets. Also, we have considered the delay time for all the models to be fixed, which can vary for products. So, it would be of interest to find the delay time for all models. We leave these extensions for our future work.

# References

Aggrawal, D., Singh, O., Anand, A., & Agarwal, M. (2014). Optimal introduction timing policy for a successive generational product. *International Journal of Technology Diffusion (IJTD)*, 5(1), 1–16.

Agliari, E., Burioni, R., Cassi, D., & Maria Neri, F. (2009). Word-of-mouth and dynamical inhomogeneous markets: An efficiency measure and optimal sampling policies for the pre-launch stage. *IMA Journal of Management Mathematics*, 21(1), 67–83.

Anand, A., Aggarwal, R., Singh, O., & Aggrawal, D. (2016a). Understanding diffusion process in the context of product dis-adoption. *St. Petersburg State Polytechnical University Journal Economics*, 9(2), 7–18.

Anand, A., Singh, O., Aggarwal, R., & Aggrawal, D. (2016b). Diffusion modeling based on customer's review and product satisfaction. *International Journal of Technology Diffusion (IJTD)*, 7(1), 20–31.

Bass, F. M. (1969). A new product growth for model consumer durables. *Management Science*, 15(5), 215–227.

Chandrasekaran, D., & Tellis, G. J. (2007). A critical review of marketing research on diffusion of new products. In N. K. Malhotra (Ed.), *Review of Marketing Research* (Vol. 3, pp. 39–80). Bingley: Emerald Group Publishing Limited.

Gatingon, H., & Robertson, T.S. (1985). A propositional inventory for new diffusion research. *Journal of Consumer Research*, 11(4), 849–867.

Goldenberg, J., Libai, B., & Muller, E. (2002). Riding the saddle: How cross-market communications can create a major slump in sales. *Journal of Marketing*, 66(2), 1–16.

Golder, P. N., & Tellis, G. T. (1998). Growing, growing, gone: Modeling the sales slowdown of really new consumer durables. University of Southern California Working Paper.

Helfat, C. E., & Peteraf, M. A. (2003). The dynamic resource-based view: Capability lifecycles. *Strategic Management Journal*, 24(10), 997–1010.

Hofer, C. W. (1975). Toward a contingency theory of business strategy. *Academy of Management Journal*, 18(4), 784–810.

Kapur, P. K., Bardhan, A. & Jha, P. C. (2004), An alternative formulation of innovation diffusion model. In V. K. Kapoor (Ed.), *Mathematics and Information Theory*, (pp. 17–23). New Delhi: Anamaya Publication.

Kim, T., Hong, J. S., & Lee, H. (2014). Predicting when the mass market starts to develop: The dual market model with delayed entry. *IMA Journal of Management Mathematics*, 27(3), 381–396.

Mahajan, V., & Muller, E. (1998). When is it worthwhile targeting the majority instead of the innovators in a new product launch? *Journal of Marketing Research*, 35, 488–495.

Moore, G. A. (1991), *Crossing the Chasm*. New York: Harper Business.

Moore, G. A. (1995), *Inside the Tornado*. New York: Harper Business.

Rogers, E. M. (1995), *The Diffusion of Innovations*, 4th ed. New York: The Free Press.

Singh, O., Kapur, P. K., & Sachdeva, N. (2015), Technology management in segmented markets. *Quality, Reliability, Infocom Technology and Industrial Technology Management* (pp. 78–89). New Delhi: I K International Publishing House.

Van den Bulte, C., & Lilien, G. L. (1997). Bias and systematic change in the parameter estimates of macro-level diffusion models. *Marketing Science*, 16(4), 338–353.

Wedel, M., & Kamakura, W. A. (2012). *Market Segmentation: Conceptual and Methodological Foundations* (Vol. 8). New York: Springer Science & Business Media.

Wong, H. K., & Ellis, P. D. (2007). Is market orientation affected by the product life cycle? *Journal of World Business*, 42(2), 145–156.

# chapter ten

# Kernel estimators for data analysis

**Piotr Kulczycki**
*Systems Research Institute, Polish Academy of Sciences;*
*AGH University of Science and Technology*

## Contents

## 10.1 Introduction

Perversely, one can state that contemporary data analysis has developed too vigorously, which negatively caused, in particular, an absence of due care and attention regarding formalism, mathematical justification, and ultimately compact subject methodology. Before the computer revolution of the second half of the 20th century, data analysis was conducted based on already well-established, effective mathematical apparatuses of statistics. The main trouble was then the inadequacy of the data, expressed primarily in small sample sizes, and – in consequence – statistical procedures were directed toward maximal effectiveness in the sense of gaining as much information as possible from them. The situation was diametrically reversed in the 1980s, with the spread of not only efficient numerical calculation systems, but also methods of automatic measurement. In a relatively short time, a total reversal in the conditions occurred: the data became too numerous and carried too complex information for processing by classic statistical methodology. Moreover, the absence of the ability to super-vise such excessive and complicated data sets through a statistician's intuition led to the danger of the appearance of evidently erroneous data, resulting, for example, in faults occur-ring during the measurements of particular elements. Such drastically reversed conditions of data analysis tasks caused an enormous need for totally new procedures, and the speed of progress brought about a situation in which they were based on separated, often specific, concepts without mathematical justification or attempts at the unification of methodology. Currently, it seems, the time for their modification, proving, and generalization has arrived.

The subject of this chapter is the presentation of a coherent concept of establishing the methodology of kernel estimators for the three main tasks of data analysis: identification/detection of atypical elements (outliers), clustering, and classification. The application of a uniform apparatus for all three basic problems facilitates comprehension of the material and, in consequence, creation of individualized modifications, and also in the latter phase of the designing of a personal computer application. The use of nonparametric kernel esti-mators frees the results from data distribution – this concerns not only the shape of their grouping, but also the possibility of their partition in separate incoherent parts. The meth-odology investigated in this chapter is practically parameter free, i.e., it is not required from the user, the calculation of parameter values, although it is possible to optionally modify them in order to achieve the specific desired properties.

This chapter is constructed as follows. After this introduction, Section 10.2 presents an outline of the methodology of kernel estimators. This will be applied in Section 10.3 to the identification of atypical element task, to clustering in Section 10.4, and in Section 10.5 – classification. In the framework of the final summary, in Section 10.5, an exam-ple application of the investigated material in the creation of a mobile phone operator's marketing support strategy is presented.

## 10.2   *Methodology of kernel estimators*

Let the $n$-dimensional random variable $X : \Omega \to \mathbf{R}^n$, with a distribution having the density $f$, be given. Its kernel estimator $\hat{f} : \mathbf{R}^n \to [0, \infty)$ is calculated on the basis of the $m$-elements simple random sample

$$x_1, x_2, \ldots, x_m \in \mathbf{R}^n \tag{10.1}$$

experimentally obtained from the variable $X$, and it is defined in a basic form by

$$\hat{f}(x) = \frac{1}{mh^n} \sum_{i=1}^{m} K\left(\frac{x - x_i}{h}\right), \tag{10.2}$$

where the measurable function $K : \mathbf{R}^n \to [0, \infty)$, symmetrical with respect to zero and having a weak global maximum at this point, fulfils the condition $\int_{\mathbf{R}^n} K(x) \, dx = 1$ and is called a kernel, whereas the positive coefficient $h$ is referred to as a smoothing parameter. For details, see the classic monographs (Kulczycki 2005; Silverman 1986; Wand and Jones 1995). Notably, a kernel estimator enables the identification of density for practically any distribution, especially with no assumptions regarding its membership of a fixed class; unusual, complex, or multimodal distributions are treated here as a typical unimodal case. The form of the kernel $K$ and the value of the smoothing parameter $h$ are commonly provided based on the mean integrated square error criterion.

Thus, the selection of the kernel form is practically meaningless from a statistical point of view and, in consequence, the user should above all take into account properties of the desired estimator or/and computational aspects, useful for the application problem being worked out; for details, see the literature (Kulczycki 2005, Section 3.1.3; Wand and Jones 1995, Sections 2.7 and 4.5).

For the one-dimensional case (i.e., when $n = 1$), the normal (Gauss) kernel

$$K_j(x) = \frac{1}{\sqrt{2\pi}} \exp\left(-\frac{x^2}{2}\right) \tag{10.3}$$

is generally held as basic. For special purposes, other types can be proposed; here, the uniform kernel

$$K_j(x) = \begin{cases} \dfrac{1}{2} & \text{for } x \in [-1,1] \\[2mm] 0 & \text{for } x \notin [-1,1] \end{cases} \tag{10.4}$$

will be used henceforth – it has bounded support and assumes a finite number of values, which will be taken advantage of later in this chapter.

In the multidimensional case (i.e., when $n > 1$), a so-called product kernel will be applied hereinafter.[*] The main idea here is the division of particular variables with the multidimensional kernel then becoming a product of $n$ one-dimensional kernels for specific coordinates. Thus, the kernel estimator (10.2) is then given as

$$\hat{f}(x) = \frac{1}{m \prod\limits_{j=1}^{n} h_j} \sum_{i=1}^{m} K_1\left(\frac{x_1 - x_{i,1}}{h_1}\right) K_2\left(\frac{x_2 - x_{i,2}}{h_2}\right) \cdots K_n\left(\frac{x_n - x_{i,n}}{h_n}\right), \tag{10.5}$$

where $K_j$ ($j = 1, 2, \ldots, n$) denotes one-dimensional kernels, e.g., normal (10.2) or uniform (10.3), $h_j$ ($j = 1, 2, \ldots, n$) are smoothing parameters individualized for particular coordinates, while assigning to coordinates

$$x = \begin{bmatrix} x_1 \\ x_2 \\ \vdots \\ x_n \end{bmatrix} \quad \text{and} \quad x_i = \begin{bmatrix} x_{i,1} \\ x_{i,2} \\ \vdots \\ x_{i,n} \end{bmatrix} \quad \text{for } i = 1, 2, \ldots, m. \tag{10.6}$$

The above kernels fulfill the additional requirements of the particular procedures used henceforth.

The value of the smoothing parameter is highly significant for the estimation quality, and many advantageous algorithms for calculating it on the basis of a random sample have been proposed.

First, consider the one-dimensional case. In specific conditions, e.g., during initial research or a numerous random sample (10.1) with relatively regular distribution, the

---

[*] For description of another – radius – type, see the monographs by Kulczycki (2005, Section 3.1.3) and Wand and Jones (1995, Section 4.5), where it is called spherically symmetric. This notion will not be used in this text.

approximate method (Kulczycki 2005, Section 3.1.5; Wand and Jones 1995, Section 3.2.1) is sufficient, according to which

$$h = \left( \frac{8\sqrt{\pi}}{3} \frac{W(K)}{U(K)^2} \frac{1}{m} \right)^{1/5} \hat{\sigma}, \tag{10.7}$$

where $W(K) = \int_{-\infty}^{\infty} K(x)^2 \, dx$ and $U(K) = \int_{-\infty}^{\infty} x^2 K(x) \, dx$, while $\hat{\sigma}$ denotes the estimator of a standard deviation:

$$\hat{\sigma} = \sqrt{\frac{1}{m-1} \sum_{i=1}^{m} (x_i - \hat{E})^2} \quad \text{with } \hat{E} = \frac{1}{m} \sum_{i=1}^{m} x_i. \tag{10.8}$$

The functional values occurring in formula (10.7) are, respectively, for normal kernel (10.3)

$$W(K) = \frac{1}{2\sqrt{\pi}}, \quad U(K) = 1 \tag{10.9}$$

and for uniform (10.4)

$$W(K) = \frac{1}{2}, \quad U(K) = \frac{1}{3}. \tag{10.10}$$

For specific cases, the more sophisticated yet effective plug-in method (Kulczycki 2005, Section 3.1.5; Wand and Jones 1995, Section 3.6.1) can be recommended. Its concept consists of the calculation of the smoothing parameter using the approximate method described above, and after $r$ steps improving the result, one obtains a value close to optimal. On the basis of simulation research carried out for the needs of the material worked out in this chapter, $r = 2$ can be proposed. In this case, the plug-in method consists of the application of the following steps:

$$d_8 = \frac{105}{32\sqrt{\pi}\hat{\sigma}^9}, \tag{10.11}$$

where $\hat{\sigma}$ is given by formula (10.8), and subsequently,

$$g_{\mathrm{II}} = \left( \frac{-2\tilde{K}^{(6)}(0)}{mU(\tilde{K})d_8} \right)^{1/9} \tag{10.12}$$

$$g_{\mathrm{I}} = \left( \frac{-2\tilde{K}^{(4)}(0)}{mU(\tilde{K})d_6(g_{\mathrm{II}})} \right)^{1/7}; \tag{10.13}$$

finally

$$h = \left( \frac{V(K)}{mU(K)^2 d_4(g_{\mathrm{I}})} \right)^{1/5}, \tag{10.14}$$

while

$$d_p(g) = \frac{1}{m^2 g^{p+1}} \sum_{i=1}^{m} \sum_{j=1}^{m} \tilde{K}^{(p)} \left( \frac{x_i - x_j}{g} \right) \quad \text{for } p = 4,6. \tag{10.15}$$

The kernel $K$, applied in estimator (10.2), is used only in the last step (10.14). In other steps, represented by formulas (10.12), (10.13), and (10.15), the different kernel $\tilde{K}$ may be used. Generally, a normal kernel (10.3) is assumed; the quantities occurring in formulas (10.12), (10.13), and (10.15) are then given by dependence (10.9) and also

$$K^{(6)}(x) = \frac{1}{\sqrt{2\pi}}(x^6 - 15x^4 + 45x^2 - 15)\exp\left(-\frac{1}{2}x^2\right), \quad K^{(6)}(0) = -\frac{15}{\sqrt{2\pi}} \tag{10.16}$$

$$K^{(4)}(x) = \frac{1}{\sqrt{2\pi}}(x^4 - 6x^2 + 3)\exp\left(-\frac{1}{2}x^2\right), \quad K^{(4)}(0) = \frac{3}{\sqrt{2\pi}}. \tag{10.17}$$

For the multidimensional case, thanks to using a product kernel, the methods presented can be simply applied $n$ times, sequentially for each coordinate.

Finally, it is worth noting that too small a value of the smoothing parameter $h$ implies the appearance of an excessive number of local extremes of the estimator $\hat{f}$, whereas too large causes its overflattening – this property will be actively used in later considerations.

In practical applications of kernel estimators, one can also use specific concepts, generally improving the estimator properties, and others optionally fitting the model to a considered reality.[*] In the first group, a so-called modification of the smoothing parameter – presented in Section 10.2.1 (Kulczycki 2005, Section 3.1.6; Silverman 1986, Section 5.3.1) – will be used henceforth, while in Section 10.2.2, the support boundary (Kulczycki 2005, Section 3.1.7; Silverman 1986, Section 2.10), belonging to the second group, is presented.

## 10.2.1   Modification of smoothing parameter

For the kernel estimator definitions (10.2) and (10.5), the impact of the smoothing parameter on individual kernels is identical. Positive results can be achieved by individualizing this effect, obtained by mapping the positive modifying parameters $s_1, s_2, \ldots, s_m$, to successive kernels, which are given by

$$s_i = \left(\frac{\hat{f}_*(x_i)}{\sqrt[m]{\prod_{i=1}^m \hat{f}_*(x_i)}}\right)^{-c} \quad \text{for } i = 1, 2, \ldots, m, \tag{10.18}$$

where $c \in [0, \infty)$, $\hat{f}_*$ means the kernel estimator without modification, and finally defining the kernel estimator with modification of the smoothing parameter as

$$\hat{f}(x) = \frac{1}{mh^n}\sum_{i=1}^m \frac{1}{s_i^n} K\left(\frac{x - x_i}{hs_i}\right). \tag{10.19}$$

If the product kernel is used, the counterpart of definition (10.5) becomes

$$\hat{f}(x) = \frac{1}{m\prod_{j=1}^n h_j}\sum_{i=1}^m \frac{1}{s_i} K_1\left(\frac{x_1 - x_{i,1}}{s_i h_1}\right) K_2\left(\frac{x_2 - x_{i,2}}{s_i h_2}\right)\cdots K_n\left(\frac{x_n - x_{i,n}}{s_i h_n}\right). \tag{10.20}$$

---

[*] According to the experience of the author and his research team, it is worth maintaining sensible self-restraint in the application of specific ideas available in the subject literature, often ineffective in practice, and increasing the complexity of the procedures.

As a consequence of the above concept, in the areas in which the elements of random sample (10.1) are rare, the kernel estimators are additionally flattened, and in the regions of their concentration – additionally peaked. The parameter $c$ determines the intensity of the modification procedure – when its value is larger/smaller, it becomes more/less distinct. Using the criterion of the integrated mean square error, one can propose

$$c = 0.5. \tag{10.21}$$

For details see the monographs by Kulczycki (2005, Section 3.1.6) and Silverman (1986, Section 5.3.1).

## 10.2.2  Support boundary

For practical applications, specific coordinates of the random variable can describe diverse quantities. A number of these, in particular representing distance or time, for their correct interpretation, must belong to properly bounded subsets, e.g., nonnegative numbers. Due to omitting misinterpretations and calculational errors resulting from this, a beneficial procedure for bounding a kernel estimator's support can be applied.

First, consider the one-dimensional case (when $n = 1$) and the left boundary – i.e., the case where the condition $\hat{f}(x) = 0$ for $x < x_*$, with $x_* \in \mathbf{R}$ (mostly $x_* = 0$), is desired. A fragment of the $i$th kernel which lays outside the interval $[x_*, \infty)$ is symmetrically "reflected" with respect to the boundary $x_*$, and becomes a part of the kernel "hooked" in the element $x_i$ "reflection"; therefore, in the point $2x_* - x_i$. So, after defining the function $K_{x_*} : \mathbf{R} \to [0, \infty)$ by

$$K_{x_*}(x) = \begin{cases} K(x) & \text{when} & x \geq x_* \\ 0 & \text{when} & x < x_* \end{cases}, \tag{10.22}$$

the basic form of kernel estimator (10.2) is the following

$$\hat{f}(x) = \frac{1}{mh} \sum_{i=1}^{m} \left[ K_{x_*}\left( \frac{x - x_i}{h} \right) + K_{x_*}\left( \frac{x + x_i - 2x_*}{h} \right) \right] \tag{10.23}$$

and analogously, the formula with the modification of the smoothing parameter (10.20)

$$\hat{f}(x) = \frac{1}{mh} \sum_{i=1}^{m} \frac{1}{s_i} \left[ K_{x_*}\left( \frac{x - x_i}{hs_i} \right) + K_{x_*}\left( \frac{x + x_i - 2x_*}{hs_i} \right) \right]. \tag{10.24}$$

Cut fragments of kernels, lying outside of the assumed support, are embodied into the support directly near its boundary, so with such a small introduced change, this is acceptable in practice.

The consideration for the right boundary of the support can be followed analogously. In the multidimensional case, the concept presented may naturally be applied subsequently for every coordinate of the considered random variable. These cases, however, will not be used further in this text. For more details, see the books by Kulczycki (2005) and Silverman (1986, Section 2.10).

A broader description regarding various aspects of kernel estimators is found in the classic monographs (Kulczycki 2005; Silverman 1986; Wand and Jones 1995). In the next sections of the chapter, this methodology will be uniformly applied to three fundamental procedures of data analysis: identification of atypical elements, clustering, and classification.

## 10.3  Identification of atypical elements

The task of identifying atypical elements is one of the fundamental problems of contemporary data analysis (Aggarwal 2013), above all, in the preliminary phase of processing. The occurrence of such elements can be interpreted in two ways. The first, and more popular, associates them with gross errors handicapping some elements of the set being considered. They are then eliminated or corrected. In this case, the identification of atypical elements can be termed "detection," which is generally connoted with negative occurrences. In the second, which is less common yet more constructive, atypical elements represent unconventional phenomena, exceptional items, and new trends. They then provide exceptionally valuable information, and stimulate nontrivial behaviors and innovative thinking. In order to cover this case, it is worth replacing the notion of "detection" with the more neutral "identification," as is done throughout this text.

There is no one definition of atypical elements. The most general is that they are observations originating from a distribution other than the remaining population. However, this view does not help to recognize them in a specific data set. The above definition is most often refined by the classic notion of "outliers," to a distance-based concept, indicating those elements furthest from the majority. Here, the frequency approach will be applied, whereby atypical elements are rare, i.e., the probability of their appearance is faint. In this way, one can identify atypical observations not only on the peripheries of the population, but in the case of multimodal distributions with wide-spreading segments, and also those lying in between these segments, even if close to the center of the set.

### 10.3.1  Basic version of the procedure

Let the set be given, with elements representative for the population

$$x_1, x_2, \dots, x_m \in \mathbf{R}^n. \tag{10.25}$$

Treat these elements as realizations of the $n$-dimensional continuous random variable $X$ with distribution having density $f$ and calculate – in accordance with Section 10.2 – the kernel estimator $\hat{f}$; preferably with modification of the smoothing parameter (10.20) and normal kernel (10.3). Next, consider the set of its value for elements of set (10.25), so

$$\hat{f}_{-1}(x_1), \hat{f}_{-2}(x_2), \dots, \hat{f}_{-m}(x_m), \tag{10.26}$$

where $\hat{f}_{-i}$ means the kernel estimator $\hat{f}$ calculated excluding the $i$th element, for $i = 1, 2, \dots, m$. It is worth noting that, regardless of the dimension of the random variable $X$, the values of set (10.26) are real (one-dimensional). Particular values $\hat{f}_{-i}(x_i)$ characterize the probability of the occurrence of the element $x_i$, and therefore, the lower the value $\hat{f}_{-i}(x_i)$, the more the element $x_i$ can be interpreted as "less typical," or rather happening more rarely.

Define now the number

$$r \in (0, 1) \tag{10.27}$$

establishing sensitivity of the procedure for identifying atypical elements. This number will determine the assumed proportion of atypical elements in relation to the total population and, therefore, the ratio of the number of atypical elements to the sum of atypical and typical. One can naturally assume $r > 0.5$, as otherwise the atypical elements become typical and *vice versa*. In practice

$$r = 0.01, \ 0.05, \ 0.1 \tag{10.28}$$

is the most often used, with particular attention paid to the second option.

Let us treat set (10.26) as realizations of a real (one-dimensional) random variable and calculate the estimator for the quantile of the order $r$. The positional estimator of the second order (Parrish 1990, Kulczycki 1998) will be applied as follows, given by the formula

$$\hat{q}_r = \begin{cases} z_1 & \text{for } mr < 0.5 \\ (0.5 + i - mr)z_i + (0.5 - i + mr)z_{i+1} & \text{for } mr \geq 0.5 \end{cases}' \tag{10.29}$$

where $i = [mr + 0.5]$, and $[y]$ denotes an integral part of the number $y \in \mathbf{R}$, while $z_i$ is the $i$th value in size of set (10.26) after being sorted; thus,

$$\left\{ z_1, z_2, \ldots, z_m \right\} = \left\{ \hat{f}_{-1}(x_1), \hat{f}_{-2}(x_2), \ldots, \hat{f}_{-m}(x_m) \right\} \tag{10.30}$$

with $z_1 \leq z_2 \leq \cdots \leq z_m$. Application of the positional quantile estimator guarantees its value does not exceed beyond the support of the random variable under investigation, or rather to be more precise, thanks to the use of kernel (10.3) with positive values, the inequality $\hat{q}_r > 0$ is fulfilled.

Generally, there are no special recommendations concerning the choice of the sorting algorithm (Canaan et al. 2011) used to specify set (10.30). However, let us interpret definition (10.29), taking into account condition (10.28). So, it is enough to sort only the $i + 1$ smallest values in the set $\{z_1, z_2, \ldots, z_m\}$ and, therefore, in practice about 1%–10% of its size. One can apply a simple algorithm consisting of subsequently finding the $i + 1$ smallest elements of the set $\{z_1, z_2, \ldots, z_m\}$.

Finally, if for a given tested element

$$\tilde{x} \in \mathbf{R}^n \tag{10.31}$$

the condition

$$\hat{f}(\tilde{x}) \leq \hat{q}_r \tag{10.32}$$

is fulfilled, then this element should be considered atypical; for the opposite

$$\hat{f}(\tilde{x}) > \hat{q}_r \tag{10.33}$$

it is typical. Note that for the correctly estimated quantities $\hat{f}$ and $\hat{q}_r$, the above guarantees obtaining the proportion of the number of atypical elements to total population at the assumed level $r$.

The above procedure for identifying atypical elements, combined with the properties of kernel estimators, allows in the multidimensional case for inferences based not only on values for specific coordinates of a tested element, but above all, on the relationships between them.

## 10.3.2  Extended pattern of population

Although, from a theoretical point of view, the procedure presented in the previous section seems complete, when the values $r$ are applied in practice – see condition (10.29) – and the size $m$ is not large, the estimator of the quantile $\hat{q}_r$ is encumbered with a large error, due to the low number of elements $z_i$ being smaller than the estimated value. To counteract this, a data set will be extended by generating additional elements with distribution identical to that characterizing the subject population, based on set (10.25).

The methodology for enlarging a set representative for the investigated population is suggested using Neumann's elimination concept (Gentle 2003). This allows the generation of a sequence of random numbers of distribution with support bounded to the interval $[a, b]$, while $a < b$, characterized by the density $f$ of values limited by the positive number $c$, i.e.,

$$f(x) \leq c \quad \text{for } x \in [a,b].   \tag{10.34}$$

In the multidimensional case, the interval $[a, b]$ generalizes to the $n$-dimensional cuboid $[a_1,b_1] \times [a_2,b_2] \times \cdots \times [a_n,b_n]$, while $a_j < b_j$ for $j = 1, 2, \ldots, n$.

First, the one-dimensional case is considered. Let us generate two pseudorandom numbers $u$ and $v$ of distribution uniform to the intervals $[a, b]$ and $[0, c]$, respectively. Next, one should check that

$$v \leq f(u).   \tag{10.35}$$

If the above condition is fulfilled, then the value $u$ ought to be assumed as the desired realization of a random variable with distribution described by the density $f$, that is

$$x = u.   \tag{10.36}$$

In the opposite case, the numbers $u$ and $v$ need to be removed and the above procedure repeated until the desired number of pseudorandom numbers $x$ with density $f$ is obtained.

In the presented procedure, the density $f$ is established by the methodology of kernel estimators, described in Section 10.2. Denote its estimator as $\hat{f}$. The uniform kernel will be employed, allowing easy calculation of the support boundaries $a$ and $b$, as well as the parameter $c$ appearing in condition (10.34). Namely,

$$a = \min_{i=1,2,\ldots,m} x_i - h   \tag{10.37}$$

$$b = \max_{i=1,2,\ldots,m} x_i + h   \tag{10.38}$$

and

$$c = \max_{i=1,2,\ldots,m} \left\{ \hat{f}(x_i - h), \hat{f}(x_i + h) \right\}.   \tag{10.39}$$

The last formula results from the fact that the maximum for a kernel estimator with the uniform kernel must occur on the edge of one of the kernels. It is also worth noting that calculations of parameters (10.37)–(10.39) do not require much effort. This is thanks to the appropriate choice of kernel form, taking advantage of the kernel estimators' robustness in form.

In the multidimensional case, Neumann's elimination algorithm is similar to the previously discussed one-dimensional version. The edges of the $n$-dimensional cuboid

$[a_1,b_1] \times [a_2,b_2] \times \cdots \times [a_n,b_n]$ are calculated from formulas comparable to dependences (10.34)–(10.36) separate for particular coordinates. The kernel estimator maximum is thus located in one of the corners of one of the kernels; therefore,

$$c = \max_{i=1,2,\ldots,m} \left\{ \hat{f} \left( \begin{bmatrix} x_{i,1} \pm h \\ x_{i,2} \pm h \\ \vdots \\ x_{i,n} \pm h \end{bmatrix} \right) \right\} \quad \text{following all combinations of } \pm. \qquad (10.40)$$

The number of these combinations is finite and equal to $2^n$. Using the formula presented, $n$ particular coordinates of pseudorandom vector $u$ and the subsequent number $v$ are generated, after which, condition (10.35) is checked.

The results of empirical research show that for the properly extended set (10.25), the procedure investigated here for identifying atypical elements allows us to obtain a proportion of this type of element throughout the whole population, with great accuracy, sufficient from an applicational point of view.

### 10.3.3   Equal-sized patterns of atypical and typical elements

Let us consider set (10.25) introduced in Section 10.3.1, consisting of elements representative for an investigated population, and potentially extended as described in accordance with Section 10.3.2. In taking its subset comprising these observations $x_i$ for which condition (10.32) is fulfilled, one can treat it as a pattern of atypical elements. Denote it thus:

$$x_1^{at}, x_2^{at}, \ldots, x_{m_{at}}^{at}. \qquad (10.41)$$

Similarly, the set of observations for which the opposite inequality (10.33) is true may be considered as a pattern of typical elements:

$$x_1^{t}, x_2^{t}, \ldots, x_{m_t}^{t}. \qquad (10.42)$$

Sizes of the above patterns equal $m_{at}$ and $m_t$, respectively. Of course $m_{at} + m_t = m$; we also have

$$\frac{m_{at}}{m_{at} + m_t} \cong r. \qquad (10.43)$$

In this way, unsupervised in its nature, the problem of identifying atypical elements has been reduced to a supervised classification task, although with strongly unbalanced patterns – taking into account relation (10.43) with condition (10.28), set (10.41) is in practice around 10–100 times smaller than pattern (10.42). Classification is relatively conveniently conditioned and can use many different well-developed methods. However, most procedures work much better if patterns are of similar or even equal sizes (Kaufman and Rousseeuw 1990). Using once again the algorithm presented in Section 10.3.3, the size of the set can be increased to $m_t$, so that $m_{at} = m_t$, thus equaling patterns of atypical (10.41) and typical (10.42) elements.

### 10.3.4   Comments for Section 10.3

The conducted broad empirical research confirmed the validity of the investigated method. For any fixed value for the parameter $m$, the quality of the procedure increased together with the value $m^*$, but only within the range of 1000–10,000. One can interpret this as exhausting the information contained in the $m$-element sample (10.25). Moreover, an interesting phenomena occurred in the case $m^* = m$, i.e., when the generated – according to Section 10.3.2 – sample was of the same size as the basic set (10.25). The results were better for the further case, which may be explained by stabilization, of sorts, of results, "filtered" through the distribution calculated for set (10.25). Such a positive "initial condition" provides additional motivation for the concept of extending the population size, presented in Section 10.3.2. Over and above the investigations of the generated synthetic sets and benchmarks, the methodology worked out here was applied for making correct medical decisions regarding biochemical blood tests concerning plasma component analysis, and also in research on the dependency between the level of hemoglobin and death due to heart attack and conditions of the circulatory system. In the work by Kulczycki and Kruszewski (2017a), the procedure was designed to submit the results in fuzzy and intuitionistic fuzzy forms was investigated.

A detailed summary of the studies presented in this section can be found in the work by Kulczycki and Kruszewski (2017b).

## 10.4   Clustering

Clustering has become the second basic problem within data analysis – compared to other procedures, it is more loosely defined and at a lesser advanced stage in research (Everitt et al. 2011; Xu and Wunsch 2009). It can be found between classical data analysis, where the research objective has already been specified, and exploration data analysis, in which the aim of future investigations is unknown *a priori*, and its detection is an integral component of the research. In the first case, clustering may be applied for the purposes of classification, however, without fixed patterns, whereas the second treats it as a division of the explored data into a few groups, each comprising elements that are similar to each other but significantly differ between particular groups.

Consider a set of elements from an investigated population. The most intuitive and natural concept is the assumption that specific clusters are related to modes (local maxima) of distribution density; thus, the "valleys" become the borders of the resulting clusters (Fukunaga and Hostetler 1975). The algorithm described in this section is presented in its entirety, which can be applied without the requirement for users to conduct laborious research. Its attributes can be summarized as follows:

1. Parameter values can be numerically calculated based on optimizing criteria.
2. The algorithm requires stringent assumptions concerning the number of clusters, which enables the obtained number to better describe a real structure of data.
3. The parameter that is responsible for the cluster number is specified; it also displays how possible modifications to its value – e.g., obtained in point 1 – have an influence over increases or decreases in the number of clusters but still without determining their precise number.
4. Furthermore, the next parameter is indicated and the proportion of the number of clusters in dense versus sparse regions of data elements can influences its value; it can also be obtained by the optimization criteria and with the option for further

modification with the aim of simultaneously increasing the cluster quantity in dense regions and reducing or even eliminating them from sparse areas of data, or *vice versa*.

5. The suitable relationship between the parameters mentioned in points 3 and 4 enables reducing and even eliminating clusters in sparse regions, virtually without affecting the cluster quantity in dense areas of data.

The characteristics from point 4, and consequently point 5, are particularly worthwhile to highlight as being practically absent in other clustering methods. In applications, one should underline the consequences of points 1 and 2, as well as possibly point 3.

## 10.4.1 Procedure

Consider – as in the previous section – a data set

$$x_1, x_2, \ldots, x_m \in \mathbf{R}^n, \tag{10.44}$$

treated as realizations of the $n$-dimensional continuous random variable $X$ with distribution having density $f$. Appling the methodology presented in Section 10.2, one can create the kernel estimator $\hat{f}$. After assuming that subsequent clusters are mapped to the local maxima of the above function, the elements (10.44) can be shifted in the direction of gradient $\nabla\hat{f}$ with the suitable step. It can be carried out iteratively using the classic Newtonian method (Kinkaid and Cheney 2002), given as

$$x_i^0 = x_i \quad \text{for} \quad i = 1, 2, \ldots, m \tag{10.45}$$

$$x_i^{k+1} = x_i^k + b\frac{\nabla\hat{f}(x_i^k)}{\hat{f}(x_i^k)} \quad \text{for} \quad i = 1, 2, \ldots, m \quad \text{and} \quad k = 0, 1, \ldots, k^*, \tag{10.46}$$

whereas $b > 0$ and $k^* \in \mathbf{N}\backslash\{0\}$. Based on an optimizing criterion, one can suggest

$$b = \frac{1}{n+2}\left(\min_{j=1,2,\ldots,n} h_j\right)^2, \tag{10.47}$$

while $h_j$ denotes the smoothing parameter value of the $j$th coordinate. To the above task, the estimator with smoothing parameter modification with standard intensity (10.21) can be applied; the product kernel (10.5) is used in the multidimensional case. As a (one-dimensional) kernel, the normal kernel (10.3) is proposed because of its analytical convenience, differentiability within the entire domain, and the fact that its values are positive, which defends against division by zero in formula (10.46). In this case, the quotient on the right side of Equation (10.46) takes the convenient form

$$\frac{\nabla\hat{f}(x_i^k)}{\hat{f}(x_i^k)} = \begin{bmatrix} -\dfrac{(x_{i,1}^k - x_{i,1})}{s_1^2 h_1^2} \\[2ex] -\dfrac{(x_{i,2}^k - x_{i,2})}{s_2^2 h_2^2} \\[1ex] \vdots \\[1ex] -\dfrac{(x_{i,n}^k - x_{i,n})}{s_n^2 h_n^2} \end{bmatrix}, \tag{10.48}$$

with denotation of particular coordinates

$$x_i^k = \begin{bmatrix} x_{i,1}^k \\ x_{i,2}^k \\ \vdots \\ x_{i,n}^k \end{bmatrix}.$$

(10.49)

Next, it is assumed that algorithm (10.45) and (10.46) needs to be completed, in the event that the following inequality is fulfilled after the subsequent $k$th step

$$|D_k - D_{k-1}| \leq aD_0$$

(10.50)

in which $a > 0$ and

$$D_0 = \sum_{i=1}^{m-1}\sum_{j=i+1}^{m} d(x_i, x_j), \quad D_{k-1} = \sum_{i=1}^{m-1}\sum_{j=i+1}^{m} d(x_i^{k-1}, x_j^{k-1}), \quad D_k = \sum_{i=1}^{m-1}\sum_{j=i+1}^{m} d(x_i^k, x_j^k),$$

(10.51)

where $d$ denotes Euclidean metric in $\mathbf{R}^n$. Thus, $D_0$ and $D_{k-1}$, $D_k$ mean the sums of the distances between specific elements (10.44) during the starting of the algorithm and following the performance of the $(k-1)$th and $k$th steps, respectively. Initially, one can suggest $a = 0.001$. The possible reduction of this value has practically no influence over the results; however, growth needs validation of the potential consequences. Last, when condition (10.50) is fulfilled after the $k$th step, then

$$k^* = k$$

(10.52)

and as a result this is considered to be the final step.

A procedure now needs to be applied for the creation of clusters and for particular elements to be assigned to them. The set comprising elements of set (10.44) submitted to $k^*$ steps of algorithm (10.45) and (10.46) is considered to achieve this objective:

$$x_1^{k^*}, x_2^{k^*}, \ldots, x_m^{k^*}.$$

(10.53)

Now, define the set of their mutual distances

$$\left\{ d(x_i^{k^*}, x_j^{k^*}) \right\}_{\substack{i=1,2,\ldots,m-1 \\ j=i+1,i+2,\ldots,m}}.$$

(10.54)

Its size is

$$m_d = \frac{m(m-1)}{2}.$$

(10.55)

Considering set (10.54) as a one-dimensional random variable sample, one should calculate the auxiliary kernel estimator $\hat{f}_d$ of mutual distances (10.54). Normal kernel (10.3) is suggested once more; furthermore, the smoothing parameter modification procedure with the standard value of parameter (10.21), together with the left-sided support boundary to the interval $[0, \infty)$; see formula (10.24) for $x_* = 0$.

Finding – with appropriate accuracy – the "first" (in the sense of the lowest argument value) local minimum of the function $\hat{f_d}$ in the interval $(0, D)$, where

$$D = \max_{\substack{i=1,2,\ldots,m-1 \\ j=i+1,i+2,\ldots,m}} d(x_i, x_j), \qquad (10.56)$$

is the next task. For this objective, one can consider set (10.54) to be a random sample, estimate its standard deviation applying formula (10.8), and subsequently take the values $x$ from the set

$$\{0, 0.01 \cdot \sigma_d, 0.02 \cdot \sigma_d, \ldots, [\text{int}(100 \cdot D) - 1] \cdot \sigma_d\}, \qquad (10.57)$$

where $\text{int}(100 \cdot D)$ means an integral of the number $100 \cdot D$, until the condition

$$\hat{f_d}(x - 0.01\,\sigma_d) > \hat{f_d}(x) \quad \text{and} \quad \hat{f_d}(x) \le \hat{f_d}(x + 0.01\,\sigma_d) \qquad (10.58)$$

is fulfilled. The first (the smallest) value[*] will be treated as the smallest distance between cluster centers located in close proximity to each other and referred to as $x_d$ hereinafter.

The final step is the creation of the clusters. This is achieved through:

1. Taking an element of set (10.55) and first producing a one-element cluster including it.
2. Finding an element of set (10.55) which differs from the others in the cluster, and is nearer than $x_d$; if such an element exists, it needs to be then added to that cluster; in the event that this is not the case, go to point 4.
3. Discovering an element of set (10.55) different to elements in the cluster, and lying closer than $x_d$ to at least one of these other elements; in the event of there being such an element, one should add this to the cluster and subsequently repeat point 3.
4. Adding the attained cluster to a "cluster list" and removing its elements from set (10.55); if this set reduced in such a way remains not empty, go to point 1, otherwise, finish the algorithm.

The "cluster list" such obtained has all clusters defined during the above procedure contained within. Now, we have investigated the basic form of the clustering procedure – its potential modifications with their influence on the results is described in the following section.

## 10.4.2   Influence of the parameters values on obtained results

It should be underlined once more that the clustering procedure investigated in the previous section did not necessitate any initial, often arbitrary, assumption regarding the number of clusters – it mainly depends on the internal data structure. In the elementary form presented above, the parameters are calculated effectively based on optimizing criteria, but in practical applications, it can be beneficial to suitably modify the values of kernel estimator parameters – thus influencing, in such a manner, the cluster number and additionally, the proportion of their occurrence between dense and sparse regions.

---

[*] In the event that such a value is nonexistent, the presence of one cluster should be recognized and the procedure completed. The same applies to the irrational but formally possible situation $m = 1$, when set (1.52) is empty.

As discussed in Section 10.2, too great a value of the smoothing parameter $h$ results in the over-smoothing of the kernel estimator; while if it is too small, this causes the appearance of too many local extremes. Therefore, the result of increasing – with respect to that calculated by the criterion of the mean integrated square error – this parameter value is the occurrence of fewer clusters; conversely, decreases to this value yield more clusters. In both cases, one can emphasize that despite influencing the number of clusters, their number still solely depends on the data's internal structure. On the basis of the performed research, a change in the smoothing parameter value in the range –25% to +50% can be recommended. Results are in need of individual verification if they lie beyond this range.

The intensity of the smoothing parameter modification – as described in Section 10.2.1 – is defined by the parameter $c$; its standard value is provided by formula (10.21). Its increase sharpens the kernel estimator in the dense regions of set (10.44) and also smooths it in the sparse areas; as a consequence, if this parameter value rises, then the number of clusters in dense areas increases and simultaneously decreases in sparse regions. These effects are reversed in the event of this parameter value diminishing. On the basis of the performed research, the parameter $c$ value can be proposed to be between 0 (indicating a lack of modification) and 1.5. The validity of the obtained results needs individual verification in the case of exceeding a value of 1.5. In particular one can suggest $c = 1$ as standard.

However, growth of the cluster number in dense data regions and at the same time lowering or even eliminating clusters in sparse areas (as they frequently contain atypical elements appearing as a result of various errors) is frequently desired in practice. Combining the aforementioned considerations, it is appropriate to propose increases to both the change of the standard intensity of the smoothing parameter modification (10.21) and simultaneously, the smoothing parameter $h$ value calculated on the basis of the optimization criterion, to the value $h^*$ given as

$$h^* = \left(\frac{3}{2}\right)^{c-0.5} h. \tag{10.59}$$

The combined effect of both of these factors implies a twofold smoothing of the estimator $\hat{f}$ in the areas in which set (10.44) is sparse. At the same time, the above factors virtually cancel each other out in dense regions; hence, them having almost zero influence on the discovery of clusters in such areas. On the basis of the conducted research, a change in the parameter $c$ value in the range of 0.5–1.0 can be executed; however, increases that exceed 1.0 require individual validation. In particular, the value $c = 0.75$ can be recommended in such a case.

### 10.4.3   Comments for Section 10.4

The above clustering procedure was comprehensively tested both for generated synthetic sets and generally available benchmarks; it was also compared with other well-known methods. Confirming the total supremacy of any of these remains difficult; however, the concept proposed here enables greater opportunities to adjust specific data structures and, as a consequence, the acquired results become – from a human point of view – more justified. This property, along with possibilities of changes to standard values of parameters on the basis of clear and easily visualized interpretations, has been used actively in three projects: in the field of engineering (fuzzy controller synthesis); within management (a mobile phone operator's marketing support strategy); and in the domain of bioinformatics (grain categorization for seed production).

More details with visual aids are presented in the article by Kulczycki and Charytanowicz (2010). Applications were synthetically presented in the paper by Kulczycki et al. (2012), and also in more detail in the particular publications by Charytanowicz et al. (2016), Kulczycki and Daniel (2009), and Łukasik et al. (2008).

## 10.5   Classification

Classification constitutes the third of the basic tasks of data analysis (Duda et al. 2001). In the previously considered problems of atypical element identification (Section 10.3) and clustering (Section 10.4), the subject of processing was only a data set (10.25) or (10.44), respectively, with no other information, without additional prompts, supervision. These problems are therefore typical unsupervised tasks. Meanwhile, the classification issue involves a tested element being assigned to previously defined groups (classes) represented by patterns. This constitutes additional significant information, and thus the classification becomes a supervised task.

Such beneficial conditions of a classification task cause the available methodology to be rich and very varied. The concept presented in the following sections will be based on the Bayes approach. The classifiers gained in this way work quite well in complex real-world situations and are eagerly used by practitioners, chiefly because of their robustness, low requirements for patterns, and also their illustrativeness supporting individual modifications. Especially, in the method proposed here, there is an opportunity to attribute preferences to classes containing elements which – according to possible asymmetrical task requirements – must especially not be incorrectly attributed to others. The parameters of kernel estimators can be made more precise with the aim of successively improving classification quality. Moreover, the application of the sensitivity method borrowed from artificial neural networks allows the elimination of those pattern elements that have insignificant or even a negative effect on the correctness of results. These last two procedures will in turn be the basis for the creation of an effective adaptational structure, adjusting a classifier to nonstationary data (so-called concept drift).

### 10.5.1   Bayes classification

Assume $J$ sets containing elements from space $\mathbf{R}^n$:

$$x_1', x_2', \ldots, x_{m_1}' \tag{10.60}$$

$$x_1'', x_2'', \ldots, x_{m_2}'' \tag{10.61}$$

$$\vdots$$

$$x_1^{''\cdots'}, x_2^{''\cdots'}, \ldots, x_{m_J}^{''\cdots'}, \tag{10.62}$$

representing assumed classes. The sizes $m_1, m_2, \ldots, m_J$, need to be more or less proportional to the "contribution" of specific classes within the investigated population. The aim of classification is to map the tested element

$$\tilde{x} \in \mathbf{R}^n \tag{10.63}$$

to one of the groups represented by patterns (10.60)–(10.62). Denote as $\hat{f}_1, \hat{f}_2, \ldots, \hat{f}_J$ kernel estimators successively calculated on the basis of sets (10.60)–(10.62) treated as random samples (10.1) each time – the methodology used for this purpose is presented in Section 10.2.

According to the classic Bayes concept (Duda et al. 2001), the classified element (10.63) needs then to be attributed to that class in which the value

$$m_1 \hat{f}_1(\tilde{x}), m_2 \hat{f}_2(\tilde{x}), \ldots, m_J \hat{f}_J(\tilde{x}) \tag{10.64}$$

is the largest. By introducing the positive coefficients $z_1, z_2, \ldots, z_J$, the above can be generalized to

$$z_1 m_1 \hat{f}_1(\tilde{x}), z_2 m_2 \hat{f}_2(\tilde{x}), \ldots, z_J m_J \hat{f}_J(\tilde{x}). \tag{10.65}$$

Giving the values $z_1 = z_2 = \cdots = z_J = 1$, formula (10.65) is reducing to dependency (10.64). By applying a suitable increase to the value $z_i$, one introduces an inversely proportional decrease to the probability of mistakenly assigning the $i$th class elements to an incorrect class; however, the danger of slightly increasing the overall quantity of misclassifications then theoretically appears. Bearing this in mind, it is possible to increase the values of even a few coefficients $z_i$.

## 10.5.2   Correction of values of smoothing parameter and modification intensity

It is often presented within literature on the subject that classic techniques for the smoothing parameter value calculation – typically originating from the quadratic criterion – are mostly inappropriate for the purposes of classification. Available publications practically fail to provide a comprehensive solution to such problems, in particular for more than two classes and in multidimensional cases. A procedure is proposed in this chapter, which will suit the specific conditions pertaining to classification with the nonstationary patterns considered here – this especially refers to a successive adaptation to the appearing changes.

Let us propose the introduction of $n + 1$ multiplicative correcting coefficients, relating to the values of the intensity of modification procedure $c$ and the smoothing parameter for particular coordinates $h_1, h_2, \ldots, h_n$, with respect to those obtained applying the integrated square error criterion. These are denoted as $b_0 \geq 0, b_1, b_2, \ldots, b_n > 0$, respectively. Note that $b_0 = b_1 = \cdots = b_n = 1$ implies a lack of correction. Using, for a comprehensive search, a grid with a rather sizable discretization value, one may find the most beneficial points within the context of correct classification. The last stage is a classic optimization algorithm in the $(n + 1)$-dimensional space, with starting conditions being the points found above, whereas the performance index is

$$J(b_0, b_1, \ldots, b_n) = \# \{\text{incorrect classifications}\}, \tag{10.66}$$

where $\#$ means the number of elements within a set. The classic leave-one-out method can be applied for the calculation of the above functional value for any fixed argument. Due to this value being an integer, the modified Hook–Jeeves algorithm (Kelley 1999) was used to find a minimum. Alternate conceptions are described in the survey paper (Venter 2010).

As a result of performed research, the assumption can be made that for every coordinate, the grid should usually have nodes at the points 0.25, 0.5, …, 1.75. The functional (10.66) values are calculated for these nodes; the attained results are then sorted and the five best become starting conditions for the Hook–Jeeves procedure, in which the initial step value is proposed as 0.2. Following completion of each of the above five executions, the values of functional (10.66) for the obtained end points are calculated, and that which has the smallest value is the sought-after vector of the parameters $b_0, b_1, b_2, \ldots, b_n$.

It is worthy of note that in this procedure it is not necessary to correct the classification parameters; however, doing so would enhance the classification quality and, moreover, would allow applying an easy and convenient formula (10.7) to calculate smoothing parameter values.

## 10.5.3  Reduction to pattern sizes

In reality, it is possible that some elements of sets (10.60)–(10.62) representing patterns of classes have irrelevant or even negative influence over classification quality. Therefore, their removal may result in both a reduction in incorrect results and also a decrease in calculation times. To achieve the objectives of this procedure, the generalization of the kernel estimator definition given as follows will consist in the addition of the nonnegative coefficients $w_1, w_2, \ldots, w_m$, normed by

$$\sum_{i=1}^{m} w_i = m, \tag{10.67}$$

subsequently assigned to specific random sample (10.1) elements. Then the initial definition for kernel estimator (10.2) becomes

$$\hat{f}(x) = \frac{1}{mh^n} \sum_{i=1}^{m} w_i K\left(\frac{x - x_i}{h}\right). \tag{10.68}$$

Formulas (10.2), (10.5), (10.19), (10.20), and (10.23), (10.24) can be changed analogously. The coefficients $w_i$ describe the weight (significance) of the $i$th pattern element with respect to classification quality. It should be noted that if $w_i \equiv 1$, definition (10.68) is then reduced to basic form (10.2).

The procedure for reducing pattern sets (10.60)–(10.62) consists of two stages. The first consists of the weights $w_i$ calculation; the second is the removal of such random sample elements which have the lowest respective weights. To realize the former of these two stages, separate neural networks can be built for each class. For simplicity of the forthcoming notations, let the index $j = 1, 2, \ldots, J$ which characterizes specific classes, be fixed. The constructed network has three layers and is unidirectional: with $m$ inputs which are related to subsequent elements of the pattern; a hidden layer of a size equal to the integral of the number $\sqrt{m}$; with one output neuron. This network learns through the use of a data set consisting of values of specific kernels for consecutive pattern elements, while the output is the kernel estimator value for the considered pattern element. Network learning is achieved through backward propagation of errors, with a momentum factor. At the completion of the procedure, the network is subjected to an analysis of sensitivity on learning data; for details see the book by Zurada (1992). The essence of this method constitutes the establishment – after network learning – of the influence of the subsequent inputs $u_i$ on the output $y$; this is represented by the real coefficients

$$S_i = \frac{\partial y(x_1, x_2, \ldots, x_m)}{\partial x_i} \quad \text{for } i = 1, 2, \ldots, m. \tag{10.69}$$

Define the coefficients $S_i^{(p)}$, which aggregate information originating from consecutive iterations of the previous stage (with $p = 1, 2, \ldots, P$) and characterize the sensitivity of successive learning data. This results in the coefficient $\bar{S}_i$ defined as

$$\bar{S}_i = \sqrt{\frac{\displaystyle\sum_{p=1}^{P}(S_i^{(p)})^2}{P}} \quad \text{for } i = 1,2,\dots,m, \tag{10.70}$$

which will be used to calculate the coefficients $w_i$. Thus, first let

$$\tilde{w}_i = \left(1 - \frac{\bar{S}_i}{\displaystyle\sum_{j=1}^{m}\bar{S}_j}\right) \quad \text{for } i = 1,2,\dots,m, \tag{10.71}$$

and then norm to

$$w_i = m\frac{\tilde{w}_i}{\displaystyle\sum_{i=1}^{m}\tilde{w}_i} \quad \text{for } i = 1,2,\dots,m, \tag{10.72}$$

in order to guarantee condition (10.67). The form of definition (10.71) is due to the network created here being the most sensitive to redundant and atypical elements, which suggests – as a consequence of the kernel estimator (10.68) form – a requirement to assign to them the suitably smaller values $\tilde{w}_i$, and consequently, $w_i$; these coefficients characterize the significance of specific pattern elements with respect to the classification quality.

The natural requirement that those elements for which $w_i < 1$ needed to be removed from the pattern has been confirmed by the performed research [observe that the mean value of coefficients $w_i$ equals 1 due to normalization being introduced by formula (10.72)]. Increases to this value resulted in a significant drop in classification accuracy because of losses of nonredundant, valuable information carried in the pattern. Conversely, decreases of such a threshold led to a substantial fall in the reduction of pattern size; however, its effect over classification accuracy was barely noticeable in the proximity of value 1, although a significant reduction results in a significant increase in the number of errors.

### 10.5.4   Structure for nonstationary patterns (concept drift)

The material of previous Sections 10.5.2 and 10.5.3 will also be used to create the classification structure for the nonstationary case, that is when all or some patterns of classes undergo significant changes, considering the investigated task. A block diagram of the calculation procedure is presented in Figure 10.1. Blocks symbolizing operations performed on all elements of patterns (10.60)–(10.62) are jointly drawn with a continuous line; a dashed line denotes operations on particular classes, while a dotted line is used for separate operations for each element of those patterns.

First, one should fix the reference sizes of patterns (10.60)–(10.62), hereinafter denoted by $m_1^*, m_2^*, \dots, m_j^*$. The patterns of these sizes will be the subject of a basic reduction procedure, described in Section 10.5.3. The sizes of patterns available at the beginning of the algorithm must not be smaller than the above referential values. These values can, however, be modified during the procedure's operation, with the natural condition that their potential growth does not increase the number of elements newly provided for the patterns. For preliminary research, $m_1^* = m_2^* = \dots = m_j^* = 25 \cdot 2^n$ can be proposed. Lowering

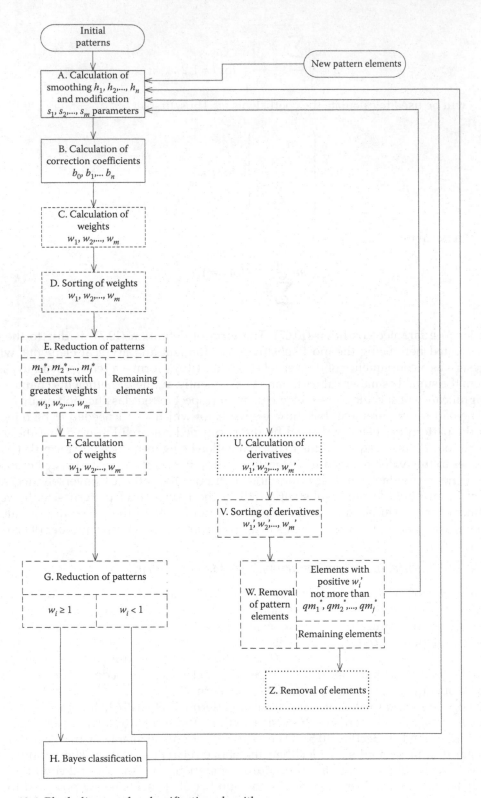

*Figure 10.1* Block diagram for classification algorithm.

these values may worsen the classification quality, whereas an increase results in an excessive calculation time.

The elements of initial patterns (10.60)–(10.62) are provided as introductory data. Based on these – according to the procedures presented in Section 10.2 – the value of the parameter $h$ is calculated (for the parameter $c$ it is given by formula (10.21)). Figure 10.1 shows this action in block A. Next, corrections in the parameters $h$ and $c$ values are made by taking the coefficients $b_0, b_1, \ldots, b_n$, as described in Section 10.5.2 (block B in Figure 10.1).

The subsequent procedure, shown by block C, is the calculation of the parameters $w_i$ values mapped to particular elements of patterns, separately for each class, as in Section 10.5.3. Following this, within each class, the values of the parameter $w_i$ are sorted (block D), and then – in block E – the appropriate $m_1^*, m_2^*, \ldots, m_j^*$ elements of the largest values $w_i$ are designated to the classification phase itself. The remaining undergo further treatment, denoted in block U, which is presented in the following sections, after Bayes classification has been dealt with.

The reduced patterns separately go through a procedure newly calculating the values of parameters $w_i$, presented in Section 10.5.3 and depicted in block F. In turn, as block G in Figure 10.1 denotes, these pattern elements for which $w_i \geq 1$ are submitted to further stages of the classification procedure, while those with $w_i < 1$ are sent to block A for further processing in the next steps of the algorithm after adding new elements of patterns. The final, and also the principal part of the procedure worked out here is Bayes classification, presented at the beginning of this section and marked by block H. Obviously, many tested elements (10.64) can be subjected to classification separately. After the procedure has been finished, elements of patterns which have undergone classification are sent to the beginning of the algorithm to block A, for further processing in the next steps, following the addition of new elements of patterns.

Now – as mentioned two paragraphs earlier, in the last sentence – it remains to consider those pattern elements, whose values $w_i$ were not counted among the $m_1^*, m_2^*, \ldots, m_j^*$ largest for particular patterns. Thus, within block U, the derivative $w_i'$ is calculated[*] for each of them. If the element is "too new" and does not possess the $k - 1$ previous values $w_i$, then the gaps are filled with zeros (because the values $w_i$ generally oscillate around unity, such behavior significantly increases the derivative value, and in consequence, ensures against premature elimination of this element). Next, for each separate class, the elements $w_i'$ are sorted (block V). As marked in block W, the respective

$$qm_1^*, qm_2^*, \ldots, qm_j^* \quad \text{for } i = 1, 2, \ldots, m, \tag{10.73}$$

elements of each pattern with the largest derivative values, on the additional requirement that the value is positive, go back to block A for further calculations carried out after the addition of new elements. If the number of elements with positive derivative is less than $qm_1^*, qm_2^*, \ldots, qm_j^*$, then the number of elements returning may be smaller (including even zero). The remaining elements are permanently eliminated from the procedure, as shown in block Z. In the above notation, $q$ is a positive constant influencing the proportion of

---

[*] As the task considered here does not require the differences between subsequent values $t_1, t_2, \ldots, t_k$ to be equal, it is therefore advantageous to apply interpolation methods. In the procedure worked out here, favorable results were achieved using a classic method based on Newton's interpolation polynomial. Detailed formulas, as well as a treatment of other related concepts are found in the survey paper (Venter 2010). A backward derivative, after taking into consideration the last three values, can be assumed as standard, i.e., a useful compromise between stability of results and possibility to react to changes (the derivative has then two degrees of freedom).

patterns' elements with little, but successively increasing meaning. As a standard value $q = 0.2$ is proposed, or more generally $q \in [0.1, 0.25]$ depending on the size/speed of changes. An increase in this parameter value allows more effective conforming to pattern changes, although this potentially increases the calculation time, while lowering it may significantly worsen adaptation. In the general case, this parameter can be different for particular patterns – then formula (10.73) takes the form $q_1 m_1^*, q_2 m_2^*, ..., q_j m_j^*$, where $q_1, q_2, ..., q_j$ are positive.

The above procedure is repeated following the addition of new elements (block A in Figure 10.1). Besides these elements – as has been mentioned earlier – for particular patterns $m_1^*, m_2^*, ..., m_j^*$ elements of the greatest values $w_i$ are taken, respectively, as well as up to $q m_1^*, q m_2^*, ..., q m_j^*$ (or in the generalized case $q_1 m_1^*, q_2 m_2^*, ..., q_j m_j^*$) elements of the greatest derivative $w_i'$, so successively increasing its significance, most often due to the nonstationarity of patterns.

### 10.5.5 Comments for Section 10.5

The properties and functioning of the concepts presented in this chapter have been checked numerically in detail and subsequently confirmed. First, consider the influence of the coefficients $z_1, z_2, ..., z_J$, introduced by formula (10.65). Above various natural costs of individual classification errors, in the case of concept drift considered in Section 10.5.4, the nonstationary patterns are more difficult conditioned in a natural manner, which indicates the need to increase the value of the coefficients $z_i$ proportional to the changes of speed to particular classes. A value of 1.25 is initially suggested. The correction of the modification intensity and smoothing parameter values, described in Section 10.5.3 caused a 10%– 20% decrease in the number of classification errors. Moreover, following the application of the reducing pattern procedure, as discussed in Section 10.5.4, there was a decrease in the number of misclassifications by an additional 10%–20%, with a concurrent reduction in the pattern sizes by around 40%. The joint occurrence of these effects is worthy of particular note; at a same time as the reduction of pattern sizes – implying a significant calculation speed increase – there is also an enhanced quality of classification. Finally, the nonstationary structure described in Section 10.5.5 was also intensively examined for cases where changes successively grew, occurred in a stepped manner, and were periodical. This was proven to be especially beneficial for such consecutive growth.

The broader description of the stationary case, also including classification of interval information, can be found in the article by Kulczycki and Kowalski (2011). The nonstationary structure was presented in the paper by Kulczycki and Kowalski (2015a), and for the case interval information, in the publication by Kulczycki and Kowalski (2015b). The task to submit the result in fuzzy and intuitionistic fuzzy forms is being currently investigated.

## 10.6   Example practical application and final comments

The three procedures presented in this chapter were successfully applied to the creation of the methodologically uniform concept of a mobile phone operator's marketing support strategy.

Due to the high dynamic growth which is prevalent within the market of mobile phone networks, there is a natural necessity for such companies to permanently focus their strategies with the aim of satisfying their clients' different requirements, while simultaneously maximizing income. However, the natural difficulty in exercising control over such activity can give rise to a lack of coherence with regard to serving particular clients, thus resulting in their possible defection to competitors. In order to counter this, it is necessary to ensure that services remain uniform for comparable clients. The results of investigations concerning long-term business clients conducted on the network of a Polish mobile phone operator are presented in the following text.

There exists an enormous range of characteristic quantities that can be applied in practice to describe specific subscribers. These are primarily mean monthly income per SIM card, subscription length, number of active SIM cards, and possibly others, appropriately adapted to the current market specifics. Therefore, each $m$-element within a database characterizes successive business clients, and comprises $n$ their features (attributes) easily available for an operator. For the previous example we have

$$x_i = \begin{bmatrix} x_{i,1} \\ x_{i,2} \\ x_{i,3} \\ \vdots \\ x_{i,n} \end{bmatrix} \quad \text{for } i = 1, 2, \ldots, m, \qquad (10.74)$$

where $x_{i,1}$ denotes the average monthly income per SIM card of the $i$th client, $x_{i,2}$ is its length of subscription, $x_{i,3}$ is the number of active SIM cards, and possibly others $x_{i,4}, x_{i,5}, \ldots, x_{i,n}$ in accordance with the current market situation.

Firstly, atypical elements within set (10.74) were removed, according to the procedure presented in Section 10.3 (with $r = 0.1$). The regularity of the data structure was thus enhanced; it is worthy of note that this was achieved through the cancellation of only elements that had only negligible importance for the further results of the investigated procedure.

Secondly, the data set was submitted to clustering by the procedure described in Section 10.4. The consequence was the partitioning of the data set which consisted of particular clients, into separate groups each composed of similar members. The results, achieved for ordinary values of the modification intensity $c$ and smoothing parameters $h$, showed too great a number of small-sized clusters lying in low density areas of data – mostly containing irrelevant, unusual clients – and an excessively large main cluster containing more than half of all elements. Taking into account the properties of the used algorithm, this value was raised to $c = 1$. As a consequence, the desired effects – the significant lowering of the number of "peripheral" clusters as well as the splitting of the main cluster – were thus attained. The number of clusters was then satisfactory and changes to the smoothing parameters $h$ value became redundant. At this point, the data set comprising 1639 elements was partitioned into 26 clusters with sizes 488, 413, 247, 128, 54, 41, 34, 34, 33, 28, 26, 21, 20, 14, 13, 12, 10, two containing four elements, three of three elements, two of two elements, and two of one element. Note that four groups can be clearly distinguished – the first includes two large clusters of 488 and 413 elements, the following contains two medium clusters with 247 and 128 elements, the next nine are small and have 20–54, and finally there are 13 each with fewer than 20 elements. It is now appropriate to eliminate the last of these clusters, although, those including key or prestige clients (14, 13, 12, and 10 elements) were excluded from removal. Finally, for further analysis, 17 clusters remained.

Then, in the case of each of the clusters found in this manner, an optimal scheme – with regard to anticipated operator profit – was defined for the treatment of subscribers belonging to this group. Elements of preference theory (Fodor and Rubens 1994) and fuzzy logic were applied due to the usually imprecise nature of expert evaluation of such problems; however, details of this operation lie beyond the remit of this chapter – details can be found in the publication by Kulczycki and Daniel (2009).

It is worthy of note that, of the above calculations, none must be performed during client negotiations; instead they should merely be updated (every 1–6 months in practice).

The client with whom negotiations are conducted can be characterized through the use of – in accordance with formula (10.74) – an $n$-dimensional vector the specific coordinates of which represent the respective features of this client. Such data can be obtained from the operator database archive – if the client has previously been the subscriber – or alternatively, from historic invoices issued by a rival network should they be attempting to poach a client. Attributing the client to an appropriate subscriber group during negotiations – according to clusters defined earlier – was performed by applying Bayes classification presented in Section 10.5. Because the marketing strategies regarding specific clusters have been previously established, the above action completes the procedure of supporting the marketing strategy with regard to business clients, which was the objective of the project presented above.

The comments summarizing the above application example are symptomatic and can usefully be treated as recapitulation of all the material presented in this chapter.

Thus, use in the above concept of a marketing support strategy for a mobile phone operator methodologically uniform apparatus of kernel estimators makes the analysis and creation of a useful computer application significantly easy. In turn, its nonparametric character freed the concept from difficult to foresee – often nonstandard – distribution of data appearing in contemporary complex tasks. In particular, there are no restrictions on the shape of their grouping, and even in the number of separate parts that are divided. The values of each (excluding the easy-to-interpret participation of atypical elements $r$) parameter are set on the basis of optimization criteria, after which, they can be appropriately matched to individual preferences. In this text all the necessary formulas are given, apart from standard procedures used in Sections 10.5.2–10.5.4.

Currently, the fundamental challenge of kernel estimators is large sets of high-dimensional data. Thanks to averaging the properties of this type of estimator, quite satisfactory results can be obtained for the first of these aspects even by natural sampling of data set elements. With its fixed size $m$ it is worth using classic random sampling (Vitter 1985), and in the case of streaming data – the algorithm presented in the paper by Aggarwal (2006). For multidimensionality one can apply classic reduction using the statistical method PCA (Jolliffe 2001) or a refined approach based on calculated intelligence (Kulczycki and Łukasik 2014). More sophisticated methods, also with the presence of categorical features, are currently the subject of intensive research by the author and his team.

## Acknowledgments

The work was supported in parts by the Systems Research Institute of the Polish Academy of Sciences in Warsaw, and the Faculty of Physics and Applied Computer Science of the AGH University of Science and Technology in Cracow, Poland.

I thank my close associates – former Ph.D.-students – Małgorzata Charytanowicz, D.Sc., Ph.D., Karina Daniel, Ph.D., Piotr A. Kowalski, Ph.D., Damian Kruszewski, Ph.D., Szymon Łukasik, Ph.D., coauthors of the common publications (Kulczycki and Charytanowicz 2010; Kulczycki et al 2012; Kulczycki and Daniel 2009; Kulczycki and Kowalski 2011, 2015a,

2015b; Kulczycki and Kruszewski 2017a, 2017b; Kulczycki and Łukasik 2014). With their consent, this text also contains results of our joint research.

## *References*

Aggarwal C.C., *Outlier Analysis.* Springer, New York, 2013.

Aggarwal C.C., On biased reservoir sampling in the presence of stream evolution. In *Proceedings of the 32nd International Conference on Very Large Data Bases,* Seoul, 12–15 September 2006, U. Dayal, K.-Y. Whang, D.B. Lomet, G. Alonso, G.M. Lohman, M.L. Kersten, S.K. Cha, Y.-K. Kim (eds.), VLDB Endowment, 2006.

Canaan C., Garai M.S., Daya M., Popular sorting algorithms. *World Applied Programming,* vol. 1, pp. 62–71, 2011.

Charytanowicz M., Niewczas J., Kulczycki P., Kowalski P.A., Lukasik S., Discrimination of wheat grain varieties using X-ray images. In: *Information Technologies in Medicine,* Pietka E., Badura P., Kawa J., Wieclawek W. (eds.), Springer, Berlin, 2016, pp. 39–50.

Duda R.O., Hart P.E., Storck D.G., *Pattern Classification.* Wiley, New York, 2001.

Everitt B.S., Landau S., Leese M., Stahl D., *Cluster Analysis.* Wiley, New York, 2011.

Fodor J., Roubens M., *Fuzzy Preference Modelling and Multicriteria Decision Support.* Kluwer, Dordrecht, 1994.

Fukunaga K., Hostetler L.D., The estimation of the gradient of a density function, with applications in pattern recognition. *IEEE Transactions on Information Theory,* vol. 21, pp. 32–40, 1975.

Gentle J.E., *Random Number Generation and Monte Carlo Methods.* Springer, New York, 2003.

Jolliffe I.T., *Principal Component Analysis.* Springer, New York, 2001.

Kaufman L., Rousseeuw P.J., *Finding Groups in Data: An Introduction to Cluster Analysis.* Wiley, New York, 1990.

Kelley C.T., *Iterative Methods for Optimization.* SIAM, Philadelphia, 1999.

Kinkaid, D., Cheney, W., *Numerical Analysis.* Brooks/Cole, Pacific Grove, 2002.

Kulczycki P., *Wykrywanie uszkodzeń w systemach zautomatyzowanych metodami statystycznymi.* Alfa, Warsaw, 1998.

Kulczycki P., *Estymatory jądrowe w analizie systemowej.* WNT, Warsaw, 2005.

Kulczycki P., Charytanowicz M., A complete gradient clustering algorithm formed with kernel estimators. *International Journal of Applied Mathematics and Computer Science,* vol. 20, pp. 123–134, 2010.

Kulczycki P., Charytanowicz M., Kowalski P.A., Łukasik S., The complete gradient clustering algorithm: Properties in practical applications. *Journal of Applied Statistics,* vol. 39, pp. 1211–1224, 2012.

Kulczycki P., Daniel K., Metoda wspomagania strategii marketingowej operatora telefonii komórkowej. *Przegląd Statystyczny,* vol. 56, no. 2, pp. 116–134, 2009; Errata: vol. 56, no. 3–4, p. 3, 2009.

Kulczycki P., Kowalski P.A., Bayes classification of imprecise information of interval type. *Control and Cybernetics,* vol. 40, pp. 101–123, 2011.

Kulczycki P., Kowalski P.A., Bayes classification for nonstationary patterns. *International Journal of Computational Methods,* vol. 12, ID 1550008 (19 pages), 2015a.

Kulczycki P., Kowalski P.A., Classification of interval information with data drift. In: *Modeling and Using Context,* Christiansen H., Stojanovic I., Papadopoulos G.A. (eds.), Springer, Berlin, 2015b, pp. 495–500.

Kulczycki P., Kruszewski D., Detection of atypical elements with fuzzy and intuitionistic fuzzy evaluations. In: *Trends in Advanced Intelligent Control, Optimization and Automation,* Mitkowski W., Kacprzyk J., Oprzedkiewicz K., Skruch P. (eds.), Springer, Cham, 2017a, pp. 774–786.

Kulczycki P., Kruszewski D., Identification of atypical elements by transforming task to supervised form with fuzzy and intuitionistic fuzzy evaluations. *Applied Soft Computing,* vol. 60, no. 11, pp. 623–633, 2017b.

Kulczycki P., Łukasik S., An algorithm for reducing dimension and size of sample for data exploration procedures. *International Journal of Applied Mathematics and Computer Science,* vol. 24, pp. 133–149, 2014.

Łukasik S., Kowalski P.A., Charytanowicz M., Kulczycki P., Fuzzy models synthesis with kernel-density-based clustering algorithm. In: *Fifth International Conference on Fuzzy Systems and Knowledge Discovery*, J. Ma, Y. Yin, J. Yu, S. Zhou (eds.), IEEE Computer Society, Los Alamitos, vol. 3, pp. 449–453, 2008.

Parrish R., Comparison of quantile estimators in normal sampling. *Biometrics*, vol. 46, pp. 247–257, 1990.

Silverman, B.W., *Density Estimation for Statistics and Data Analysis*. Chapman and Hall, London, 1986.

Venter G., Review of optimization techniques. In: *Encyclopedia of Aerospace Engineering*, Blockley R., Shyy W. (eds.), Wiley, New York, 2010, pp. 5229–5238.

Vitter J.S., Random sampling with reservoir. *ACM Transactions on Mathematical Software*, vol. 11, pp. 37–57, 1985.

Wand M., Jones M., *Kernel Smoothing*. Chapman and Hall, London, 1995.

Xu R., Wunsch D., *Clustering*. Wiley, New York, 2009.

Zurada J., *Introduction to Artificial Neural Neural Systems*. West Publishing, St. Paul, 1992.

# chapter eleven

# A new technique for constructing exact tolerance limits on future outcomes under parametric uncertainty

**N.A. Nechval**
*University of Latvia*

**K.N. Nechval**
*Transport and Telecommunication Institute*

**G. Berzins**
*University of Latvia*

## Contents

## 11.1   Introduction

Statistical tolerance (prediction) limits are another tool for making statistical inference on an unknown population. As opposed to a confidence limit that provides information concerning an unknown population parameter, a tolerance limit provides information on the entire population. In this chapter, two types of statistical tolerance limits are defined: (1) $\gamma$-content tolerance limit with expected $(1 - \alpha)$-confidence, and (2) $(1 - \alpha)$-expectation tolerance limit.

To be specific, let $\gamma$ denote a proportion between 0 and 1. Then one-sided $\gamma$-content tolerance limit with expected $(1 - \alpha)$-confidence is determined to capture a proportion $\gamma$ or more of the population, with a given expected confidence level $1 - \alpha$. For example, an upper $\gamma$-content tolerance limit with expected $(1 - \alpha)$-confidence for a univariate population is such that with the given expected confidence level $1 - \alpha$, a specified proportion $\gamma$ or more of the population will fall below the limit. A lower $\gamma$-content tolerance limit with expected $(1 - \alpha)$-confidence satisfies similar conditions.

An upper $(1 - \alpha)$-expectation tolerance limit is determined so that the expected proportion of the population falling below the limit is $(1 - \alpha)$. A lower $(1 - \alpha)$-expectation tolerance limit satisfies similar conditions.

It is often desirable to have statistical tolerance limits available for the distributions used to describe time-to-failure data in reliability problems. For example, one might wish to know if at least a certain proportion, say $\gamma$, of a manufactured product will operate at least $T$ hours. This question cannot usually be answered exactly, but it may be possible to determine a lower statistical tolerance limit $L(X_1, ..., X_n)$, based on a preliminary random sample $(X_1, ..., X_n)$, such that one can say with a certain expected confidence $(1 - \alpha)$ that at least $100\gamma\%$ of the product will operate longer than $L(X_1, ..., X_n)$. Then reliability statements can be made based on $L(X_1, ..., X_n)$, or, decisions can be reached by comparing $L(X_1, ..., X_n)$ to $T$. Statistical tolerance limits of the types mentioned above are considered in this chapter.

The problem can be stated more formally as follows. Let $X_1, ..., X_n$ represent an experimental random sample from a distribution with a probability density function $f_\theta(x)$ (distribution function $F_\theta(x)$, survival function $\bar{F}_\theta(x) = 1 - F_\theta(x)$) and $S$ be any statistic (say, sufficient statistic or maximum likelihood estimator) obtained from the experimental random sample $X_1, ..., X_n$, and let a random variable $Y$ (in a future random sample $Y_1, ..., Y_m$) have the same distribution with the probability density function $f_\theta(y)$ (distribution function $F_\theta(y)$, survival function $\bar{F}_\theta(y) = 1 - F_\theta(y)$), where a parameter $\theta$ (in general, vector) is common to both distributions and it is assumed that some or all numerical values of components of the parametric vector $\theta$ are unspecified. On the basis of the experimental random sample $X_1, ..., X_n$ we wish to make a prediction about a future outcome of $Y_k$ ($k$th-order statistic, $1 \leq k \leq m$, in a future random sample of $m$-ordered observations $Y_1 \leq ... \leq Y_m$), usually in the form of one-sided statistical tolerance limits on future outcomes of $Y_k$ (lower $\gamma$-content tolerance limit $L_k$ with expected $[1 - \alpha]$-confidence [for a specified proportion $\gamma$] and upper $\gamma$-content tolerance limit $U_k$ with expected $[1 - \alpha]$-confidence [for a specified proportion $\gamma$]). That is, if $L_k$ and $U_k$ are functions of $S$, then $L_k(S)$ is a lower $\gamma$-content tolerance limit with expected $(1 - \alpha)$-confidence on future outcomes of the $k$th-order statistic $Y_k$ if

$$E_\theta\left\{\Pr\left(\int_{L_k(S)}^{\infty} g_\theta(y_k)dy_k \geq \gamma\right)\right\} = E_\theta\left\{\Pr\left(\bar{G}_\theta(L_k(S)) \geq \gamma\right)\right\} = 1 - \alpha, \tag{11.1}$$

and $U_k(S)$ is an upper $\gamma$-content tolerance limit with expected $(1 - \alpha)$-confidence on future outcomes of the $k$th-order statistic $Y_k$ if

$$E_\theta\left\{\Pr\left(\int_{0}^{U_k(S)} g_\theta(y_k)dy_k \geq \gamma\right)\right\} = E_\theta\left\{\Pr\left(G_\theta(U_k(S)) \geq \gamma\right)\right\} = 1 - \alpha, \tag{11.2}$$

where

$$g_\theta(y_k) = \frac{1}{B(k, m-k+1)}[F_\theta(y_k)]^{k-1}[1 - F_\theta(y_k)^{m-k}f_\theta(y_k) \tag{11.3}$$

is a probability density function of the $k$th-order statistic $Y_k$,

$$B(a,b) = \int_0^1 t^{a-1}(1-t)^{b-1} dt = \frac{\Gamma(a)\Gamma(b)}{\Gamma(a+b)} \tag{11.4}$$

is the beta-function, and $\Gamma(a) = \int_0^\infty t^{a-1}e^{-t}dt$ is the gamma function,

$$G_\theta(y_k) = \Pr(Y_k \le y_k) = \sum_{i=k}^m \binom{m}{i} [F_\theta(y_k)]^i [1 - F_\theta(y_k)]^{m-i}$$

$$= \sum_{i=k}^m \binom{m}{i} \left[1 - \bar{F}_\theta(y_k)\right]^i \left[\bar{F}_\theta(y_k)\right]^{m-i} = \sum_{i=k}^m \binom{m}{i} \sum_{j=0}^i \binom{i}{j} (-1)^j \left[\bar{F}_\theta(y_k)\right]^{m-i+j}$$

$$= \int_0^{F_\theta(y_k)} \varphi(t \mid k, m-k+1) dt \tag{11.5}$$

is a probability distribution function of the $k$th-order statistic $Y_k$,

$$\varphi(t \mid a,b) = \frac{1}{B(a,b)} t^{a-1}(1-t)^{b-1}, \quad t \in (0,1), \tag{11.6}$$

is a probability density function of the beta-distribution with the shape parameters $a$ and $b$,

$$\bar{G}_\theta(y_k) = 1 - G_\theta(y_k) = \Pr(Y_k > y_k) = \sum_{i=0}^{k-1} \binom{m}{i} [F_\theta(y_k)]^i [1 - F_\theta(y_k)]^{m-i}$$

$$= \sum_{i=0}^{k-1} \binom{m}{i} \left[1 - \bar{F}_\theta(y_k)\right]^i \left[\bar{F}_\theta(y_k)\right]^{m-i} = \sum_{i=0}^{k-1} \binom{m}{i} \sum_{j=0}^i \binom{i}{j} (-1)^j \left[\bar{F}_\theta(y_k)\right]^{m-i+j}$$

$$= \int_{F_\theta(y_k)}^1 \varphi(t \mid k, m-k+1) dt. \tag{11.7}$$

It can be shown that

$$\frac{dG_\theta(y_k)}{dy_k} = g_\theta(y_k). \tag{11.8}$$

Indeed, taking into account (11.5), we have

$$\frac{dG_\theta(y_k)}{dy_k} = \frac{d}{dy_k} \sum_{i=k}^{m} \binom{m}{i} [F_\theta(y_k)]^i [1 - F_\theta(y_k)]^{m-i}$$

$$= \sum_{i=k}^{m} \binom{m}{i} \left( i[F_\theta(y_k)]^{i-1} [1 - F_\theta(y_k)]^{m-i} F_\theta'(y_k) - (m-i)[F_\theta(y_k)]^i [1 - F_\theta(y_k)]^{m-i-1} F_\theta'(y_k) \right)$$

$$= \sum_{i=k}^{m} \frac{m!}{(i-1)!(m-i)!} [F_\theta(y_k)]^{i-1} [1 - F_\theta(y_k)]^{m-i} f_\theta(y_k)$$

$$- \sum_{i=k}^{m-1} \frac{m!}{i!(m-i-1)!} [F_\theta(y_k)]^i [1 - F_\theta(y_k)]^{m-i-1} f_\theta(y_k)$$

$$= \sum_{i=k}^{m} \frac{m!}{(i-1)!(m-i)!} [F_\theta(y_k)]^{i-1} [1 - F_\theta(y_k)]^{m-i} f_\theta(y_k)$$

$$- \sum_{i=k+1}^{m} \frac{m!}{(j-1)!(m-j)!} [F_\theta(y_k)]^{j-1} [1 - F_\theta(y_k)]^{m-j} f_\theta(y_k)$$

$$= \frac{m!}{(k-1)!(m-k)!} [F_\theta(y_k)]^{k-1} [1 - F_\theta(y_k)]^{m-k} f_\theta(y_k) = g_\theta(y_k), \tag{11.9}$$

where $j = i + 1$;

$$\frac{dG_\theta(y_k)}{dy_k} = \frac{d}{dy_k} \int_0^{F_\theta(y_k)} \varphi(t \mid k, m-k+1) dt = \frac{d}{dy_k} \int_0^{F_\theta(y_k)} \frac{1}{B(k, m-k+1)} t^{k-1} (1-t)^{m-k} dt$$
$$\tag{11.10}$$

$$= \frac{1}{B(k, m-k+1)} [F_\theta(y_k)]^{k-1} [1 - F_\theta(y_k)]^{m-k} F_\theta'(y_k) = g_\theta(y_k).$$

The problem considered in this chapter is to find $L_k$ (S) (lower statistical $\gamma$-content tolerance limit $L_k$ with expected $[1 - \alpha]$-confidence on future outcomes of the $k$th-order statistic $Y_k$) satisfying (11.1) and $U_k(S)$ (upper statistical $\gamma$-content tolerance limit with expected $[1 - \alpha]$-confidence on future outcomes of the $k$th-order statistic $Y_k$) satisfying (11.2) on the basis of the experimental random sample $X_1, \ldots, X_n$ when some or all numerical values of components of the parametric vector $\theta$ are unspecified.

Thus, the logical purpose for a tolerance limit must be the prediction of future outcomes for some production process. The coverage value $\gamma$ is the percentage of the future process outcomes to be captured by the prediction, and the confidence level $(1 - \alpha)$ is the proportion of the time we hope to capture that percentage $\gamma$.

The common distributions used in life testing problems are the normal, exponential, Weibull, and gamma distributions (Mendenhall 1958). Tolerance limits for the normal distribution have been considered in (Guttman 1957; Wald and Wolfowitz 1946; Wallis 1951), and others.

Tolerance (prediction) limits enjoy a fairly rich history in the literature and have a very important role in engineering and manufacturing applications. Patel (1986) provides a review (which was fairly comprehensive at the time of publication) of tolerance intervals (limits) for many distributions as well as a discussion of their relation with confidence intervals (limits) for percentiles. Dunsmore (1978) and Guenther, Patil, and Uppuluri (1976) discuss two-parameter exponential tolerance intervals (limits) and the estimation procedure in greater detail. Engelhardt and Bain (1978) discuss how to modify the formulas when dealing with type II censored data. Guenther (1972) and Hahn and Meeker (1991) discuss how one-sided tolerance limits can be used to obtain approximate two-sided tolerance intervals by applying Bonferroni's inequality. In Nechval et al. (2011, 2016a–c), the exact statistical tolerance and prediction limits are discussed under parametric uncertainty of underlying models.

In contrast to other statistical limits commonly used for statistical inference, the tolerance limits (especially for the order statistics) are used relatively rarely. One reason is that the theoretical concept and computational complexity of the tolerance limits is significantly more difficult than that of the standard confidence and prediction limits. Thus, it becomes necessary to use the innovative approaches that will allow one to construct tolerance limits on future order statistics for many populations.

In this chapter, new approaches to constructing lower and upper statistical $\gamma$-content tolerance limits with expected $(1 - \alpha)$-confidence as well as $(1 - \alpha)$-expectation tolerance limits on order statistics in future samples are proposed. For illustration, a two-parameter Weibull distribution is considered.

## 11.2   Two-parameter Weibull distribution

In this chapter, the two-parameter Weibull distribution with the pdf (probability density function),

$$f_\theta(x) = \frac{\delta}{\beta}\left(\frac{x}{\beta}\right)^{\delta-1} \exp\left[-\left(\frac{x}{\beta}\right)^{\delta}\right], \quad x > 0, \ \beta > 0, \delta > 0, \tag{11.11}$$

and cdf (cumulative distribution function),

$$F_\theta(x) = 1 - \exp\left[-\left(\frac{x}{\beta}\right)^{\delta}\right], \quad x > 0, \beta > 0, \delta > 0, \tag{11.12}$$

indexed by scale and shape parameters $\beta$ and $\delta$ is used as the underlying distribution of a random variable $X$ in a sample of the lifetime data, where $\theta = (\beta, \delta)$.

The Weibull distribution is widely used in reliability and survival analysis due to its flexible shape and ability to model a wide range of failure rates. It can be derived theoretically as a form of extreme value distribution, governing the time to occurrence of the "weakest link" of many competing failure processes. Its special case with shape parameter $\delta = 2$ is the Rayleigh distribution, which is commonly used for modeling the magnitude of radial error when $x$ and $y$ coordinate errors are independent normal variables with zero mean and the same standard deviation while the case $\delta = 1$ corresponds to the widely used exponential distribution. For illustration, probability density functions of the two-parameter Weibull distribution for selected values of $\beta$ and $\delta$ are shown in Figure 11.1.

Let $X$ follow a Weibull distribution with scale parameter $\beta$ and shape parameter $\delta$. We consider both parameters $\beta, \delta$ to be unknown. Let $(X_1, ..., X_n)$ be a random sample from

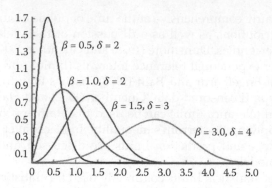

*Figure 11.1* The Weibull probability density functions for selected values of $\beta$ and $\delta$.

the two-parameter Weibull distribution (11.11), and let $\hat{\beta}$, $\delta$ be the maximum likelihood estimates of $\beta$, $\delta$, respectively, computed on the basis of $(X_1, ..., X_n)$:

$$\hat{\beta} = \left( \sum_{i=1}^{n} x_i^{\delta} \Big/ n \right)^{1/\hat{\delta}},$$

(11.13)

and

$$\hat{\delta} = \left[ \left( \sum_{i=1}^{n} x_i^{\hat{\delta}} \ln x_i \right) \left( \sum_{i=1}^{n} x_i^{\hat{\delta}} \right)^{-1} - \frac{1}{n} \sum_{i=1}^{n} \ln x_i \right]^{-1}.$$

(11.14)

In terms of the Weibull variates, we have that

$$V_1 = \left( \frac{\hat{\beta}}{\beta} \right)^{\delta}, V_2 = \frac{\delta}{\hat{\delta}}, V_2 = \frac{\delta}{\hat{\delta}},$$

(11.15)

are pivotal quantities. Furthermore, let

$$Z_i = (X_i / \hat{\beta})^{\hat{\delta}}, \quad i = 1, ..., n.$$

(11.16)

It is readily verified that any $n - 2$ of the $Z_i$'s, say $Z_1, ..., Z_{n-2}$ form a set of $n - 2$ functionally independent ancillary statistics. The appropriate conditional approach is to consider the distributions of $V_1$, $V_2$, $V_3$ conditional on the observed value of $Z^{(n)} = (Z_1, ..., Z_n)$. (For purposes of symmetry of notation we include all of $Z_1, ..., Z_n$ in expressions stated here; it can be shown that $Z_n$, $Z_{n-1}$, can be determined as functions of $Z_1, ..., Z_{n-2}$ only.)

> *Theorem 1* (Joint pdf of the pivotal quantities $V_1$, $V_2$ from the two-parameter Weibull distribution). Let $(X_1, ..., X_n)$ be a random sample of $n$ observations from the two-parameter Weibull distribution (11.11). Then the joint pdf of the pivotal quantities

$$V_1 = \left( \frac{\hat{\beta}}{\beta} \right)^{\delta}, V_2 = \frac{\delta}{\hat{\delta}},$$

(11.17)

conditional on fixed

$$\mathbf{z}^{(n)} = (z_i, \ldots, z_n), \tag{11.18}$$

where

$$Z_i = \left( \frac{X_i}{\hat{\beta}} \right)^{\hat{\delta}}, \quad i = 1, \ldots, n, \tag{11.19}$$

are ancillary statistics, any $n - 2$ of which form a functionally independent set, $\hat{\beta}$ and $\hat{\delta}$ are the maximum likelihood estimates for $\beta$ and $\delta$, respectively, based on a random sample of $n$ observations $(X_1, \ldots, X_n)$ from the two-parameter Weibull distribution (11.11), is given by

$$f_n(v_1, v_2 \mid \mathbf{z}^{(n)}) = \left( \frac{1}{\Gamma(n)} \left[ \sum_{i=1}^{n} z_i^{v_2} \right]^n v_1^{n-1} \exp\left( -v_1 \sum_{i=1}^{n} z_i^{v_2} \right) \right) \left( \frac{1}{\vartheta(\mathbf{z}^{(n)})} v_2^{n-2} \prod_{i=1}^{n} z_i^{v_2} \left[ \sum_{i=1}^{n} z_i^{v_2} \right]^{-n} \right)$$

$$= f_n(v_1 \mid \mathbf{z}^{(n)}, v_2) f_n(v_2 \mid \mathbf{z}^{(n)}), \quad v_1 \in (0, \infty), v_2 \in (0, \infty),$$

$$\tag{11.20}$$

where

$$f_n(v_1 \mid \mathbf{z}^{(n)}, v_2) = \frac{1}{\Gamma(n)} \left[ \sum_{i=1}^{n} z_i^{v_2} \right]^n v_1^{n-1} \exp\left( -v_1 \sum_{i=1}^{n} z_i^{v_2} \right), v_1 \in (0, \infty) \tag{11.21}$$

$$f_n(v_2 \mid \mathbf{z}^{(n)}) = \frac{1}{\vartheta(\mathbf{z}^{(n)})} v_2^{n-2} \prod_{i=1}^{n} z_i^{v_2} \left( \sum_{i=1}^{n} z_i^{v_2} \right)^{-n}, \quad v_2 \in (0, \infty), \tag{11.22}$$

$$\vartheta(\mathbf{z}^{(n)}) = \int_0^{\infty} v_2^{n-2} \prod_{i=1}^{n} z_i^{v_2} \left( \sum_{i=1}^{n} z_i^{v_2} \right)^{-n} dv_2 \tag{11.23}$$

is the normalizing constant.

*Proof.* The joint density of $X_1, \ldots, X_n$ is given by

$$f_\theta(x_1, \ldots, x_n) = \prod_{i=1}^{n} \frac{\delta}{\beta} \left( \frac{x_i}{\beta} \right)^{\delta-1} \exp\left( -\left( \frac{x_i}{\beta} \right)^{\delta} \right). \tag{11.24}$$

Using the invariant embedding technique (Nechval and Vasermanis 2004; Nechval et al. 2008, 2010), we transform (11.24) to

$$f_\theta(x_1,\ldots,x_n)\,d\hat\beta\,d\hat\delta = \prod_{i=1}^{n}\frac{\delta}{x_i}\left(\frac{x_i}{\beta}\right)^{\delta}\exp\left(-\sum_{i=1}^{n}\left(\frac{x_i}{\beta}\right)^{\delta}\right)d\hat\beta\,d\hat\delta$$

$$= -\hat\beta\hat\delta^{n}\prod_{i=1}^{n}\frac{1}{x_i}\left(\frac{\delta}{\hat\delta}\right)^{n-2}\prod_{i=1}^{n}\left(\frac{x_i}{\hat\beta}\right)^{\delta}\left(\frac{\hat\beta}{\beta}\right)^{\delta(n-1)}$$

$$\times \exp\left(-\left(\frac{\hat\beta}{\beta}\right)^{\delta}\sum_{i=1}^{n}\left(\frac{x_i}{\hat\beta}\right)^{\delta}\right)\left(\frac{\delta}{\beta}\left(\frac{\hat\beta}{\beta}\right)^{\delta-1}d\hat\beta\right)\left(-\frac{\delta}{\hat\delta^2}d\hat\delta\right)$$

$$= -\hat\beta\hat\delta^{n}\prod_{i=1}^{n}\frac{1}{x_i}\left(\frac{\delta}{\hat\delta}\right)^{n-2}\prod_{i=1}^{n}\left(\frac{x_i}{\hat\beta}\right)^{\hat\delta\left(\frac{\delta}{\hat\delta}\right)}\left(\frac{\hat\beta}{\beta}\right)^{\delta(n-1)}$$

$$\times \exp\left(-\left(\frac{\hat\beta}{\beta}\right)^{\delta}\sum_{i=1}^{n}\left(\frac{x_i}{\hat\beta}\right)^{\hat\delta\left(\frac{\delta}{\hat\delta}\right)}\right)\left(\frac{\delta}{\beta}\left(\frac{\hat\beta}{\beta}\right)^{\delta-1}d\hat\beta\right)\left(-\frac{\delta}{\hat\delta^2}d\hat\delta\right)$$

$$= -\hat\beta\hat\delta^{n}\prod_{i=1}^{n}\frac{1}{x_i}\left(\frac{\delta}{\hat\delta}\right)^{n-2}\prod_{i=1}^{n}\left(\frac{x_i}{\hat\beta}\right)^{\hat\delta\left(\frac{\delta}{\hat\delta}\right)}\left(\frac{\hat\beta}{\beta}\right)^{\delta(n-1)}$$

$$\times \exp\left(-\left(\frac{\hat\beta}{\beta}\right)^{\delta}\sum_{i=1}^{n}\left(\frac{x_i}{\hat\beta}\right)^{\hat\delta\left(\frac{\delta}{\hat\delta}\right)}\right)d\left(\frac{\hat\beta}{\beta}\right)^{\delta}d\left(\frac{\delta}{\hat\delta}\right)$$

$$= -\hat\beta\hat\delta^{n}\prod_{i=1}^{n}\frac{1}{x_i}v_2^{n-2}\prod_{i=1}^{n}z_i^{v_2}v_1^{n-1}\exp\left(-v_1\sum_{i=1}^{n}z_i^{v_2}\right)dv_1dv_2$$

$$= -\hat\beta\hat\delta^{n}\prod_{i=1}^{n}\frac{1}{x_i}\Gamma(n)v_2^{n-2}\prod_{i=1}^{n}z_i^{v_2}\left[\sum_{i=1}^{n}z_i^{v_2}\right]^{-n}$$

$$\times \frac{1}{\Gamma(n)}\left[\sum_{i=1}^{n}z_i^{v_2}\right]^{n}v_1^{n-1}\exp\left(-v_1\sum_{i=1}^{n}z_i^{v_2}\right)dv_1dv_2. \tag{11.25}$$

Normalizing (11.25),

$$\frac{-\hat\beta\hat\delta^{n}\displaystyle\prod_{i=1}^{n}\frac{1}{x_i}\Gamma(n)v_2^{n-2}\prod_{i=1}^{n}z_i^{v_2}\left[\sum_{i=1}^{n}z_i^{v_2}\right]^{-n}\frac{1}{\Gamma(n)}\left[\sum_{i=1}^{n}z_i^{v_2}\right]^{n}v_1^{n-1}\exp\left(-v_1\sum_{i=1}^{n}z_i^{v_2}\right)}{\displaystyle\int_0^\infty\int_0^\infty -\hat\beta\hat\delta^{n}\prod_{i=1}^{n}\frac{1}{x_i}\Gamma(n)v_2^{n-2}\prod_{i=1}^{n}z_i^{v_2}\left[\sum_{i=1}^{n}z_i^{v_2}\right]^{-n}\frac{1}{\Gamma(n)}\left[\sum_{i=1}^{n}z_i^{v_2}\right]^{n}v_1^{n-1}\exp\left(-v_1\sum_{i=1}^{n}z_i^{v_2}\right)dv_1\,dv_2}$$

$$
= \frac{v_2^{n-2} \prod\limits_{i=1}^{n} z_i^{v_2} \left[ \sum\limits_{i=1}^{n} z_i^{v_2} \right]^{-n} \frac{1}{\Gamma(n)} \left[ \sum\limits_{i=1}^{n} z_i^{v_2} \right]^{n} v_1^{n-1} \exp\left( -v_1 \sum\limits_{i=1}^{n} z_i^{v_2} \right)}{\int\limits_{0}^{\infty} v_2^{n-2} \prod\limits_{i=1}^{n} z_i^{v_2} \left[ \sum\limits_{i=1}^{n} z_i^{v_2} \right]^{-n} dv_2}
$$

$$
= \frac{1}{\Gamma(n)} \left[ \sum\limits_{i=1}^{n} z_i^{v_2} \right]^{n} v_1^{n-1} \exp\left( -v_1 \sum\limits_{i=1}^{n} z_i^{v_2} \right) \frac{v_2^{n-2} \prod\limits_{i=1}^{n} z_i^{v_2} \left[ \sum\limits_{i=1}^{n} z_i^{v_2} \right]^{-n}}{\vartheta(\mathbf{z}^{(n)})} = f_n(v_1, v_2 \mid \mathbf{z}^{(n)}), \quad (11.26)
$$

we obtain (11.20). This ends the proof.

*Corollary 1.1* If a pivotal quantity is given by

$$
W = V_1 \left[ \sum_{i=1}^{n} Z_i^{V_2} \right] = \left( \frac{\widehat{\beta}}{\beta} \right)^{\widehat{\delta}\frac{\delta}{\widehat{\delta}}} \sum_{i=1}^{n} Z_i^{V_2} = \left( \frac{\widehat{\beta}^{\widehat{\delta}}}{\beta^{\widehat{\delta}}} \right)^{V_2} \sum_{i=1}^{n} Z_i^{V_2} = V_3^{V_2} \sum_{i=1}^{n} Z_i^{V_2}, \tag{11.27}
$$

it follows from (11.21) that

$$
W \sim g_n(w) = \frac{1}{\Gamma(n)} w^{n-1} \exp(-w), \quad w \in (0, \infty) \tag{11.28}
$$

where $g_n(w)$ is a probability density function of the gamma distribution

$$
f(w \mid a, b) = \frac{1}{\Gamma(a)b} \left( \frac{w}{b} \right)^{a-1} \exp(-w/b), \quad w \in (0, \infty), \tag{11.29}
$$

with the shape parameter $a = n$ and scale parameter $b = 1$.

## 11.3   Lower statistical γ-content tolerance limit with expected (1 − α)-confidence

*Theorem 2.* Let $X_1, \ldots, X_n$ be the $n$ observations of the experimental random sample of size $n$ from a two-parameter Weibull distribution defined by the probability density function (11.11). Then a lower statistical γ-content tolerance limit with expected $(1 - \alpha)$-confidence, $L_k \equiv L_k (S)$, on future outcomes of the $k$th-order statistic $Y_k$ from a set of $m$ future ordered observations $Y_1 \leq \cdots \leq Y_m$ also from the distribution (11.11), which satisfies (11.1) is given by

$$
L_k = \eta_{L_k}^{1/\widehat{\delta}} \widehat{\beta}, \tag{11.30}
$$

where

$$\eta_{L_k} = \arg\left( \int_0^\infty \int_0^{\frac{\ln(1-q_{1-\gamma})^{-1}\sum_{i=1}^n z_i^{v_2}}{[\eta_{L_k}]^{v_2}}} \frac{1}{\Gamma(n)} w^{n-1} \exp(-w) \frac{1}{\vartheta(\mathbf{z}^{(n)})} v_2^{n-2} \prod_{i=1}^n z_i^{v_2} \left( \sum_{i=1}^n z_i^{v_2} \right)^{-n} dw\, dv_2 = 1 - \alpha \right)$$

(11.31)

is a tolerance factor; the maximum likelihood estimates $\hat{\beta}$ and $\hat{\delta}$ of the parameters $\beta$ and $\delta$ are determined from (11.13) and (11.14), respectively; the ancillary statistics $Z_i$, $i = 1, \ldots, n$, are given by (11.19); $q_{1-\gamma}$ is a quantile of the beta-distribution satisfying

$$\int_0^{q_{1-\gamma}} \varphi(t \mid k, m-k+1)dt = \int_0^{q_{1-\gamma}} \frac{1}{B(k, m-k+1)} t^{k-1}(1-t)^{m-k}\, dt = 1 - \gamma.$$

(11.32)

*Proof.* Taking into account (11.2), (11.7), (11.27), and (11.28), the following probability transformation can be carried out:

$$\Pr\left( \int_{L_k}^\infty g_\theta(y_k)dy_k \geq \gamma \right) = \Pr\left( \bar{G}_\theta(L_k) \geq \gamma \right)$$

$$= \Pr\left( 1 - G_\theta(L_k) \geq \gamma \right) = \Pr\left( 1 - \int_0^{F_\theta(L_k)} \varphi(t \mid k, m-k+1) \geq \gamma \right)$$

$$= \Pr\left( \int_0^{F_\theta(L_k)} \varphi(t \mid k, m-k+1)dt \leq 1 - \gamma \right) = \Pr\left( F_\theta(L_k) \leq q_{1-\gamma} \right)$$

$$= \Pr\left( 1 - \exp\left[ -\left( \frac{L_k}{\beta} \right)^\delta \right] \leq q_{1-\gamma} \right) = \Pr\left( \exp\left[ -\left( \frac{L_k}{\beta} \right)^\delta \right] \geq 1 - q_{1-\gamma} \right)$$

$$= \Pr\left( -\left( \frac{L_k}{\beta} \right)^\delta \geq \ln(1-q_{1-\gamma}) \right) = \Pr\left( \left( \frac{L_k}{\beta} \right)^\delta \leq \ln(1-q_{1-\gamma})^{-1} \right)$$

$$= \Pr\left( \left( \frac{L_k}{\hat{\beta}} \frac{\hat{\beta}}{\beta} \right)^\delta \leq \ln(1-q_{1-\gamma})^{-1} \right) = \Pr\left( \left( \frac{\hat{\beta}}{\beta} \right)^\delta \left( \frac{L_k}{\hat{\beta}} \right)^{\hat{\delta}\frac{\delta}{\hat{\delta}}} \leq \ln(1-q_{1-\gamma})^{-1} \right)$$

$$= \Pr\left( V_1 \leq \frac{\ln(1-q_{1-\gamma})^{-1}}{\left[ (L_k/\hat{\beta})^{\hat{\delta}} \right]^{V_2}} \right) = \Pr\left( V_1 \sum_{i=1}^n Z_i^{V_2} \leq \frac{\ln(1-q_{1-\gamma})^{-1} \sum_{i=1}^n Z_i^{V_2}}{\left[ (L_k/\hat{\beta})^{\hat{\delta}} \right]^{V_2}} \right)$$

$$= \Pr\left( W \le \frac{\ln\left(1 - q_{1-\gamma}\right)^{-1} \displaystyle\sum_{i=1}^{n} Z_i^{V_2}}{\left[\left(L_k/\hat{\beta}\right)^{\hat{\delta}}\right]^{V_2}} \right), \tag{11.33}$$

where $q_{1-\gamma}$ is the $(1 - \gamma)$-quantile of the beta-distribution with the shape parameters $a = k$ and $b = m - k + 1$. Using pivotal quantity averaging, it follows from (11.1) and (11.33) that

$$E\left\{ \Pr\left( W \le \frac{\ln\left(1 - q_{1-\gamma}\right)^{-1} \displaystyle\sum_{i=1}^{n} Z_i^{V_2}}{\left[\left(L_k/\hat{\beta}\right)^{\hat{\delta}}\right]^{V_2}} \right) \right\} = E\left\{ \int_0^{\frac{\ln\left(1 - q_{1-\gamma}\right)^{-1} \sum_{i=1}^{n} Z_i^{V_2}}{\left[\left(L_k/\hat{\beta}\right)^{\hat{\delta}}\right]^{V_2}}} g_n(w)\,dw \right\}$$

$$= \int_0^\infty \int_0^{\frac{\ln\left(1 - q_{1-\gamma}\right)^{-1} \sum_{i=1}^{n} Z_i^{V_2}}{\left[\left(L_k/\hat{\beta}\right)^{\hat{\delta}}\right]^{V_2}}} g_n(w) f_n(v_2 \mid \mathbf{z}^{(n)})\,dw\,dv_2$$

$$= \int_0^\infty \int_0^{\frac{\ln\left(1 - q_{1-\gamma}\right)^{-1} \sum_{i=1}^{n} Z_i^{V_2}}{\left[\left(L_k/\hat{\beta}\right)^{\hat{\delta}}\right]^{V_2}}} \frac{1}{\Gamma(n)} w^{n-1} \exp(-w) \frac{1}{\vartheta(\mathbf{z}^{(n)})} v_2^{n-2} \prod_{i=1}^{n} z_i^{v_2} \left( \sum_{i=1}^{n} z_i^{v_2} \right)^{-n} dw\,dv_2 = 1 - \alpha. \tag{11.34}$$

It follows from (11.34) that

$$L_k = \arg\left( \int_0^\infty \int_0^{\frac{\ln\left(1 - q_{1-\gamma}\right)^{-1} \sum_{i=1}^{n} Z_i^{V_2}}{\left[\left(L_k/\hat{\beta}\right)^{\hat{\delta}}\right]^{V_2}}} \frac{1}{\Gamma(n)} w^{n-1} \exp(-w) \frac{1}{\vartheta(\mathbf{z}^{(n)})} v_2^{n-2} \prod_{i=1}^{n} z_i^{v_2} \left( \sum_{i=1}^{n} z_i^{v_2} \right)^{-n} dw\,dv_2 = 1 - \alpha \right). \tag{11.35}$$

Assuming that

$$\left(L_k/\hat{\beta}\right)^{\hat{\delta}} = \eta_{L_k}, \tag{11.36}$$

we have (11.30). This completes the proof.

Corollary 2.1. If $k = 1$, then

$$L_k = \eta_{L_k}^{1/\hat{\delta}} \hat{\beta}, \tag{11.37}$$

where $\eta_{L_k}$ satisfies

$$\int_0^\infty \int_0^{\frac{\ln(1-q_{1-\gamma})^{-1}\sum_{i=1}^{n} z_i^{v_2}}{[\eta_{L_k}]^{v_2}}} \frac{1}{\Gamma(n)} w^{n-1} \exp(-w) \frac{1}{\vartheta(\mathbf{z}^{(n)})} v_2^{n-2} \prod_{i=1}^{n} z_i^{v_2} \left( \sum_{i=1}^{n} z_i^{v_2} \right)^{-n} dw\, dv_2 = 1 - \alpha, \tag{11.38}$$

$q_{1-\gamma}$ is a quantile of the beta-distribution (with $k = 1$) satisfying (11.32).

Corollary 2.2. If $k = m = 1$, then

$$L_k = \eta_{L_k}^{1/\hat{\delta}} \hat{\beta}, \tag{11.39}$$

where

$$\eta_{L_k} = \arg\left( \int_0^\infty \int_0^{\frac{\ln\gamma^{-1}\sum_{i=1}^{n} z_i^{v_2}}{[\eta_{L_k}]^{v_2}}} \frac{1}{\Gamma(n)} w^{n-1} \exp(-w) \frac{1}{\vartheta(\mathbf{z}^{(n)})} v_2^{n-2} \prod_{i=1}^{n} z_i^{v_2} \left( \sum_{i=1}^{n} z_i^{v_2} \right)^{-n} dw\, dv_2 = 1 - \alpha \right). \tag{11.40}$$

## 11.4  Upper statistical γ-content tolerance limit with expected (1 − α)-confidence

*Theorem 3.* Let $X_1, \ldots, X_n$ be the $n$ observations of the experimental random sample of size $n$ from the two-parameter Weibull distribution defined by the probability density function (11.11). Then an upper statistical $\gamma$-content tolerance limit with expected $(1 - \alpha)$-confidence, $U_k \equiv U_k(S)$, on future outcomes of the $k$th-order statistic $Y_k$ from a set of $m$ future ordered observations $Y_1 \leq \cdots \leq Y_m$ also from the distribution (11.11)), which satisfies (11.2), is given by

$$U_k = \eta_{U_k}^{1/\hat{\delta}} \hat{\beta}, \tag{11.41}$$

where

$$\eta_{U_k} = \arg\left( \int_0^\infty \int_0^{\frac{\ln(1-q_\gamma)^{-1}\sum_{i=1}^{n} z_i^{v_2}}{[\eta_{U_k}]^{v_2}}} \frac{1}{\Gamma(n)} w^{n-1} \exp(-w) \frac{1}{\vartheta(\mathbf{z}^{(n)})} v_2^{n-2} \prod_{i=1}^{n} z_i^{v_2} \left( \sum_{i=1}^{n} z_i^{v_2} \right)^{-n} dw\, dv_2 = \alpha \right) \tag{11.42}$$

is a tolerance factor; the maximum likelihood estimates $\hat{\beta}$ and $\hat{\delta}$ of the parameters $\beta$ and $\delta$ are determined from (11.13) and (11.14), respectively; the ancillary statistics $Z_i$, $i = 1, \ldots, n$, are given by (11.19); $q_\gamma$ is a quantile of the beta-distribution satisfying

$$\int_0^{q_\gamma} \varphi(t \mid k, m-k+1) dt = \int_0^{q_\gamma} \frac{1}{B(k, m-k+1)} t^{k-1}(1-t)^{m-k} dt = \gamma. \tag{11.43}$$

*Proof.* Taking into account (11.2), (11.5), (11.27), and (11.28), the following probability transformation can be carried out:

$$\Pr\left(\int_0^{U_k} g_\theta(y_k) dy_k \geq \gamma\right) = \Pr\left(G_\theta(U_k) \geq \gamma\right) = \Pr\left(\int_0^{F_\theta(U_k)} \varphi(t \mid k, m-k+1) \geq \gamma\right) = \Pr\left(F_\theta(U_k) \geq q_\gamma\right)$$

$$= \Pr\left(1 - \exp\left[-\left(\frac{U_k}{\beta}\right)^\delta\right] \geq q_\gamma\right) = \Pr\left(\exp\left[-\left(\frac{U_k}{\beta}\right)^\delta\right] \leq 1 - q_\gamma\right)$$

$$= \Pr\left(-\left(\frac{U_k}{\beta}\right)^\delta \leq \ln(1-q_\gamma)\right) = \Pr\left(\left(\frac{L_k}{\beta}\right)^\delta \geq \ln(1-q_\gamma)^{-1}\right)$$

$$= \Pr\left(\left(\frac{U_k}{\hat{\beta}}\frac{\hat{\beta}}{\beta}\right)^\delta \geq \ln(1-q_\gamma)^{-1}\right) = \Pr\left(\left(\frac{\hat{\beta}}{\beta}\right)^\delta \left(\frac{U_k}{\hat{\beta}}\right)^{\delta\frac{\hat{\delta}}{\delta}} \geq \ln(1-q_\gamma)^{-1}\right)$$

$$= \Pr\left(V_1 \geq \frac{\ln(1-q_\gamma)^{-1}}{\left[\left(U_k/\hat{\beta}\right)^{\hat{\delta}}\right]^{V_2}}\right) = \Pr\left(V_1 \sum_{i=1}^n Z_i^{V_2} \geq \frac{\ln(1-q_\gamma)^{-1}\sum_{i=1}^n Z_i^{V_2}}{\left[\left(U_k/\hat{\beta}\right)^{\hat{\delta}}\right]^{V_2}}\right)$$

$$= \Pr\left(W \geq \frac{\ln(1-q_\gamma)^{-1}\sum_{i=1}^n Z_i^{V_2}}{\left[\left(U_k/\hat{\beta}\right)^{\hat{\delta}}\right]^{V_2}}\right), \tag{11.44}$$

where $q_\gamma$ is the $\gamma$ quantile of the beta-distribution (11.6) with the shape parameters $a = k$ and $b = m - k + 1$.

Using pivotal quantity averaging, it follows from (11.2) and (11.44) that

$$E\left\{\Pr\left(W \geq \frac{\ln(1-q_\gamma)^{-1} \sum_{i=1}^{n} Z_i^{V_2}}{\left[\left(U_k/\hat{\beta}\right)^{\hat{\delta}}\right]^{V_2}}\right)\right\} = E\left\{1 - \int_0^{\frac{\ln(1-q_\gamma)^{-1} \sum_{i=1}^{n} Z_i^{V_2}}{\left[\left(u_k/\hat{\beta}\right)^{\hat{\delta}}\right]^{V_2}}} g_n(w)\,dw\right\}$$

$$= 1 - \int_0^\infty \int_0^{\frac{\ln(1-q_\gamma)^{-1} \sum_{i=1}^{n} Z_i^{V_2}}{\left[\left(u_k/\hat{\beta}\right)^{\hat{\delta}}\right]^{V_2}}} g_n(w) f_n(v_2 \mid \mathbf{z}^{(n)})\,dw\,dv_2$$

$$= 1 - \int_0^\infty \int_0^{\frac{\ln(1-q_\gamma)^{-1} \sum_{i=1}^{n} Z_i^{V_2}}{\left[\left(u_k/\hat{\beta}\right)^{\hat{\delta}}\right]^{V_2}}} \frac{1}{\Gamma(n)} w^{n-1} \exp(-w) \frac{1}{\vartheta(\mathbf{z}^{(n)})} v_2^{n-2} \prod_{i=1}^{n} z_i^{v_2} \left(\sum_{i=1}^{n} z_i^{v_2}\right)^{-n} dw\,dv_2 = 1 - \alpha.$$

$$(11.45)$$

It follows from (11.45) that

$$U_k = \arg\left(\int_0^\infty \int_0^{\frac{\ln(1-q_\gamma)^{-1} \sum_{i=1}^{n} Z_i^{V_2}}{\left[\left(u_k/\hat{\beta}\right)^{\hat{\delta}}\right]^{V_2}}} \frac{1}{\Gamma(n)} w^{n-1} \exp(-w) \frac{1}{\vartheta(\mathbf{z}^{(n)})} v_2^{n-2} \prod_{i=1}^{n} z_i^{v_2} \left(\sum_{i=1}^{n} z_i^{v_2}\right)^{-n} dw\,dv_2 = \alpha\right).$$

$$(11.46)$$

Assuming that

$$\left(U_k/\hat{\beta}\right)^{\hat{\delta}} = \eta_{U_k},$$

$$(11.47)$$

we have (11.41). This completes the proof.

Corollary 3.1. If $k = 1$, then

$$U_k = \eta_{U_k}^{1/\hat{\delta}} \hat{\beta},$$

$$(11.48)$$

where $\eta_{U_k}$ satisfies

$$\int_0^\infty \int_0^{\frac{\ln(1-q_\gamma)^{-1} \sum_{i=1}^{n} Z_i^{V_2}}{[\eta_{U_k}]^{V_2}}} \frac{1}{\Gamma(n)} w^{n-1} \exp(-w) \frac{1}{\vartheta(\mathbf{z}^{(n)})} v_2^{n-2} \prod_{i=1}^{n} z_i^{v_2} \left(\sum_{i=1}^{n} z_i^{v_2}\right)^{-n} dw\,dv_2 = \alpha. \quad (11.49)$$

$q_\gamma$ is a quantile of the beta-distribution (with $k = 1$) satisfying

$$\int_0^{q_\gamma} \varphi(t \mid k, m-k+1)dt = \int_0^{q_\gamma} \frac{1}{\mathrm{B}(k, m-k+1)} t^{k-1}(1-t)^{m-k} dt = \gamma. \tag{11.50}$$

Corollary 3.2. If $k = m = 1$, then

$$U_k = \eta_{U_k}^{1/\hat{\delta}} \hat{\beta}, \tag{11.51}$$

where

$$\eta_{U_k} = \arg\left( \int_0^\infty \int_0^{\frac{\ln(1-\gamma)^{-1} \sum_{i=1}^n z_i^{v_2}}{[\eta_{U_k}]^{v_2}}} \frac{1}{\Gamma(n)} w^{n-1} \exp(-w) \frac{1}{\vartheta(\mathbf{z}^{(n)})} v_2^{n-2} \prod_{i=1}^n z_i^{v_2} \left( \sum_{i=1}^n z_i^{v_2} \right)^{-n} dw \, dv_2 = \alpha \right). \tag{11.52}$$

*Remark 1.* It will be noted that an upper statistical $\gamma$-content tolerance limit with expected $(1 - \alpha)$-confidence may be obtained from a lower statistical $\gamma$-content tolerance limit with expected $(1 - \alpha)$-confidence by replacing $1 - \alpha$ by $\alpha$, $1 - \gamma$ by $\gamma$.

## 11.5   *Lower statistical $(1 - \alpha)$-expectation tolerance limit*

*Theorem 4.* Let $X_1, \ldots, X_n$ be the $n$ observations of the experimental random sample of size $n$ from a two-parameter Weibull distribution defined by the probability density function (11.11). Then a lower statistical $(1 - \alpha)$-expectation tolerance limit, $L_k^\circ \equiv L_k^\circ(S)$, on future outcomes of the $k$th-order statistic $Y_k$ from a set of $m$ future ordered observations $Y_1 \leq \cdots \leq Y_m$ also from the distribution (11.11), which satisfies

$$E_\theta\left\{\Pr\left(Y_k > L_k^\circ(S)\right)\right\} = E_\theta\left\{ \int_{L_k^\circ(S)}^\infty g_\theta(y_k)dy_k \right\} = 1 - \alpha, \tag{11.53}$$

is given by

$$L_k^\circ = \eta_{L_k^\circ}^{1/\hat{\delta}} \hat{\beta}, \tag{11.54}$$

where $\eta_{L_k^\circ}$ is a tolerance factor satisfying

$$\sum_{l=0}^{k-1} \binom{m}{l} \sum_{j=0}^l \binom{l}{j} (-1)^j \frac{1}{\vartheta(\mathbf{z}^{(n)})} \int_0^\infty \frac{v_2^{n-2} \prod_{i=1}^n z_i^{v_2}}{\left( \sum_{i=1}^n z_i^{v_2} + \left[ \eta_{L_k^\circ} \right]^{v_2} (m-l+j) \right)^n} dv_2 = 1 - \alpha. \tag{11.55}$$

*Proof.* Taking into account (11.53), (11.7), (11.27), and (11.28), the following probability transformation can be carried out:

$$\Pr\left(Y_k > L_k^\circ\right) = \int_{L_k^\circ}^{\infty} g_\theta(y_k)\,dy_k = \bar{G}_\theta(L_k^\circ)$$

$$= \sum_{l=0}^{k-1} \binom{m}{l} \left[1 - \bar{F}_\theta(L_k^\circ)\right]^l \left[\bar{F}_\theta(L_k^\circ)\right]^{m-l} = \sum_{l=0}^{k-1} \binom{m}{l} \sum_{j=0}^{l} \binom{l}{j} (-1)^j \left[\bar{F}_\theta(L_k^\circ)\right]^{m-l+j}$$

$$= \sum_{l=0}^{k-1} \binom{m}{l} \sum_{j=0}^{l} \binom{l}{j} (-1)^j \left[\exp\left[-\left(\frac{L_k^\circ}{\beta}\right)^\delta\right]\right]^{m-l+j}$$

$$= \sum_{l=0}^{k-1} \binom{m}{l} \sum_{j=0}^{i} \binom{l}{j} (-1)^j \left[\exp\left[-\left(\frac{\beta}{\beta}\right)^\delta \left(\frac{L_k^\circ}{\hat{\beta}}\right)^{\hat{\delta}\frac{\delta}{\hat{\delta}}}\right]\right]^{m-l+j}$$

$$= \sum_{l=0}^{k-1} \binom{m}{l} \sum_{j=0}^{l} \binom{l}{j} (-1)^j \left[\exp\left[-V_1 \sum_{i=1}^{n} z_i^{V_2} \left(\frac{L_k^\circ}{\hat{\beta}}\right)^{\hat{\delta}V_2} \left(\sum_{i=1}^{n} z_i^{V_2}\right)^{-1}\right]\right]^{m-l+j}$$

$$= \sum_{l=0}^{k-1} \binom{m}{l} \sum_{j=0}^{l} \binom{l}{j} (-1)^j \left[\exp\left[-W\left[\left(L_k^\circ/\hat{\beta}\right)^{\hat{\delta}}\right]^{V_2} \left(\sum_{i=1}^{n} z_i^{V_2}\right)^{-1} (m-l+j)\right]\right].$$

$$(11.56)$$

Using pivotal quantity averaging, it follows from (11.22), (11.27), (11.28), (11.53), and (11.56) that

$$E_\theta\left\{\Pr\left(Y_k > L_k^\circ(S)\right)\right\}$$

$$= E\left\{\sum_{l=0}^{k-1} \binom{m}{l} \sum_{j=0}^{l} \binom{l}{j} (-1)^j \left[\exp\left[-W\left[\left(L_k^\circ/\hat{\beta}\right)^{\hat{\delta}}\right]^{V_2} \left(\sum_{i=1}^{n} z_i^{V_2}\right)^{-1} (m-l+j)\right]\right]\right\}$$

$$= \sum_{l=0}^{k-1} \binom{m}{l} \sum_{j=0}^{l} \binom{l}{j} (-1)^j \int_0^{\infty}\int_0^{\infty} \exp\left[-w\left[\left(L_k^\circ/\hat{\beta}\right)^{\hat{\delta}}\right]^{v_2} \left(\sum_{i=1}^{n} z_i^{v_2}\right)^{-1} (m-l+j)\right] g_n(w) f_n(v_2 \mid \mathbf{z}^{(n)})\,dw\,dv_2$$

$$= \sum_{l=0}^{k-1} \binom{m}{l} \sum_{j=0}^{l} \binom{l}{j} (-1)^j \frac{1}{\vartheta(\mathbf{z}^{(n)})} \int_0^{\infty} \frac{v_2^{n-2} \prod_{i=1}^{n} z_i^{v_2}}{\left(\sum_{i=1}^{n} z_i^{v_2} + \left[\left(L_k^\circ/\hat{\beta}\right)^{\hat{\delta}}\right]^{v_2} (m-l+j)\right)^n}\,dv_2, \qquad (11.57)$$

where

$$
L_k^\circ = \arg \left( \sum_{l=0}^{k-1} \binom{m}{l} \sum_{j=0}^{l} \binom{l}{j} (-1)^j \frac{1}{\vartheta(\mathbf{z}^{(n)})} \int_0^\infty \frac{v_2^{n-2} \prod_{i=1}^{n} z_i^{v_2}}{\left( \sum_{i=1}^{n} z_i^{v_2} + \left[ \left( L_k^\circ / \widehat{\beta} \right)^{\widehat{\delta}} \right]^{v_2} (m-l+j) \right)^n} dv_2 = 1 - \alpha \right).
$$

(11.58)

Assuming that

$$
\left( L_k^\circ / \widehat{\beta} \right)^{\widehat{\delta}} = \eta_{L_k^\circ},
$$

(11.59)

we have (11.54). This completes the proof.

Corollary 4.1. If $k = 1$, then

$$
L_k^\circ = \eta_{L_k^\circ}^{1/\widehat{\delta}} \widehat{\beta},
$$

(11.60)

where $\eta_{L_k^\circ}$ satisfies

$$
\frac{1}{\vartheta(\mathbf{z}^{(n)})} \int_0^\infty \frac{v_2^{n-2} \prod_{i=1}^{n} z_i^{v_2}}{\left( \sum_{i=1}^{n} z_i^{v_2} + \left[ \eta_{L_k^\circ} \right]^{v_2} m \right)^n} dv_2 = 1 - \alpha.
$$

(11.61)

Corollary 4.2. If $k = m = 1$, then

$$
L_k^\circ = \eta_{L_k^\circ}^{1/\widehat{\delta}} \widehat{\beta},
$$

(11.62)

where $\eta_{L_k^\circ}$ satisfies

$$
\frac{1}{\vartheta(\mathbf{z}^{(n)})} \int_0^\infty \frac{v_2^{n-2} \prod_{i=1}^{n} z_i^{v_2}}{\left( \sum_{i=1}^{n} z_i^{v_2} + \left[ \eta_{L_k^\circ} \right]^{v_2} \right)^n} dv_2 = 1 - \alpha.
$$

(11.63)

## 11.6  Upper statistical $(1 - \alpha)$-expectation tolerance limit

*Theorem 5.* Let $X_1, \ldots, X_n$ be the $n$ observations of the experimental random sample of size $n$ from a two-parameter Weibull distribution defined by the probability density function (11.11). Then an upper statistical $(1 - \alpha)$-expectation tolerance limit, $U_k^\circ \equiv U_k^\circ(S)$, on future outcomes of the $k$th-order statistic $Y_k$ from a set of $m$ future ordered observations $Y_1 \leq \equiv \leq Y_m$ also from the distribution (11.11), which satisfies

$$E_\theta \left\{ \Pr\left(Y_k \le U_k^\circ(S)\right) \right\} = E_\theta \left\{ \int_0^{U_k^\circ(S)} g_\theta(y_k) dy_k \right\} = 1 - \alpha, \tag{11.64}$$

is given by

$$U_k^\circ = \eta_{U_k^\circ}^{1/\delta} \hat{\beta}, \tag{11.65}$$

where $\eta_{U_k^\circ}$ is a tolerance factor satisfying

$$\sum_{l=0}^{k-1} \binom{m}{l} \sum_{j=0}^{l} \binom{l}{j} (-1)^j \frac{1}{\vartheta(\mathbf{z}^{(n)})} \int_0^\infty \frac{v_2^{n-2} \prod_{i=1}^{n} z_i^{v_2}}{\left( \sum_{i=1}^{n} z_i^{v_2} + \left[ \eta_{U_k^\circ} \right]^{v_2} (m-l+j) \right)^n} dv_2 = \alpha. \tag{11.66}$$

*Proof.* Taking into account (11.63), (11.7), (11.27), and (11.28), the following probability transformation can be carried out:

$$\Pr\left(Y_k \le L_k^\circ\right) = 1 - \int_{U_k^\circ}^\infty g_\theta(y_k) dy_k = 1 - \bar{G}_\theta(U_k^\circ)$$

$$= 1 - \sum_{l=0}^{k-1} \binom{m}{l} \left[ 1 - \bar{F}_\theta(U_k^\circ) \right]^l \left[ \bar{F}_\theta(U_k^\circ) \right]^{m-l} = 1 - \sum_{l=0}^{k-1} \binom{m}{l} \sum_{j=0}^{l} \binom{l}{j} (-1)^j \left[ \bar{F}_\theta(U_k^\circ) \right]^{m-l+j}$$

$$= 1 - \sum_{l=0}^{k-1} \binom{m}{l} \sum_{j=0}^{l} \binom{l}{j} (-1)^j \left[ \exp\left[ -\left( \frac{U_k^\circ}{\beta} \right)^\delta \right] \right]^{m-l+j}$$

$$= 1 - \sum_{l=0}^{k-1} \binom{m}{l} \sum_{j=0}^{l} \binom{l}{j} (-1)^j \left[ \exp\left[ -\left( \frac{\hat{\beta}}{\beta} \right)^\delta \left( \frac{L_k^\circ}{\hat{\beta}} \right)^{\hat{\delta}\frac{\delta}{\hat{\delta}}} \right] \right]^{m-l+j}$$

$$= 1 - \sum_{l=0}^{k-1} \binom{m}{l} \sum_{j=0}^{l} \binom{l}{j} (-1)^j \left[ \exp\left[ -V_1 \sum_{i=1}^{n} z_i^{V_2} \left( \frac{U_k^\circ}{\hat{\beta}} \right)^{\hat{\delta}V_2} \left( \sum_{i=1}^{n} z_i^{V_2} \right)^{-1} \right] \right]^{m-l+j}$$

$$= 1 - \sum_{l=0}^{k-1} \binom{m}{l} \sum_{j=0}^{l} \binom{l}{j} (-1)^j \left[ \exp\left[ -W\left[ \left( U_k^\circ / \hat{\beta} \right)^{\hat{\delta}} \right]^{V_2} \left( \sum_{i=1}^{n} z_i^{V_2} \right)^{-1} (m-l+j) \right] \right].$$

$$\tag{11.67}$$

Using pivotal quantity averaging, it follows from (11.22), (11.27), (11.28), (11.64), and (11.67) that

$E_\theta \left\{ \Pr\left(Y_k \leq U_k^\circ(S)\right) \right\}$

$$= E \left\{ 1 - \sum_{l=0}^{k-1} \binom{m}{l} \sum_{j=0}^{l} \binom{l}{j} (-1)^j \left[ \exp\left[ -W\left[ \left(U_k^\circ/\widehat{\beta}\right)^{\widehat{\delta}} \right]^{V_2} \left( \sum_{i=1}^{n} z_i^{V_2} \right)^{-1} (m-l+j) \right] \right] \right\}$$

$$= 1 - \sum_{l=0}^{k-1} \binom{m}{l} \sum_{j=0}^{l} \binom{l}{j} (-1)^j$$

$$\times \int_0^\infty \int_0^\infty \exp\left[ -w\left[ \left(U_k^\circ/\widehat{\beta}\right)^{\widehat{\delta}} \right]^{v_2} \left( \sum_{i=1}^{n} z_i^{v_2} \right)^{-1} (m-l+j) \right] g_n(w) f_n(v_2 \mid \mathbf{z}^{(n)}) dw\, dv_2$$

$$= 1 - \sum_{l=0}^{k-1} \binom{m}{l} \sum_{j=0}^{l} \binom{l}{j} (-1)^j \frac{1}{\vartheta(\mathbf{z}^{(n)})} \int_0^\infty \frac{v_2^{n-2} \prod_{i=1}^{n} z_i^{v_2}}{\left( \sum_{i=1}^{n} z_i^{v_2} + \left[ \left(U_k^\circ/\widehat{\beta}\right)^{\widehat{\delta}} \right]^{v_2} (m-l+j) \right)^n} dv_2,$$

$$(11.68)$$

where

$$U_k^\circ = \arg \left( \sum_{l=0}^{k-1} \binom{m}{l} \sum_{j=0}^{l} \binom{l}{j} (-1)^j \frac{1}{\vartheta(\mathbf{z}^{(n)})} \int_0^\infty \frac{v_2^{n-2} \prod_{i=1}^{n} z_i^{v_2}}{\left( \sum_{i=1}^{n} z_i^{v_2} + \left[ \left(U_k^\circ/\widehat{\beta}\right)^{\widehat{\delta}} \right]^{v_2} (m-l+j) \right)^n} dv_2 = \alpha \right).$$

$$(11.69)$$

Assuming that

$$\left(U_k^\circ/\widehat{\beta}\right)^{\widehat{\delta}} = \eta_{U_k^\circ},$$

$$(11.70)$$

we have (11.65). This completes the proof.

Corollary 5.1. If $k = 1$, then

$$U_k^\circ = \eta_{U_k^\circ}^{1/\widehat{\delta}} \widehat{\beta},$$

$$(11.71)$$

where $\eta_{U_k^\circ}$ satisfies

$$\frac{1}{\vartheta(\mathbf{z}^{(n)})} \int_0^\infty \frac{v_2^{n-2} \prod_{i=1}^{n} z_i^{v_2}}{\left( \sum_{i=1}^{n} z_i^{v_2} + \left[ \eta_{U_k^\circ} \right]^{v_2} m \right)^n} dv_2 = \alpha.$$

$$(11.72)$$

Corollary 5.2. If $k = m = 1$, then

$$U_k^\circ = \eta_{U_k^\circ}^{1/\hat{\delta}} \hat{\beta},$$  (11.73)

where $\eta_{U_k^\circ}$ satisfies

$$\frac{1}{\vartheta(\mathbf{z}^{(n)})} \int\limits_0^\infty \frac{v_2^{n-2} \prod\limits_{i=1}^n z_i^{v_2}}{\left( \sum\limits_{i=1}^n z_i^{v_2} + \left[ \eta_{U_k^\circ} \right]^{v_2} \right)^n} dv_2 = \alpha.$$  (11.74)

*Remark 2.* It will be noted that an upper statistical $(1 - \alpha)$-expectation tolerance limit may be obtained from a lower statistical $(1 - \alpha)$-expectation tolerance limit by replacing $1 - \alpha$ by $\alpha$.

## 11.7   Numerical example 1

Consider the data in an example discussed by Mann and Saunders (1969). They regard the data coming from the Weibull distribution as the results of full-scale fatigue tests on a particular type of component. The data are for a complete sample of size $n = 3$, with observations $X_1 = 45.952$, $X_2 = 54.143$, and $X_3 = 65.440$, results being expressed here in number of thousands of cycles. On the basis of these data, it is wished to obtain the lower $(1 - \alpha)$-expectation tolerance limit for the minimum $(Y_1)$ of independent lifetimes in a group of $m = 500$ components which are to be put into service.

The maximum likelihood estimates of the unknown parameters $\delta$ and $\beta$, computed on the basis of $(X_1, X_2, X_3)$, are $\hat{\delta} = 7.726$ and $\hat{\beta} = 58.706$, respectively. Taking $1 - \alpha = 0.8$ and $k = 1$, with $n = 3$ and $m = 500$, we have from (11.60) that the statistical lower $(1 - \alpha)$-expectation tolerance limit, $L_k^\circ \equiv L_k^\circ(S)$, for the minimum $(Y_1)$ of independent lifetimes in a group of $m = 500$ components which are to be put into service, is given by

$$L_k^\circ = \eta_{L_k^\circ}^{1/\hat{\delta}} \hat{\beta} = 5.527411,$$  (11.75)

where it follows from (11.61) that

$$\eta_{L_k^\circ} = \arg\left( \frac{1}{\vartheta(\mathbf{z}^{(n)})} \int\limits_0^\infty \frac{v_2^{n-2} \prod\limits_{i=1}^n z_i^{v_2}}{\left( \sum\limits_{i=1}^n z_i^{v_2} + \left[ \eta_{L_k^\circ} \right]^{v_2} m \right)^n} dv_2 = 1 - \alpha \right) = 1.18/10^8.$$  (11.76)

Lawless (1973) obtained for this example (via conditional approach in terms of a Gumbel distribution) the lower 80% prediction limit of 5.623, which is slightly larger than (11.75). The resulting lower 80% prediction limit of Mee and Kushary (1994) for this example (obtained via simulation) was 5.225, which is slightly smaller than (11.75). The Mann and Saunders (1969) result for this example was only 0.766.

Taking $\gamma = 0.8$, $1 - \alpha = 0.8$ and $k = 1$, with $n = 3$ and $m = 500$, we have from (11.37) that a lower statistical $\gamma$-content tolerance limit, $L_k \equiv L_k$ (S), with expected $(1 - \alpha)$-confidence for the minimum $(Y_1)$ of independent lifetimes in a group of $m = 500$ components which are to be put into service, is

$$L_k = \eta_{L_k}^{1/\hat{\delta}}\hat{\beta}, = 4.082282, \tag{11.77}$$

where it follows from (11.38) that

$$\eta_{L_k} = \arg\left(\int_0^\infty \int_0^{\frac{\ln(1-q_{1-\gamma})^{-1}\sum_{i=1}^n z_i^{v_2}}{[\eta_{L_k}]^{v_2}}} \frac{1}{\Gamma(n)} w^{n-1}\exp(-w)\frac{1}{\vartheta(\mathbf{z}^{(n)})}v_2^{n-2}\prod_{i=1}^n z_i^{v_2}\left(\sum_{i=1}^n z_i^{v_2}\right)^{-n} dw\,dv_2 = 1-\alpha\right)$$

$$= 1.135 / 10^9, \tag{11.78}$$

$q_{1-\gamma}$ is a quantile of the beta-distribution (with $k = 1$) satisfying (11.32).

## 11.8 Numerical example 2

To investigate the performance of a logic circuit for a small electronic calculator, a circuit manufacturer puts $n = 5$ of the circuits on life test without replacement under specified environmental conditions, and the failures are observed after $X_1 = 830$, $X_2 = 1020$, $X_3 = 1175$, $X_4 = 1424$, and $X_5 = 1603$ hours. A buyer tells the circuit manufacturer that he wants to place three orders ($l = 3$) for the same type of logic circuits to be shipped to three different destinations. The buyer wants to select a random sample of $q = 5$ logic circuits from each shipment to be tested. An order is accepted only if all of five logic circuits in each selected sample meet the warranty lifetime. What warranty lifetime should the manufacturer offer so that all of five logic circuits in each selected sample meet the warranty with probability of 0.95?

In order to find this warranty lifetime, the manufacturer wishes to use a random sample of size $n = 5$ given above and to calculate the lower statistical simultaneous tolerance limit $L_{k=1}(S)$ (warranty lifetime) which is expected to capture a certain proportion, say, $\gamma = 0.975$ or more of the population of selected items ($m = lq = 15$), with a given confidence level $1 - \alpha = 0.95$. This lower statistical simultaneous tolerance limit is such that one can say with a certain confidence $1 - \alpha$ that at least $100\gamma\%$ of the military carriers in each sample selected by the buyer for testing will operate longer than $L_1(S)$.

*Goodness-of-fit testing.* Let us assume that $(X_1, ..., X_n)$ is a random sample from the two-parameter Weibull distribution (11.11), and let $\hat{\beta}$, $\hat{\delta}$, be the maximum likelihood estimates of $\beta$, $\delta$, respectively, computed on the basis of $(X_1, ..., X_n)$:

$$\hat{\delta} = \left[\left(\sum_{i=1}^n x_i^{\hat{\delta}}\ln x_i\right)\left(\sum_{i=1}^n x_i^{\hat{\delta}}\right)^{-1} - \frac{1}{n}\sum_{i=1}^n \ln x_i\right]^{-1} = 4.977351, \tag{11.79}$$

and

$$\hat{\beta} = \left( \sum_{i=1}^{n} x_i^{\hat{\delta}} \Big/ n \right)^{1/\hat{\delta}} = 1321.323. \tag{11.80}$$

We assess the statistical significance of departures from the Weibull model (11.11) by performing the Anderson–Darling goodness-of-fit test. There are many goodness-of-fit tests, for example: Kolmogorov–Smirnov and Anderson–Darling tests. The Anderson–Darling test is more sensitive to deviations in the tails of a distribution than the older Kolmogorov–Smirnov test. The Anderson–Darling test statistic value is determined by (e.g., D'Agostino and Stephens 1986)

$$A^2 = -\left[ \sum_{i=1}^{n} (2i-1)\left( \ln F_\theta(x_i) + \ln\left(1 - F_\theta(x_{n+1-i})\right)\right) \right] \Big/ n - n, \tag{11.81}$$

where $F_\theta(x)$ is the cumulative distribution function of $X$,

$$\theta = (\beta = \hat{\beta} = 1321.323, \ \delta = \hat{\delta} = 4.977351), \tag{11.82}$$

and $n = 5$ is the number of observations. The result from (11.81) needs to be modified for small sampling values. For the Weibull distribution, the modification of $A^2$ is

$$A_{\text{mod}}^2 = A^2 \left( 1 + \frac{0.2}{\sqrt{n}} \right). \tag{11.83}$$

The $A_{\text{mod}}^2$ value must then be compared with critical values, $A_\alpha^2$, which depend on the significance level $\alpha$ and the distribution type. As an example, for the Weibull distribution the determined $A_{\text{mod}}^2$ value has to be less than the critical value $A_\alpha^2$ for acceptance of goodness-of-fit test. For this example, $\alpha = 0.05$, $A_{\alpha=0.05}^2 = 0.757$,

$$A^2 = -\left[ \sum_{i=1}^{5} (2i-1)\left( \ln F_\theta(x_i) + \ln\left(1 - F_\theta(x_{n+1-i})\right)\right) \right] \Big/ 5 - 5 = 0.202335, \tag{11.84}$$

$$A_{\text{mod}}^2 = A^2 \left( 1 + \frac{0.2}{\sqrt{n}} \right) = 0.220432 < A_{\alpha=0.05}^2 = 0.757. \tag{11.85}$$

Since the test statistic is less than the critical value, we do not reject the null hypothesis at the significance level $\alpha = 0.05$. Thus, there is not evidence to rule out the Weibull lifetime model (11.11).

Now the lower one-sided simultaneous $\gamma$-content tolerance limit at the confidence level $1 - \alpha$, $L_1 \equiv L_1 (S)$ (on the order statistic $Y_1$ from a set of $m = 15$ future ordered observations $Y_1 \le \ldots \le Y_m$) is given by (11.37)

$$L_k = \eta_{L_k}^{1/\hat{\delta}} \hat{\beta} = 328.7676 \cong 329, \tag{11.86}$$

where it follows from (11.38) that

$$\eta_{L_k} = \arg \left( \int_0^\infty \int_0^{\frac{\ln(1-q_{1-\gamma})^{-1} \sum_{i=1}^n z_i^{v_2}}{[\eta_{L_k}]^{v_2}}} \frac{1}{\Gamma(n)} w^{n-1} \exp(-w) \frac{1}{\vartheta(\mathbf{z}^{(n)})} v_2^{n-2} \prod_{i=1}^n z_i^{v_2} \left( \sum_{i=1}^n z_i^{v_2} \right)^{-n} dw \, dv_2 = 1 - \alpha \right),$$

$$= 0.0009842,$$  (11.87)

$q_{1-\gamma}$ is a quantile of the beta-distribution (with $k = 1$) satisfying (11.32).

*Statistical inference.* Thus, the manufacturer has 95% assurance that no failures will occur in the proportion $\gamma = 0.975$ or more of the population of selected logic circuits ($m = 15$) before $L_k = 329$ hours.

## 11.9  Conclusion

The new technique (based on probability transformation and pivotal quantity averaging) given and illustrated in this chapter is offered as a conceptually simple, efficient, and useful method for constructing exact statistical tolerance limits on future outcomes under parametric uncertainty of underlying models. It is based also on the idea of invariant embedding of a sufficient statistic in the underlying model in order to construct pivotal quantities and to eliminate the unknown parameters from the problem via pivotal quantity averaging. Using the proposed technique, the exact statistical tolerance limits on future order statistics (under parametric uncertainty of underlying models)) associated with sampling from corresponding distributions can be found easily and quickly making tables, simulation, and special computer programs unnecessary.

We consider the one-sided statistical tolerance limits defined as follows: (1) one-sided statistical tolerance limit that covers at least $100\gamma\%$ of the measurements with expected $100(1 - \alpha)\%$ confidence and (2) one-sided statistical tolerance limit determined so that the expected proportion of the measurements covered by this limit is $(1 - \alpha)$. For example, such tolerance limits are required when planning life tests, engineers may need to predict the number of failures that will occur by the end of the test or to predict the amount of time that it will take for a specified number of units to fail. The methodology described in this chapter is illustrated for the two-parameter Weibull distribution. Applications to other log-location-scale distributions could follow directly. Finally, we give two numerical examples.

It should be noted that the results obtained in this chapter (Sections 11.3–11.8) via the proposed technique are new.

## References

D'Agostino, R.B. and M.A. Stephens, *Goodness-of-Fit Techniques*. New York: Marcel Dekker, 1986.

Dunsmore, J.R., "Some approximations for tolerance factors for the two parameter exponential distribution," *Technometrics*, vol. 20, pp. 317–318, 1978.

Engelhardt, M. and L.J. Bain, "Tolerance limits and confidence limits on reliability for the two-parameter exponential distribution," *Technometrics*, vol. 20, pp. 37–39, 1978.

Guenther, W.C., "Tolerance intervals for univariate distributions," *Naval Research Logistics Quarterly*, vol. 19, pp. 309–333, 1972.

Guenther, W.C., S.A. Patil and V.R.R. Uppuluri, "One-sided $\beta$-content tolerance factors for the two parameter exponential distribution," *Technometrics*, vol. 18, pp. 333–340, 1976.

Guttman, I., "On the power of optimum tolerance regions when sampling from normal distributions," *Annals of Mathematical Statistics*, vol. XXVIII, pp. 773–778, 1957.

Hahn, G.J. and W.Q. Meeker, *Statistical Intervals: A Guide for Practitioners*. New York: John Wiley & Sons, 1991.

Lawless, J.F., "On estimation of the safe life when the underlying life distribution is Weibull," *Technometrics*, vol. 15, 857–865, 1973.

Mann, N.R. and S.C. Saunders, "On evaluation of warranty assurance when life has a Weibull distribution," *Biometrika*, vol. 56, pp. 615–625, 1969.

Mee, R.W. and D. Kushary, "Prediction limits for the Weibull distribution utilizing simulation," *Computational Statistics & Data Analysis*, vol. 17, 327–336, 1994.

Mendenhall, V. "A bibliography on life testing and related topics," *Biometrika*, vol. XLV, pp. 521–543, 1958.

Nechval, N.A. and E.K. Vasermanis, *Improved Decisions in Statistics*. Riga: Izglitibas soli, 2004.

Nechval, N.A., G. Berzins, M. Purgailis and K.N. Nechval, "Improved estimation of state of stochastic systems via invariant embedding technique," *WSEAS Transactions on Mathematics*, vol. 7, pp. 141–159, 2008.

Nechval, N.A., M. Purgailis, G. Berzins, K. Cikste, J. Krasts and K.N. Nechval, "Invariant embedding technique and its applications for improvement or optimization of statistical decisions," in Al-Begain, K., Fiems, D., Knottenbelt, W. (Eds.), *Analytical and Stochastic Modeling Techniques and Applications*, (LNCS) (vol. 6148, pp. 306–320). Berlin: Springer-Verlag, 2010.

Nechval, N.A., K.N. Nechval and M. Purgailis, "Statistical inferences for future outcomes with applications to maintenance and reliability," in *Lecture Notes in Engineering and Computer Science: Proceedings of the World Congress on Engineering*, WCE 2011, 6–8 July, 2011 (pp. 865–871). London, UK, 2011.

Nechval, N.A. and K.N. Nechval, "Tolerance limits on order statistics in future samples coming from the two-parameter exponential distribution,"*American Journal of Theoretical and Applied Statistics*, vol. 5, pp. 1–6, 2016a.

Nechval, N.A., K.N. Nechval, S.P. Prisyazhnyuk and V.F. Strelchonok, "Tolerance limits on order statistics in future samples coming from the Pareto distribution," *Automatic Control and Computer Sciences*, vol. 50, pp. 423–431, 2016b.

Nechval, N.A., K.N. Nechval and V.F. Strelchonok, "A new approach to constructing tolerance limits on order statistics in future samples coming from a normal distribution," *Advances in Image and Video Processing (AIVP)*, vol. 4, pp. 47–61, 2016c.

Patel, J.K., "Tolerance limits: A review," *Communications in Statistics: Theory and Methodology*, vol. 15, pp. 2719–2762, 1986.

Wald, A. and J. Wolfowitz, "Tolerance limits for a normal distribution," *Annals of Mathematical Statistics*, vol. XVII, pp. 208–215, 1946.

Wallis, W.A., "Tolerance intervals for linear regression," in Neyman, J. (Ed.), *Second Berkeley Symposium on Mathematical Statistics and Probability* (pp. 43–51) Berkeley: University of California Press, 1951.

# chapter twelve

# Design of neural network–based PID controller for biped robot while ascending and descending the staircase

*Ravi Kumar Mandava and Pandu R. Vundavilli*
IIT Bhubaneswar

## Contents

## 12.1  Introduction

Compared to industrial manipulators, legged robots are having much more interaction with the ground and it is a tough job to control the robot in an effective manner. Further, the mechanism, structure and balancing of the two-legged robot is complex in nature when compared with other legged robots. Over the past few decades, people are working on the stability and controlling aspects of the biped robot on various terrains. Generating the stable gait for the biped robot while walking on various terrains is a difficult task and was taken up by many researchers. Presently, researchers are utilizing zero moment point (ZMP) [1] based control algorithms to control the gait of the two-legged robot. Some other researchers had tried to optimize the parameters of the ZMP-based controller after utilizing nontraditional algorithms [2,3]. It is important to note that the conventional PID controllers were widely deployed in both industrial as well as nonindustrial applications due to its ease, simple design, and cost effectiveness. Based on the demand for the usage of PID controller in various applications, the real-time tuning/adaption of the controller gains (i.e., $K_p$, $K_d$, and $K_i$) in an online manner is a challenging task. Moreover, the tuning methods such as Zeigler–Nicholas [4] and Cohen [5] methods were already proved that it is not possible to use them in highly nonlinear, uncertain, and coupled robotic applications.

The advancements in the computational techniques motivated several researchers to use evolutionary and nature-inspired optimization algorithms, such as genetic algorithms (GAs) [6], particle swarm optimization (PSO) [7,8], ant colony optimization (ACO) [9], cuckoo search algorithm (CSA) [10] and bacterial forging optimization (BFO) [11], for tuning the gains of the conventional PID controller. The evolutionary and nature-inspired optimization algorithms provide a fixed set of optimal tuning parameters (i.e., gains). This fixed set of gains most of the time could not provide optimum performance when used in uncertain and nonlinear processes. Therefore, an efficient and effective online tuning mechanism is required for controlling the joints of the robot in a systematic manner. It is also important to mention that in order to improve the performance of the conventional PID controller, few researchers had tried with fuzzy logic technique for obtaining the optimal gains of the PID controller in an online manner. The PID controller was successfully implemented in various industrial applications, such as induction machine, ship plant [12–14], micro grid [15], hybrid electrical vehicle system [16] and AVR system [17], to control them in a more effective manner.

Based on the above literature, it is evident that the performance of the conventional PID controller can be improved by optimizing the critical parameters of the control algorithm using evolutionary and nature-inspired algorithms. The evolutionary algorithms, such as GA and PSO, were widely used in optimization of the gains of the PID controller for various industrial applications. Along with the nature-inspired optimization algorithms, recently Mehrabain and Lucas [18] established an IWO-based optimization technique for improving the convergence and performance of the system [19,20]. In IWO, the weeds were reproduced without matting and the fitter plants that produce a greater number of seeds might have led to the improvement in the convergence and the performance of the algorithm. Further, the application of neural network in the area of dynamical control system was also tested by a few researchers. Cembrano et al. [21] discussed the concept of neural network in learning inverse kinematics, inverse dynamics and visual positioning of the robot in real-time control applications. Ghorbani et al. [22] designed a general regression neural network (GRNN) feedback controller for stabilizing the biped robot at the upright position and satisfied all the constraints in between the foot link and the ground. The obtained controller action resulted in reducing energy consumption and minimizing the torque required at the ankle joint. A genetic algorithm was used to train the network, and the stability of the controller was analyzed by using the Lyapunov exponent. Later on, Jiang [23] developed a dynamic trajectory tracking control algorithm for an industrial manipulator after using a PD controller and a neural network (NN)–based controller. It was observed that after an increased learning time, the torque consumed by the NN-based controller was seen to more when compared with the PD controller. Saeed [24] also proposed a NN-based PID controller to improve the performance and robustness of the robotic manipulator. They added an uncertainty rejecting property to the NN-based PID controller.

The motivation of the present work is to preserve the favorable characteristics of the conventional PID controller for moving the joints of the biped robot in a systematic manner. The contributions of the present chapter are as follows:

- In this research work, the authors initiated a torque-based PID controller to control each joint of the biped robot in a systematic manner and to reduce the error between the two consecutive intervals of various joints.
- Optimal tuning of the PID controller occurred using the MCIWO algorithm, instead of the time-consuming manual tuning process. A NN tool has also been developed to tune the gains of the PID controller.

- Further, the authors implemented a cosine and chaotic variables to the standard invasive weed optimization (i.e., modified chaotic invasive weed optimization) algorithm to evolve the structure of NN automatically and to generate the gains in an adaptive manner. With the best of the authors' knowledge, none of the researchers had used the MCIWO algorithm to evolve the structure of the neural network in control applications.

## 12.2   Kinematics and dynamics of the biped robot

In the present research, an 18-Degrees of Freedom (DOF) biped robot (Figure 12.1) has been chosen to carry out the simulations and experimental validation. Initially, the coordinate frames are assigned at various joints and the D–H parameters [25,26] are extracted for each joint of the biped robot. Alongside, a systematic study has been conducted to study the influence of swing foot trajectory of the biped robot after using quadratic, cubic, and fifth-ordered polynomial trajectories. According to the above study, the swing foot trajectory is having its influence on the balance of the robot while ascending and descending the staircase. In the present study, cubic polynomial trajectories are assumed to be used by the swing foot and hip joints of the robot to generate a smooth gait on the terrain. Further, the concept of inverse kinematics has been used to derive the gaits for the lower and upper limbs of the biped robot. The included angles between the lower and upper limbs of the swing leg in the sagittal plane (i.e., $\theta_3$ and $\theta_4$), in the frontal plane (i.e., $\theta_2$ and $\theta_5$), the stand leg in the sagittal plane (i.e., $\theta_9$ and $\theta_{10}$) and the frontal plane (i.e., $\theta_8$ and $\theta_{11}$) are given as follows:

$$\theta_4 = \sin^{-1}\left( \frac{H_1 l_3 \sin\psi + L_1 \left(l_4 + l_3 \cos\psi\right)}{\left(l_4 + l_3 \cos\psi\right)^2 + \left(l_3 \sin\psi\right)^2} \right) \tag{12.1}$$

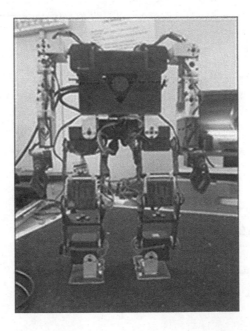

*Figure 12.1* Structure of the biped robot.

where $H_1 = l_4 \cos(\theta_4) + l_3 \cos(\theta_3)$, $L_1 = l_4 \sin(\theta_4) + l_3 \sin(\theta_3)$, $\psi = \theta_4 - \theta_3 = \text{arcos}((H_1^2 + L_1^2 - l_4^2 - l_3^2)/2\,l_4 l_3)$. The angle $\theta_3$ can be calculated after using the equation $\theta_3 = \theta_4 - \psi$.

$$\theta_{10} = \sin^{-1}\left( \frac{H_2 l_9 \sin\psi + L_2 \left(l_{10} + l_9 \cos\psi\right)}{\left(l_{10} + l_9 \cos\psi\right)^2 + \left(l_9 \sin\psi\right)^2} \right) \tag{12.2}$$

where $H_2 = l_9 \cos(\theta_9) + l_{10} \cos(\theta_{10})$, $L_2 = l_9 \sin(\theta_9) + l_{10} \sin(\theta_{10})$, $\psi = \theta_{10} - \theta_9$. The angle $\theta_9$ can be calculated by using $\theta_9 = \theta_{10} - \psi$.

Further, the following mathematical expressions are used to calculate the joint angles in the frontal plane:

$$\theta_2 = \theta_8 = \tan^{-1}\left(f_w/H_1\right) \tag{12.3}$$

$$\theta_5 = \theta_{11} = \tan^{-1}\left((0.5 f_w)/H_2\right) \tag{12.4}$$

### 12.2.1   *Dynamic balance margin while ascending the staircase*

The gait generated by the biped robot should be dynamically balanced while ascending the staircase (Figure 12.2). For attaining dynamically balanced gait for the biped robot, the zero moment point should lie inside the foot support polygon. The position of ZMP in X- and Y-directions are shown in Figure 12.3, and it is calculated by using the subsequent equations:

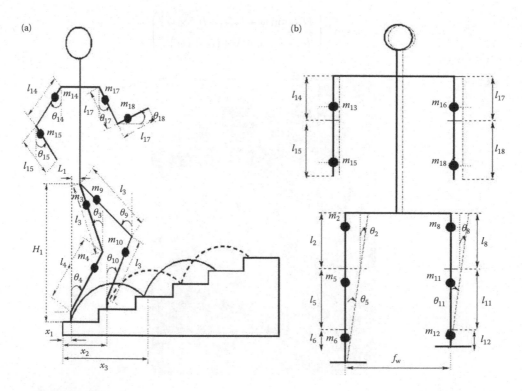

*Figure 12.2* Biped robot walking on ascending the staircase. (a) Sagittal view and (b) frontal view.

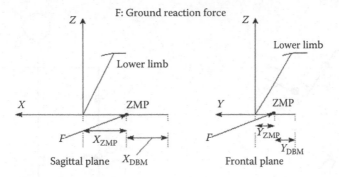

*Figure 12.3* ZMP and DBM in both sagittal and frontal planes.

$$x_{ZMP} = \frac{\sum_{i=1}^{n}\left(I_i\dot{\omega}_i - m_i\ddot{x}_i z_i + m_i x_i(g - \ddot{z}_i)\right)}{\sum_{i=1}^{n}\left(m_i(\ddot{z}_i - g)\right)} \qquad (12.5)$$

$$y_{ZMP} = \frac{\sum_{i=1}^{n}\left(I_i\dot{\omega}_i - m_i\ddot{y}_i z_i + m_i y_i(g - \ddot{z}_i)\right)}{\sum_{i=1}^{n}\left(m_i(\ddot{z}_i - g)\right)} \qquad (12.6)$$

where $\dot{\omega}_i$, $I_i$, and $m_i$ represent the angular acceleration (rad/s²), mass moment of inertia (kg m²) and mass (kg) of the link $i$, $g$ is the acceleration due to gravity (m/s²), $\ddot{z}_i$ and $\ddot{x}_i$ denote the acceleration (m/s²) of the link in $z$- and $x$-directions, respectively, and $(x_i, y_i, z_i)$ indicates the coordinates of the $i^{th}$ lumped mass.

After determining the position of ZMP, the dynamic balance margin (DBM) of the biped robot in X- and Y-directions (Figure 12.3) are calculated by using Equations (12.7) and (12.8), respectively. It is important to note that for generating the dynamically balanced gait, the ZMP should lie inside the foot support polygon. If the ZMP moves outside the polygon, we need to move the links of the biped robot in such a way that they push the ZMP inside the foot support polygon. Further, the dynamic balance margin has been defined as the distance between the point where the ZMP is acting and to the end of the foot support polygon:

$$x_{DBM} = \left(\frac{f_s}{2} - |x_{ZMP}|\right) \qquad (12.7)$$

$$y_{DBM} = \left(\frac{f_w}{2} - |y_{ZMP}|\right) \qquad (12.8)$$

## 12.2.2 Dynamic balance margin while descending the staircase

The schematic diagram showing the structure of the biped robot while descending the staircase in the sagittal and the frontal views, are given in Figure 12.4a,b, respectively. The mathematical model used for the gait generation of the biped robot while descending the staircase is similar to that of ascending the staircase. Further, the DBM of the biped robot during descending the staircase can also be calculated by implementing a

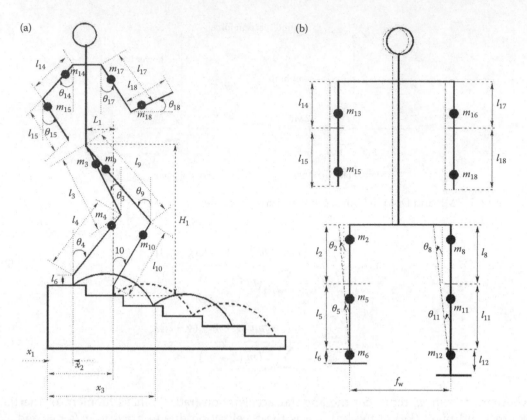

**Figure 12.4** Biped robot walking on descending the staircase. (a) Sagittal view and (b) frontal view.

similar procedure as that of the ascending case. However, there is a small difference in the descending case with the acceleration due to gravity "*g*" gacting in the direction opposite to that of the movement of the robot.

### 12.2.3 *Design of torque-based PID controllers for the biped robot*

Once the gait is generated, then suitable torque-based PID controllers are developed for each joint of the biped robot while ascending and descending the staircase. The dynamics of the biped robot which is used in the design of the controller has been derived by using the Lagrange–Euler formulation (refer to Equation 12.9). The theoretical torques required at different joints of the biped robot are calculated by using the following equation:

$$\tau_{i,the} = \sum_{j=1}^{n} M_{ij}(q)\ddot{q}_j + \sum_{j=1}^{n} \sum_{k=1}^{n} C_{ijk}\dot{q}_j\dot{q}_k + G_i \quad i,j,k = 1,2,\ldots,n \tag{12.9}$$

where $\tau_{i,the}$, $q$, $\dot{q}_j$, and $\bar{q}_j$ represent the theoretical torque required, displacement of the joint in (m), velocity of the joint in (m/s), and acceleration of the joint in (m/s²). Further, the expanded form of inertia ($M_{i,j}$), centrifugal/coriolis ($h_{i,j,k}$), and gravity ($G_i$) terms are as follows:

$$M_{ij} = \sum_{p=\max(i,j)}^{n} Tr\left[d_{pj}I_p d_{pi}^T\right] \quad i,j = 1,2,\ldots,n \tag{12.10}$$

$$h_{ijk} = \sum_{p=\max(i,j,k)}^{n} Tr\left[\frac{\partial(d_{pk})}{\partial q_p} I_p d_{pi}^T\right] \quad i,j = 1,2,\ldots,n \tag{12.11}$$

$$G_i = -\sum_{p=i}^{n} m_p g d_{pi}\, {}_e^p \bar{r}_p \quad i,j = 1,2,\ldots,n \tag{12.12}$$

where ${}_e^p \bar{r}_p$, $I_p$, and $g$ denote the mass center (m), mass moment of inertia (kg m/s$^2$) tensor of $p^{th}$ link and acceleration due to gravity in (m/s$^2$), respectively. It is important to note that the acceleration of the joint plays a significant role in controlling each link of the biped robot. By rearranging Equation (12.9), the expression related to the acceleration of link $i$ is given as follows:

$$\ddot{q}_j = \sum_{j=1}^{n} M_{ij}(q)^{-1}\left[-\sum_{j=1}^{n}\sum_{k=1}^{n} C_{ijk}\dot{q}_j\dot{q}_k - G_i\right] + \left(\sum_{j=1}^{n} M_{ij}(q)^{-1} * \tau_{i,the}\right) \quad i,j,k = 1,2,\ldots,n$$

$$\tag{12.13}$$

now considering the term

$$\sum_{j=1}^{n} M_{ij}(q)^{-1} * \tau_{i,the} = \hat{\tau} \tag{12.14}$$

and substituting Equation (12.14) into Equation (12.13), it can be written as

$$\ddot{q}_j = \sum_{j=1}^{n} M_{ij}(q)^{-1}\left[-\sum_{j=1}^{n}\sum_{k=1}^{n} C_{ijk}\dot{q}_j\dot{q}_k - G_i\right] + \hat{\tau} \quad i,j,k = 1,2,\ldots,n \tag{12.15}$$

In real-time applications, the theoretical torque required and acceleration of each link are not suitable to estimate the value of actual torque and acceleration. Based on the above reason the actual torque required at different joints of the biped robot is calculated by using the following expression:

$$\tau_{act} = K_p e + K_d \dot{e} + K_i \int e\, dt \tag{12.16}$$

where $\tau_{act}$ represents the actual torques required at different joints of the biped robot; the terms $K_p$, $K_d$, and $K_i$ denote proportional, derivative, and integral gains of the PID controller, respectively; and $e$ indicates the value of error (i.e., difference between the desired and actual value, $e(\theta_i) = \theta_{if} - \theta_{is}$) related to each joint. After including the terms $e$ and $\dot{e}$ the above equation can be written as follows:

$$\tau_{i,act} = K_{pi}\left(\theta_{if} - \theta_{is}\right) - K_{di}\dot{\theta}_{is} + K_{ii} \int e(\theta_{is})\, dt \quad i = 1,2,\ldots,n \tag{12.17}$$

where $\theta_{if}$ and $\theta_{is}$ represent the final and initial angular positions at different joints of the biped robot, respectively. Therefore, the final control equation that represents the acceleration of the link is

$$\ddot{q}_j = \sum_{j=1}^{n} M_{ij}(q)^{-1}\left[-\sum_{j=1}^{n}\sum_{k=1}^{n} C_{ijk}\dot{q}_j\dot{q}_k - G_i\right] + K_{pi}\left(\theta_{if} - \theta_{is}\right) - K_{di}\dot{\theta}_{is} + K_{ii} \int e(\theta_{is})\, dt$$

$$\tag{12.18}$$

## 12.3   MCIWO-based PID controller

Mehrabain and Lucas developed a novel stochastic IWO algorithm [17] inspired from the colonizing behavior of weeds, which is a common phenomenon in agriculture. So far, researchers have used this algorithm in many engineering problems because it is not only robust in nature but also simple and easy to implement. In a standard IWO algorithm, weeds that indicate the feasible solution to the problem are considered as the set of weeds. Initially, a finite number of weeds are dispersed randomly over the search space. Depending on the fitness of the problem in a colony, each weed produces new seeds. The new seeds are distributed randomly over the search space by using a normally distributed random number with mean equal to zero. The above process will continue until the maximum number of weeds is reached. Note that only the weeds with better fitness survive and produce new seeds, and others are eliminated. The above process continues until the maximum number of iterations is reached or hopefully, the weed with best fitness is closest to the optimal solution. The flowchart shown in Figure 12.5 explains the operation of the MCIWO algorithm. The step-by-step procedure used for the implementation of the algorithm is explained as follows:

1. *Initialize a population*
   The population of the initial solution is being dispersed randomly over the $N$-dimensional search space. It is important to note that each position of the weed signifies one possible solution to the problem.
2. *Reproduction*
   After growing, the individual weeds are allowed to reproduce new seeds depending on their own, the lowest and highest fitness values in the colony. Based on the fitness of the weed, the number of seeds produced by the weed varies from lower to higher in a linear manner. The weed with better fitness will produce more seeds, and worst fitness will produce less seeds. The goodness of this algorithm is that, all the worst and best weeds in the solution space will contribute in the reproduction process, and the weed that is giving the worst fitness value also shares some useful information in the evolution process. The number of seeds ($S$) produced by each weed will be given by the following equation:

$$S = Floor\left[ S_{min} + \frac{f - f_{min}}{f_{max} - f_{min}} \times S_{max} \right] \tag{12.19}$$

where $f_{min}$ and $f_{max}$ denote the minimum and maximum fitness value in the colony, respectively, and $S_{min}$ and $S_{max}$ represent minimum and maximum number of seeds produced by each plant, respectively.

3. *Spatial dispersal*
   The randomly generated seeds are distributed around the parent weed with a certain value of variance and mean equal to zero. Moreover, the standard deviation ($\sigma$) of the random function will be reduced nonlinearly from a previously mentioned initial value ($\sigma_{initial}$) to a final value ($\sigma_{final}$) in each generation. The equation governing this process is as follows:

$$\sigma_{Gen} = \frac{(Gen_{max} - Gen)^n}{(Gen_{max})^n} (\sigma_{initial} - \sigma_{final}) + \sigma_{final} \tag{12.20}$$

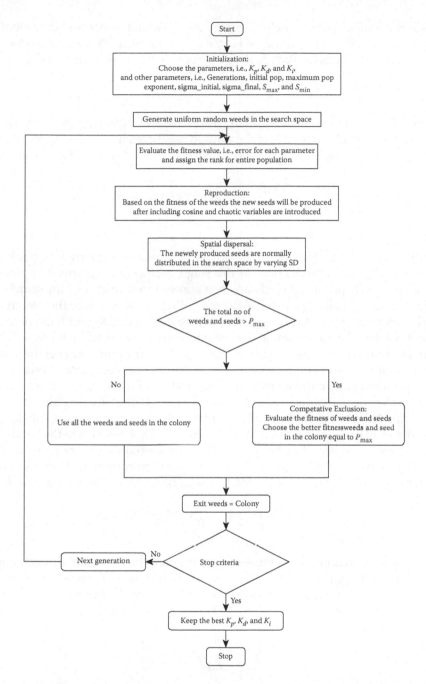

*Figure 12.5* Flow chart showing the step-by-step procedure of the MCIWO algorithm.

where $\sigma_{initial}$ and $\sigma_{final}$ indicate the initial and final standard deviation, respectively, $Gen_{max}$ and "$n$" represent the maximum number of generations and modulation index, respectively.

In order to improve the performance of the algorithm in the present research, the authors introduced two new terms, namely, chaotic [27] and cosine [28,29] variable in spatial dispersal section. The first variable, i.e., chaotic random variable, is used to distribute the seeds equally. This will help in enhancing the search space and in minimizing the chances of the solution being trapped in the local optimum. The chaotic random number considered in the present study is attained from the Chebyshev map and is

$$X_{k+1} = \cos\left(k \cos^{-1}(X_k)\right) \tag{12.21}$$

Further, the cosine variable assists in enhancing and exploring the search space in a better manner, and it will utilize the unused resources in a search space. After introducing the cosine variable, Equation (12.20) can be modified and is as follows:

$$\sigma_{Gen} = \frac{(Gen_{\max} - Gen)^n}{(Gen_{\max})^n} \times |\cos(Gen)| \times (\sigma_{\text{initial}} - \sigma_{\text{final}}) + \sigma_{\text{final}} \tag{12.22}$$

4. *Competitive exclusion*

After passing several iterations, the number of weeds in a colony will reach its maximum ($P_{\max}$) by fast reproduction. At this stage, each weed is allowed to produce new seeds. The newly produced seeds are then allowed to spread over the search space by using a chaotic random number. After spreading the seeds over the search area, the weeds occupy their position and a rank will be assigned along with the parent weeds. Once it reaches the maximum allowable population, the weeds with lower fitness are eliminated, and the weeds with better fitness will join the population in the next generation. This process will continue until the maximum numbers of iterations are reached. In the present chapter, the gains (i.e., $K_p$, $K_d$, and $K_i$) of the various PID controllers that are used to control the individual joints of the biped robot are considered as weeds of the MCIWO algorithm. In this study, the PID controllers are designed only to control 12 joints out of 18 joints of the biped robot. The rest six DOF belongs to the hands and they are not considered here. Further, each PID controller requires three gains for its operation. Therefore, in total, 36 gain values are required to control all 12 joints of the biped robot. One such population of the MCIWO algorithm looks like the one that follows:

$$\underbrace{866.54}_{K_{p1}}, \underbrace{400.25}_{K_{d1}}, \underbrace{958.32}_{K_{i1}}, \dots, \underbrace{758.35}_{K_{p36}}, \underbrace{550.96}_{K_{d36}}, \underbrace{688.78}_{K_{i36}}$$

It is important to note that a fitness value needs to be assigned for each population of the MCIWO algorithm. Here, the average angular error between the start and end points of the interval for various joints is considered as the fitness of each population. The fitness function ($f$) of the MCIWO PID controller is as follows:

$$f = \frac{1}{b} \sum_{j=1}^{b} \sqrt{\frac{1}{2} \sum_{k=1}^{p} \left(\alpha_{ijkf} - \alpha_{ijks}\right)^2} \tag{12.23}$$

where $b$ denotes the number of intervals considered in one step, and $p$ indicates the number of joints, respectively, for which the controllers are designed.

## 12.4 MCIWO-NN–based PID controller

In the present research work, the authors used a feed-forward hierarchical type of NN (Figure 12.6) that consists of input, hidden, and output layers that are used to tune the gains

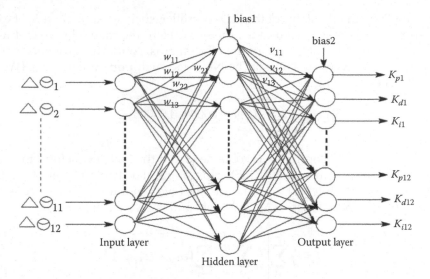

***Figure 12.6*** Structure of the neural network.

of the PID controller. The architecture of the NN consists of 12 input neurons that indicate the angular error of each joint at different instances of time. The output layer consists of 36 neurons that represent the proportional, integral, and derivative gains of the PID controllers that are used to control each joint of the biped robot. The number of neurons in the hidden layer has been decided with the help of a systematic study. It is important to note that the performance of a NN depends on the connecting weights between the neurons of input-hidden and hidden-output layers. During the parametric study, the connecting weights, the bias value of the network and the coefficients of transfer function values of individual layers are optimized with the help of the MCIWO algorithm. A batch mode of training has been implemented to train the NN. Once the structure of the NN is optimized, it can be used to predict the gains of the PID controller in a more adaptive manner. The operating principle of a MCIWO-NN is shown in Figure 12.7.

Let us consider that the change in the angular position of joints, such as $\Delta\theta_1$, $\Delta\theta_2$, $\Delta\theta_3$, $\Delta\theta_4$, $\Delta\theta_5$, $\Delta\theta_6$, $\Delta\theta_7$, $\Delta\theta_8$, $\Delta\theta_9$, $\Delta\theta_{10}$, $\Delta\theta_{11}$, $\Delta\theta_{12}$, of the biped robot at equal intervals of time as

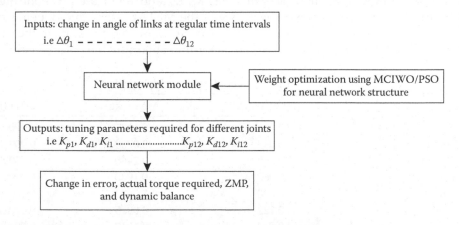

***Figure 12.7*** Flow chart showing the structure of the MCIWO-NN algorithm.

the input to the NN and the gains of the PID controller, such as, $K_p$, $K_d$, and $K_i$ of each joint of the biped robot, are taken as outputs of the NN. Here the main objective of using the MCIWO is to optimize the structure of the NN—that is, connecting weights, bias values and coefficients of transfer function of the NN. One such population of MCIWO in this case looks like the following:

$$\underbrace{0.25}_{w_{11}},\underbrace{0.4}_{w_{12}},\underbrace{0.72}_{w_{13}},\ldots,\underbrace{0.34}_{v_{1434}},\underbrace{0.00009}_{b_1},\underbrace{0.00005}_{b_2},\underbrace{2}_{c_{11}},\underbrace{5}_{c_{12}},\underbrace{7}_{c_{13}}$$

Once the gains of the PID controller are obtained from the NN, it is helpful to control the joint motors. The root mean square (RMS) error of angular displacement of all the joints between the end of each interval ($\alpha_{ijkf}$) and the start of each interval ($\alpha_{ijks}$) is considered as the fitness ($f$) of each population of MCIWO-NN and is given as follows:

$$f = \frac{1}{d}\sum_{i=1}^{d}\left[\frac{1}{b}\sum_{j=1}^{b}\sqrt{\frac{1}{2}\sum_{k=1}^{p}\left(\alpha_{ijkf}-\alpha_{ijks}\right)^2}\right] \tag{12.24}$$

where $d$ represents the number of training scenarios and other terms have their usual meaning.

## 12.5   Results and discussion

In the present research, two types of controllers, namely, optimal PID controller (i.e., MCIWO-tuned PID controller) and adaptive PID controller (i.e., MCIWO-NN-tuned PID controller) are developed to control the motion of a biped robot while ascending and descending the staircase. Once the controllers are developed, initially their performances are tested in computer simulations. The parameters (i.e., length, mass, and inertia) related to the biped robot that is used in the present study are given in Table 12.1. Further, the results related to the present research work are discussed in the subsequent subsections.

### 12.5.1   Ascending the staircase

The results related to the optimal and adaptive PID controllers are presented here. While designing the optimal controller, the gains (i.e., $K_p$, $K_d$, and $K_i$) of the PID controller are varied within their range. As the performance of the MCIWO algorithm depends on the values of its parameters, a parametric study has been conducted to estimate the optimal parameters of the MCIWO algorithm that are helpful in obtaining the optimal gains of the

*Table 12.1* Parameters related to the biped robot

| Link | Length (m) | Mass (kg) | Inertia (kg m²) |
|---|---|---|---|
| Lower limb of the leg | 0.093 | 0.1190 | 0.00007440 |
| Upper limb of the leg | 0.093 | 0.0700 | 0.00012600 |
| Ankle to foot | 0.033 | 0.2460 | 0.00003300 |
| Upper arm | 0.060 | 0.1930 | 0.00008569 |
| Lower arm | 0.060 | 0.0592 | 0.00012000 |
| Trunk | 0.122 | 0.0975 | 0.00017700 |
| Pelvis | 0.037 | 0.1940 | 0.00671000 |

PID controller. After conducting the study, the optimal values related to the parameters of MCIWO algorithm are seen to be as follows: final value of standard deviation ($\sigma_{final}$) = 0.00001; initial standard deviation ($\sigma_{initial}$) = 3%; maximum number of seeds ($S_{max}$) = 5; minimum number of seeds ($S_{min}$) = 0; initial population size (npop$_i$) = 10; final population size (npop$_f$) = 25; nonlinear modulation index ($n$) = 2; and generations (Gen) = 30. Further, a systematic study has also been conducted in the MCIWO-NN approach to obtain the suitable transfer functions for hidden and output layers, and to find the optimal number of neurons in the hidden layer of the network. From this study, it has been observed that the log-sigmoid and tan-sigmoid transfer functions are seen to exhibit better performance with the hidden and output layers, respectively. Further, the number of neurons in the hidden layer that produce better fitness is found to be equal to 14. The numbers of neurons in the input, hidden, and output layers are kept equal to 12, 14, and 36, respectively. Therefore, the numbers of connecting weights used in the network are seen to be equal to 672 ($12 \times 14 + 14 \times 36$). Finally, the number of variables that represent the NN architecture are found to equal 675 that includes connection weights, one bias value of the network and two coefficients of transfer functions.

For solving the above problem, the weights of the network are varying in the range of 0.0–1.0, the bias values are varied in the range of 0.0–0.0001 and the coefficients are transfer functions that are varied in the range of 1–10. In addition to the above study, here also a parametric study is conducted to determine the optimal parameters of the MCIWO algorithm that evolve the NN architecture. The optimal MCIWO parameters obtained through the study are as follows: final value of standard deviation ($\sigma_{final}$) = 0.00001; initial standard deviation ($\sigma_{initial}$) = 4%; modulation index ($n$) = 4; final population ($n$pop$_f$) = 10; initial population ($n$pop$_i$) = 5; maximum number of seeds ($S_{max}$) = 4; minimum number of seeds ($S_{min}$) = 0; and maximum number of generations (*Gen*) = 50. Once the optimal parameters of MCIWO and MCIWO-NN algorithms are identified, a comparative study has been conducted in computer simulations on the said algorithms in terms of variation of error, estimation of the torque required at each joint, variation of ZMP and DBM in X- and Y-directions of the biped robot.

The variations of error at different joints of the biped robot using MCIWO and MCIWO-NN algorithms are shown in Figure 12.8. It can be observed that the magnitude of error reaches zero at the end of every interval. Further, the enlarged views show that the MCIWO-NN (i.e., adaptive PID controller) converges faster than the MCIWO (i.e., optimal controller) controller. It might be due to the reason that the trained NN could have predicted the values of gains of the controller with respect to the magnitude of error in the joint angles, whereas in the MCIWO approach the gain values are constant and fixed, and are not varying whenever there is a change in the magnitude of input signal to a particular joint. This adaptiveness of the MCIWO-NN PID controller helped in converging the error faster than the MCIWO PID controller.

Figure 12.9 shows the torque required at different joints of the biped robot while ascending the staircase. It can be observed that the adaptive (MCIWO-NN) PID controller requires less torque compared to optimized (MCIWO) PID controller. The reason for the better performance is the same as the one discussed earlier. Moreover, Figure 12.10 shows the variation of ZMP in X- and Y-directions of the biped robot while ascending the staircase. It can be observed that the ZMP of the MCIWO-NN–based PID controller is close to the center of the foot when compared to the MCIWO-based PID controller. It means that the adaptive PID controller has produced more dynamically balanced gaits than the optimal PID controller.

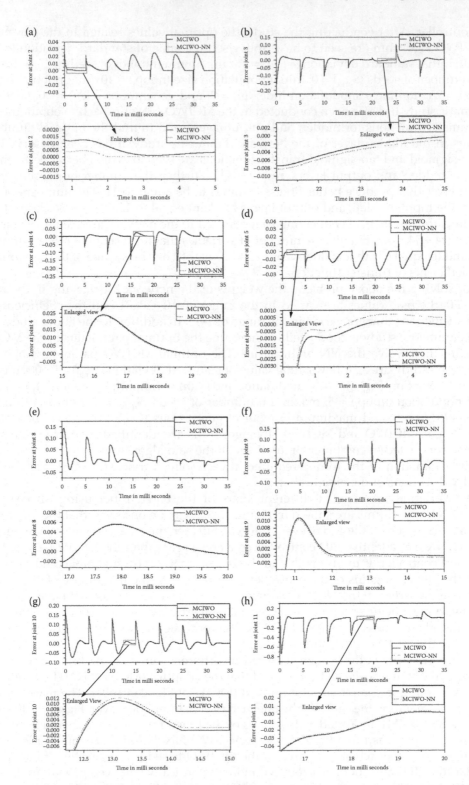

***Figure 12.8*** Variation of error at different joints of the biped robot. (a) Joint 2, (b) Joint 3, (c) Joint 4, (d) Joint 5, (e) Joint 8, (f) Joint 9, (g) Joint 10 and (h) Joint 11.

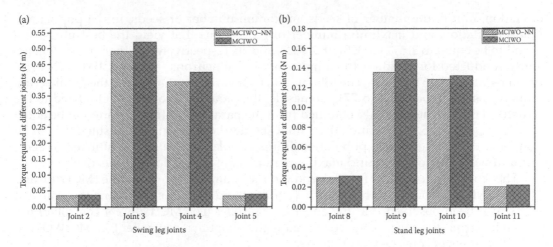

**Figure 12.9** Torque required at different joints of the biped robot while walking on ascending the staircase. (a) Swing leg and (b) stand leg.

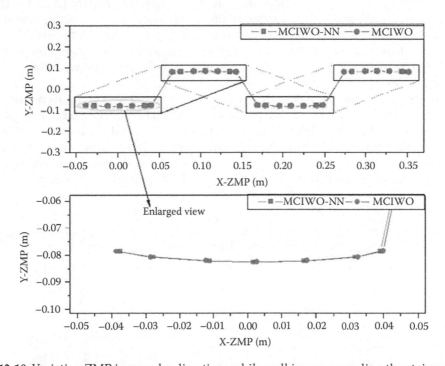

**Figure 12.10** Variation ZMP in $x$- and $y$-directions while walking on ascending the staircase.

## 12.5.2 Descending the staircase

Like designing the controller for ascending the staircase, in this case also a similar study is conducted for determining the optimal gain values and optimal structure of NN in MCIWO and MCIWO-NN algorithms, respectively. The optimal values of the MCIWO parameters, such as modulation index, initial standard deviation, final standard deviation, minimum number of seeds, maximum number of seeds, initial population, final population, and maximum number of generations that yield the best fitness are found to be equal to 2, 5%, 0.00001, 0, 2, 5, 25, and 40, respectively. Further, the systematic study conducted for deciding the number of hidden neurons in the MCIWO-NN algorithm resulted in 16 neurons. The total numbers of variables that define the architecture of NNs is seen to be equal to 771. Alongside, the parameters of MCIWO related to the structural optimization of NN obtained from the parametric study is found to be equal to 0.00001, 4%, 2, 15, 5, 2, 0, and 70 for final standard deviation, initial standard deviation, modulation index, final population, initial population, maximum number of seeds, minimum number of seeds, and maximum number of generations, respectively.

The convergence error for descending the staircase for both the MCIWO and MCIWO-NN PID controllers has shown a similar trend as that of ascending the staircase. It has also been observed that the adaptive controller (i.e., MCIWO-NN controller) has shown better performance when compared with the optimal controller (i.e., MCIWO controller) in terms of convergence of the error.

Similarly, the average torque required at various joints of the biped robot (Figure 12.11) to execute the generated gait is seen to be less with the adaptive PID controller when compared with the optimal PID controller. The variation of X- and Y-ZMP for the biped robot while descending the staircase is shown in Figure 12.12. Further, the comparison of DBMs of the biped robot while ascending and descending the staircase using MCIWO and MCIWO-NN PID controllers are also shown in Figure 12.13. It has been observed that the DBM for ascending the staircase is more when compared with descending the staircase. This observation is exactly falling in line with the experience of human beings.

In this case also, the MCIWO-NN PID controller resulted in more dynamically balanced gaits when compared with the MCIWO-based PID controller. This may be due to the reason that the adaptive PID controller has produced a gait that pushed the ZMP more toward the center of the stance foot. During computer simulations, the MCIWO-NN–based PID controller is found to perform better than the MCIWO-based PID controller. The optimal gait data obtained from the MCIWO-NN PID controller for both the ascending

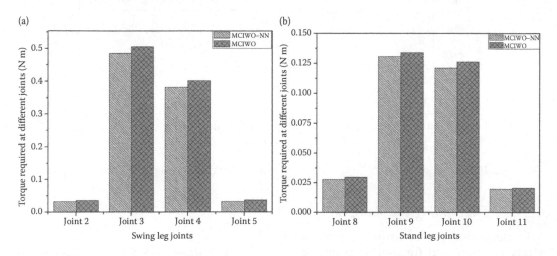

*Figure 12.11* Torque required at different joints of the biped robot while walking on descending the staircase. (a) Swing leg and (b) stand leg.

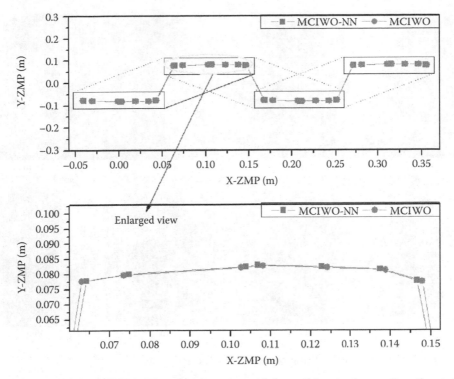

*Figure 12.12* Variation of ZMP in X- and Y-directions while walking on descending the stair-case.

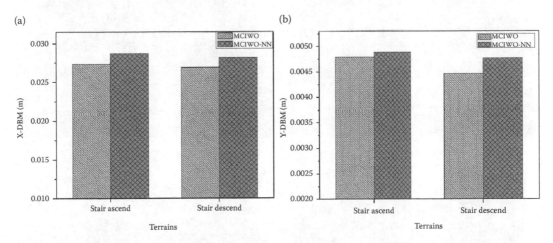

*Figure 12.13* DBM on ascending and descending the staircase. (a) X-DBM and (b) Y-DBM.

and descending case are fed to the real robot. The schematic diagrams showing the real-time walking of the biped robot while ascending and descending the staircase are shown in Figures 12.14 and 12.15, respectively. From this it can be observed that the biped robot has successfully negotiated the staircase with the help of the gait obtained by the adaptive PID controller.

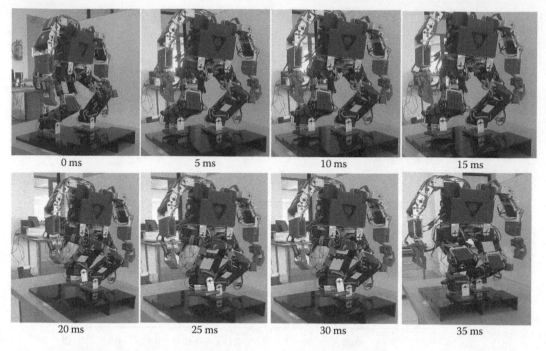

*Figure 12.14*  Real-time robot walking on ascending the staircase.

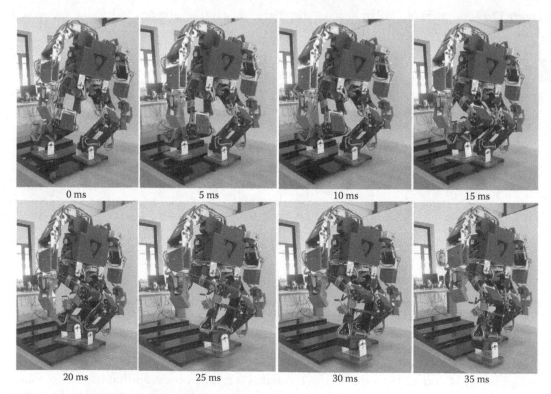

*Figure 12.15*  Real-time robot walking on descending the staircase.

## 12.6   Conclusions

In this research work, an attempt is made to develop a torque-based PID controller for the biped robot while ascending and descending the staircase. A metaheuristic optimization algorithm, that is, a MCIWO algorithm, has been used to find the optimal gains of the PID controller. Further, the same MCIWO algorithm has also been used to optimize the structure of the NN. It has been observed that in both the ascending and descending cases, the MCIWO-NN–based PID controller is found to perform better than the MCIWO-based PID controller. This may be due to the nature of the NN that is able to produce adaptive gains for the PID controller when there is a change in the value of magnitude of error for each joint. Further, the developed controllers are successfully tested in computer simulations. Finally, the gait obtained by the adaptive (i.e., MCIWO-NN controller) PID controller is tested on a real biped robot.

## References

1. Juricic D, Vukobratovic M., *Mathematical Modeling of Biped Walking Systems*, ASME Publications. Vol. 72-WA/BHF13, 1972.
2. Vundavilli PR, Sahu SK, Pratihar DK, "Dynamically balanced ascending and descending gaits of a two-legged robot", *Int J Humanoid Robotics*, Vol. 4, No. 4, pp. 717–751, 2007.
3. Vundavilli PR. Sahu SK, Pratihar DK, "Online dynamically balanced ascending and descending gait generations of a biped robot using soft computing", *Int J Humanoid Robotics* Vol. 4 No.4, pp. 777–814, 2007.
4. Ziegler JG, Nichols NB, "Optimum settings for automatic controllers", *Trans ASME*, Vol. 64, pp. 759–768, 1942.
5. Astrom KJ, Wittenmark B, *Adaptive Control*, New York: Addison-Wesley; 1995.
6. Jaung J-G, Haung M-T, Liu W-K, "PID control using presearched genetic algorithms for a MIMO system", *IEEE Trans Syst Man Cybern* Part C: Appl Rev, Vol. 38, No. 5, pp. 716–727, 2008.
7. Gaing Z-L, "A particle swarm optimization approach for optimum design of PID controller in AVR system", *IEEE Trans Energy Convers*, Vol. 19, No. 2, pp. 384–391, 2004.
8. Mandava RK, Manas KS, Vundavilli PR, "Optimization of PID controller parameters for 3-DOF planar manipulator using GA and PSO", In Bennett A. (ed) *New Developments in Expert Systems Research*, Computer Science, Technology and Applications, pp.67–88, Publisher: Nova Science Publishers, 2015.
9. Muhammet Ü, Ayça Ak, Vedat T, Hasan E, "Optimization of PID controllers using ant colony and genetic algorithms", *Stud Comput Intell*, Vol. 449, pp. 5–68, 2013, Springer-Verlag, Berlin, Heidelberg.
10. Gandomi AH, Yang X-S, Alavi AH, "Cuckoo search algorithm: A metaheuristic approach to solve structural optimization problems", *Eng Comput*, Vol. 29, pp. 17–35, 2013.
11. Latha K, Rajinikanth V, "2DOF PID controller tuning for unstable systems using bacterial foraging algorithm", In Panigrahi B.K., Das S., Suganthan P.N., Nanda P.K. (eds) *Swarm, Evolutionary, and Memetic Computing*. SEMCCO 2012. Lecture Notes in Computer Science, vol. 7677. Springer, Berlin, Heidelberg, 2012.
12. Zhao Z-Y, Tomizuka M, Isaka S, "Fuzzy gain scheduling of PID controllers", IEEE Trans Syst Man Cybern, Vol. 23, No. 5, pp. 1392–1398, 1993.
13. Hazzab A, Bousserhane IK, Zerbo M, Sicard P, "Real-time implementation of fuzzy gain scheduling of PI controller for induction motor machine control", *Neural Process Lett*, Vol. 24, pp. 203–215, 2006.
14. Yu K-W, Hsu J-H, "Fuzzy gain scheduling PID control design based on particle swarm optimization method", *Proceedings Second International Conference on Innovative Computing, Information and Control*, ICICIC'07, 337. Kumamoto; 2007.

15. Chaiyatham T, Ngamroo I, "Alleviation of power fluctuation in a microgrid by electrolyzer based on optimal fuzzy gain scheduling PID control", *IEEJ Trans Electr Electron Eng*, Vol. 9, pp. 158–164, 2014.
16. Syed FU, Kuang ML, Smith M, Okubo S, Ying H, "Fuzzy gain scheduling proportional-integral control for improving engine power and speed behavior in a hybrid electric vehicle", *IEEE Trans Veh Technol*, Vol. 58, No. 1, pp. 69–84, 2009.
17. Devaraj D, Selvabala B. "Real-coded genetic algorithm and fuzzy logic approach for real-time tuning of proportional-integral-derivative controller in automatic voltage regulator", *IET Gener Transm Distrib*, Vol. 3, No. 7, pp. 641–649, 2009.
18. Mehrabian AR, Lucas C, "A novel numerical optimization algorithm inspired from weed colonization", *Ecol Inform*, Vol. 1, No. 4, pp. 355–366, 2006.
19. Khalilpour M, Razmjooy N, Hosseini H, Moallem P, "Optimal control of DC motor using invasive weed optimization (IWO) algorithm", Majlesi Conference on Electrical Engineering, August 2011-Majlesi New Town, Isfahan, Iran.
20. Chen Z, Wang S, Deng Z, and Zhang X, "Tuning of auto-disturbance rejection controller based on the invasive weed optimization", Sixth International Conference on Bio-Inspired Computing: Theories and Applications, 2011.
21. Cembrano G, Torras C, and Wells G, "Neural networks for robot control" IFAC Artificial Intelligence in Real Time Control, Valtmcia, Spain, 1994.
22. Ghorbani R, Wu Q, Wang G, "Nearly optimal neural network stabilization of bipedal standing using genetic algorithm", *Eng Appl Artif Intell*, Vol. 20, pp. 473–480, 2007.
23. Jiang ZH, "Trajectory control of robot manipulators using a neural network controller", In Jimenez A., Al Hadithi B.M. (eds) *Robot Manipulators Trends and Development*, pp. 361–377, Publisher: InTech, Chapters published March 01, 2010.
24. Pezeshki S, Badalkhani S and Javadi A, "Performance analysis of a neuro-PID controller applied to a robot manipulator", *Int J Advanced Robotic Sys*, Vol. 9, pp. 1–10, 2012.
25. Mandava RK, Vundavilli PR, "Forward and inverse kinematic based full body gait generation of biped robot", IEEE International Conference on Electrical, Electronics, and Optimization Techniques, pp. 3301–3305, 2016.
26. Mandava RK, Vundavilli PR, "Whole body motion generation of 18-DOF biped robot on flat surface during SSP & DSP", *Int J Model Ident Contr*, 2017, Indescience publications (In press).
27. Ghasemi M, Ghavidel S, Aghaei J, Gitizadeh M, Falah H, "Application of chaos-based chaotic invasive weed optimization techniques for environmental OPF problems in the power system", *Elsevier Chaos, Solitons Fractals*, Vol. 69, pp. 271–284, 2014.
28. Basak A, Pal S, Das S and Abraham A, "A modified invasive weed optimization algorithm for time-modulated linear antenna array synthesis", *International conference on IEEE Congress on Evolutionary Computation*, Barcelona, Spain, pp. 1–10, 18–23 July 2010.
29. Roy GG, Das S, Chakraborty P, and Suganthan PN, "Design of non-uniform circular antenna arrays using a modified invasive weed optimization algorithm", *IEEE Trans Antennas Propag*, Vol. 59, No. 1, pp. 110–118, 2011.

## chapter thirteen

# Modeling fertility in Murrah bulls with intelligent algorithms

**Adesh Kumar Sharma, Ravinder Malhotra, and Atish Kumar Chakravarty**
ICAR-National Dairy Research Institute

## Contents

## 13.1   Introduction

India is popularly known for its best buffalo germplasm throughout the world, being responsible for more than 57% of the world buffalo population. Buffalo is considered as the major dairy animal and backbone of the Indian dairy industry. The country with a 108.7 million (Anon., 2014) buffalo population ranks first in the world. India ranks first in milk production, achieving an annual output of 155.5 million tonnes with per capita availability of 337 g/day in 2015–2016 (Anon., 2017). Buffaloes contribute 53% (82.41 million tonnes) to the total milk production in India.

The average productivity (5.76 kg/day/animal) of buffaloes is more than that of indigenous cattle (3.41 kg/day/animal) in the country (Anon., 2017). Besides this, buffaloes contribute significantly toward meat production, draft power, and dung for manure and fuel. Thus, the buffalo species is the most important and indispensable component of the livestock sector in the country.

Murrah is one of the best milch breeds of buffalo. The population of Murrah buffaloes is 48.25 million in India, out of which 11.68 million is pure and 36.56 million buffaloes are graded. Murrah buffaloes contribute 44.39% to the total buffalo population in India. Murrah buffaloes are known for high milk production and a higher fat percentage, which is almost twice than that of cow milk. Haryana state (especially Bhiwani, Jhajjar, Jind, and Rohtak districts) is the home tract of Murrah buffaloes, but the graded Murrah buffaloes are found throughout the country owing to their higher milk production potential coupled with adaptation to wide ecological conditions and feed conversion efficiency. The Murrah buffalo is basically the center of attraction for dairying among the various buffalo breeds available in India (Mir et al., 2015). Hence, the Murrah breed of buffalo has been appropriately named as the "black gold" of dairy animals in India. Also, several countries including Bangladesh, Brazil, Bulgaria, Egypt, etc., have used Murrah as an improver breed for upgrading their native buffaloes.

Until today, in many countries, the major attention for improvement of the dairy animals is through increasing the milk production. Although the result has been found satisfactory, over the years, it has been observed that increasing milk of the dairy animal deteriorates the reproduction performance as milk production traits are negatively associated with the fertility of the animals (Berry et al., 2011). Under these constraints, an assessment is required to compare the predictable fertility in relation to the milk production. The fertility of the breeding bull in the herd may be assessed based on total pregnancy corresponding to the total number of inseminations.

Accordingly, various studies have been carried out to predict fertility in dairy animals using classical regression analysis techniques (De Haas et al., 2007; Patil et al., 2014; Cook and Green, 2016; Utt, 2016; Eriksson et al., 2017). Conventional regression techniques are based on the assumption of a specific parametric function such as, linear, quadratic, etc., to fit the data that could be rather rigid for modeling any type of relationship (Piles et al., 2013). Alternatively, nonparametric methods like emerging machine learning (ML) algorithms could be applied for the intelligent analysis of such traits (González-Recio et al., 2014) as they do not involve prior knowledge of any parametric function. However, they can adapt intricate relationships between dependent and independent variables as well as complex dependencies among explanatory variables. Also, they are quite flexible and can learn arbitrarily complex patterns when sufficient data are presented. ML algorithms can realize how to perform important tasks by generalizing from examples, i.e., automatic predictions from instances of desired behavior or past observations. Thus, ML is the study of intelligent computer algorithms that improve automatically through incidence (Ramón et al., 2012; Shalev-Shwartz and Ben-David, 2014). These learning methods have found several applications in performance modeling and evaluation in animal sciences (Caraviello et al., 2006; Shahinfar et al., 2012; González-Recio et al., 2014; Murphy et al., 2014; Shahinfar et al., 2014; Hempstalk et al., 2015; Fenlon et al., 2016; Borchers et al., 2017). However, the majority of the studies in this area of research have been conducted outside India. Very few studies have recently been carried out in India including pioneering work by the authors at this institute (Sharma et al., 2006; Sharma et al., 2007, 2013; Panchal et al., 2016, 2017). Nevertheless, a bull's fertility prediction using ML techniques has not been attempted especially in Murrah buffaloes. Hence, in this chapter, the authors have investigated various emerging ML algorithms to predict the fertility in Murrah bulls being maintained at the ICAR-National Dairy Research Institute (NDRI), Karnal, India.

*Table 13.1* Summary statistics of the Murrah breeding bulls' fertility data set

| Variable | Mean | SD | SE | Minimum | Maximum | Range |
|---|---|---|---|---|---|---|
| Birth weight | 35.8723 | 5.3674 | 0.7829 | 25 | 50 | 25 |
| Weight (3-m) | 68.3996 | 17.9533 | 2.6188 | 43 | 116 | 73 |
| Weight (6-m) | 111.2570 | 19.4178 | 2.8324 | 76 | 170 | 94 |
| Weight (9-m) | 159.0518 | 36.7451 | 5.3598 | 110 | 270 | 160 |
| Weight (12-m) | 272.8443 | 52.2760 | 7.6252 | 172 | 363 | 191 |
| Weight (24-m) | 382.3406 | 51.6531 | 7.5344 | 264 | 493 | 229 |
| Age at first calving | 3.2955 | 0.7420 | 0.1082 | 2.1 | 4.9 | 2.8 |
| Post-thaw motility | 49.0425 | 5.3809 | 0.7849 | 40 | 60 | 20 |
| Conception rate | 62.4381 | 16.9481 | 2.4721 | 30.43 | 94.74 | 64.31 |

SD: standard deviation; SE: standard error.

## 13.2   Materials and methods

### 13.2.1   Data

The following information pertaining to Murrah bulls was collected from the NDRI archives for the period, July 1993–2016: animal number, date of birth, service sire number, dam number, date of first artificial insemination, date of successful artificial insemination, birth weight (kg); 3rd, 6th, 9th, 12th, and 24th months' body weights (kg); age at first calving (years); post-thaw motility (%); and bull conception rate (%). The data comprised 3200 records of artificial inseminations in buffaloes and 55 breeding bulls' records. Missing data were imputed through the multiple imputation (MI) technique (Berglund and Heeringa, 2014) implemented using advanced programming tools under the SAS 9.3 software. Consequently, the summary statistics computed from the data are presented in Table 13.1.

### 13.2.2   Machine learning algorithms

ML is the domain of computational intelligence, which constructs computer programs that automatically improve with experience (Witten and Frank, 2005; Smola and Vishwanathan, 2008; Daumé III, 2012; Shalev-Shwartz and Ben-David, 2014). ML paves the way to the treatment of real-life problems related to data analysis, sometimes overlooked by researchers, i.e., nonlinearity, pattern recognition, classification, adaptivity, optimization, missing variables, massive data sets, data management, causality, representation of knowledge, parallelization, etc. However, there is no systematic approach that can be employed *a priori* to identify the most fitting ML method for a specific task. Generally, several prominent ML algorithms are investigated on a new application. Hence, this chapter investigates the intelligent ML algorithms based upon supervised learning methods (Du and Swamy, 2014) such as neural networks (NNs), support vector regression (SVR), decision tree (DT), random forest (RF), and linear model (LM) for regression, with different settings for modeling and simulation experiments to efficiently predicting breeding bulls' fertility in Murrah buffaloes. The open-source R programming language has been used. A brief explanation about each of the aforementioned ML techniques with special focus on corresponding R programming tools is given in the following sections. However, a complete mathematical pedagogy for these ML techniques (in perspective with the R) can be found in Hastie et al. (2009) and Kuhn and Johnson (2013).

### 13.2.2.1 Neural network models

NN models are well known for their adaptive capabilities to "learn" relationships among variables. NNs are intelligent techniques for model fitting, which do not depend upon conventional assumptions necessary for traditional regression models (Kominakis et al., 2002; Sharma et al., 2007). The NN models are capable to efficiently analyze multivariate response data. An NN model is similar to a nonlinear regression model, with the exception that the former can encompass numerous model parameters. Thus, NN models can approximate any continuous function (Hornik, 1991; Li, 2008).

NN models are processing devices (algorithms) that are founded upon the neuronal structure of the mammalian cerebral cortex (e.g., human brain and nervous system) but on much smaller scales. These models comprise a large amount of highly interconnected processing elements (called artificial neurons or neurons in the following discussion) working parallel to solve nonlinear and complex problems. Unlike conventional number crunching algorithms, NN models work on the principle: learn by example. An NN model performs nonparametric nonlinear regression through the learning process. This adaptive process involves adjustments to synaptic weights that exist between the neurons just like the human brain.

An NN model is defined as a computing system made up of a number of simple, highly interconnected processing elements, which process information by their dynamic state response to external inputs (Hecht-Nielsen, 1990). NN models are suitable for intelligent predictive analytics. They allow complex nonlinear relationships between the response variable and its predictors.

An NN comprises a network of interconnected neurons organized into layers (Figure 13.1).

The inputs form the left-most layer, while outputs form the right-most layer. There may be intermediate layers called hidden layers, and artificial neurons contained in these layers are known as *hidden neurons*. The coefficients attached to these inputs (i.e., predictors) are termed as *synaptic weights*, which are generally random numbers of very small magnitude (Haykin, 2005). The outputs (i.e., forecasts) are obtained by a linear combination of the inputs. These synaptic weights are computed, in the NN framework, by means of a learning algorithm that minimizes a cost function such as sum of squared error (SSE) or mean sum of squared error (MSE).

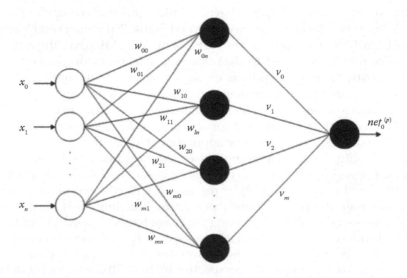

*Figure 13.1* A feed-forward NN model.

Consider the input dimension is given by $n(n \in Z_+)$ while the number of hidden neurons is denoted by $m(m \in Z_+)$. $Z_+$ signifies the set of positive integers. The training pairs are represented by $D = \left\{ x^{(p)}, t^{(p)} \right\}$, where $x^{(p)}$ and $t^{(p)}$ denote input and corresponding target patterns; $p = 1, 2, ..., P$; $P \in Z_+$, is the number of training patterns; and the index $p$ is always assumed to be present implicitly. The matrix $\mathbf{w}$ denotes the input to the hidden neurons connection strength, $w_{ij}$ is the $(i, j)$th element of the matrix $\mathbf{w}$ representing the connection strength (synaptic weights) between the $j$th input and the $i$th hidden layer neuron. With this nomenclature, the net input to the $i$th hidden layer neuron is given by

$$net_i = \sum_{j=1}^{n} w_{ij} x_j + \theta_i^{(1)} = \mathbf{w}_i \cdot \mathbf{x} + \theta_i^{(1)} \tag{13.1}$$

where $\theta_i^{(1)}$ is the bias of the $i$th hidden layer neuron. The output from the $i$th hidden layer neuron is given by

$$h_i(\mathbf{x}) = f^{(1)}(net_i) \tag{13.2}$$

where $f^{(1)}(\cdot)$ is a nonlinear transfer function.

The transfer function (also known as activation function or squash function) computes the output from a summation of the weighted inputs of a neuron. Generally, the transfer functions employed on neurons in the hidden layer are nonlinear that introduce nonlinearities into the network. The selection of transfer functions may strongly affect complexity and performance of NN models. Generally, the sigmoidal functions are used as transfer functions. The net input to the output neuron may be defined similarly as Equation (13.1) as follows:

$$net = \sum_{i=1}^{m} v_i h_i + \theta^{(2)} = \mathbf{v} \cdot \mathbf{h} + \theta^{(2)} \tag{13.3}$$

where $v_i$ represents the synaptic (or connection) strength between the $i$th hidden layer neuron and the output neuron, while $\theta^{(2)}$ is the bias of the output neuron.

Introducing a bias neuron $x_0$ with input value as +1, Equation (13.1) can be rewritten as

$$net_i = \sum_{j=0}^{n} w_{ij} x_j = \mathbf{W}_i \cdot \mathbf{x} \tag{13.4}$$

where $w_{i0} = W_{i0} \equiv \theta_i^{(1)}$ and $\mathbf{W}_i$ is the weight vector $\mathbf{w}_i$ (associated with the $i$th hidden neuron) augmented by the 0th column corresponding to the bias. Similarly, introducing an auxiliary hidden neuron $(i = 0)$ such that $h_0 = +1$, allows us to redefine Equation (13.3) as

$$net = \sum_{i=0}^{m} v_i h_i = \mathbf{V} \cdot \mathbf{h} \tag{13.5}$$

where $v_0 \equiv \theta^{(2)}$.

The equation for the network output neuron is given by

$$net_o = f^{(2)}(net) = net \tag{13.6}$$

where $f^{(2)}(\cdot)$ is a linear function.

The notations are illustrated diagrammatically in Figure 13.1. This diagram depicts an $n$-input, $m$-hidden neuron, and one-output feed-forward NN model. Such an NN model is trained to fit a data set $D$ by minimizing an error function (or performance function) as

$$F = \frac{1}{P} \sum_{p=1}^{P} \left( net_o^{(p)} - t^{(p)} \right)^2 \tag{13.7}$$

This function is minimized using any standard optimization method like Broyden–Fletcher–Goldfarb–Shanno (BFGS) optimization technique, etc.

The NN discovers knowledge from complicated or imprecise data, which is employed to find patterns and reveal trends that are too complex to be observed either by human beings or classical statistical techniques. A substantially trained NN acts as an expert system to analyze data in the specific domain of information for which it was trained.

### 13.2.2.2  *NN model building with R programming tools*

The Comprehensive R Archival Network (CRAN) comprises some packages for developing NN models, i.e., *nnet*, *neuralnet*, etc. The *nnet* and *neuralnet* packages have been investigated in this chapter. The functions in the *nnet* package facilitate development and validation of feed-forward multi-layer perceptrons (MLPs). The *nnet* functions offer sufficient dexterity to develop precision models by varying parameters' settings while learning. Training of NN models with *nnet* and *neuralnet* packages is accomplished by using the backpropagation learning algorithm equipped with the BFGS optimization method (Venables and Ripley, 2002); and resilient backpropagation (*rprop*) with or without weight backtracking (Riedmiller and Braun, 1993) or the modified globally convergent version by Anastasiadis et al. (2005), respectively. The later package supports adaptable attuning by means of custom-choice of error and transfer function. Furthermore, the computation of generalized weights (Intrator and Intrator, 1993) is employed.

### 13.2.2.3  *Support vector regression models*

The support vector machines (SVMs) are widely used due to many attractive features and promising empirical performance (Gunn, 1998; James et al., 2013). Initially, SVMs were developed to solve classification problems, which have been further extended to handle regression problems. In case of regression, support vector algorithms are called support vector regression (SVR). Among the various types of SVR, the most commonly used is $\varepsilon$-SVR (Vapnik, 1995). Hence, *eps-regression* method supported by the *e1071*-package under the CRAN (Kowalczyk, 2014) has been employed to predict fertility of Murrah breeding bulls.

The goal of the $\varepsilon$-SVR model is to find a function $f(x)$ that has at most $\varepsilon$ deviation from the actually obtained target values for all the training patterns; and simultaneously, is as flat as possible (Smola and Scholkopf, 2004). That is, if the errors are within the $\varepsilon$-insensitive band (known as $\varepsilon$-tube), these are ignored. Moreover, in order to ascertain robustness of learning, the input patterns should not strictly lie on or within the $\varepsilon$-tube. Instead, the points falling outside the $\varepsilon$-tube are penalized and slack variables are introduced to curtail such instances.

Suppose, there are training data set $\left\{ x^{(1)}, t^{(1)} \right\}, \ldots, \left\{ x^{(P)}, t^{(P)} \right\} \subset \aleph \times \Re$ where $\aleph$ stands for input space of the input patterns, e.g., $\Re^d$. In $\varepsilon$-SV regression, the errors are ignored as long as these are less than $\varepsilon$; however, any deviation beyond this point is not accepted.

First, let us consider the instance of linear functions $f$, which take the form

$$f(x) = \langle w, x \rangle + b \text{ with } w \in \aleph, b \in \Re \qquad (13.8)$$

where $\langle \cdot, \cdot \rangle$ represents the dot product in $\aleph$. In context with Equation (13.8), the flatness signifies to determine the smallest value for $w$. This can be minimized with the help of Euclidean norm, i.e., $\|w\|^2$. Formally, express this problem in the form of a convex optimization problem:

$$\text{minimize} = \frac{1}{2}\|w\|^2$$

$$\text{subject to} \begin{cases} t^{(p)} - \langle w, x^{(p)} \rangle - b \leq \varepsilon \\ \langle w, x^{(p)} \rangle + b - t^{(p)} \leq \varepsilon \end{cases} \qquad (13.9)$$

It is implicitly assumed in Equation (13.9) that a function such as $f$ actually exists, which approximates all pairs $\{x^{(p)}, t^{(p)}\}$ with $\varepsilon$ precision, i.e., the convex optimization problem is feasible. However, at times, this may not be the case, or we may tolerate some errors. Slack variables, $\xi_p$ and $\xi_p^*$ are introduced to manage the infeasible constraints of the optimization problem under consideration. Thus, the optimization problem (Equation 13.9) is reformulated:

$$\text{minimize} = \frac{1}{2}\|w\|^2 + c\sum_{p=1}^{P}\left(\xi_p + \xi_p^*\right)$$

$$\text{subject to} \begin{cases} t^{(p)} - \langle w, x^{(p)} \rangle - b \leq \varepsilon + \xi_p \\ \langle w, x^{(p)} \rangle + b - t^{(p)} \leq \varepsilon + \xi_p^* \\ \xi_p, \xi_p^* \geq 0 \end{cases} \qquad (13.10)$$

The constant $c > 0$ resolves the trade-off between the flatness of $f$ and the extent to which deviations beyond $\varepsilon$ can be permissible. The problem formulation Equation (13.10) is virtually dealing with a so-called $\varepsilon$-insensitive loss function, $|\xi|_\varepsilon$ defined as

$$|\xi|_\varepsilon := \begin{cases} 0 & \text{if } |\xi| \leq \varepsilon \\ |\xi| - \varepsilon & \text{otherwise} \end{cases} \qquad (13.11)$$

### 13.2.2.4 Decision tree models

DT builds regression models like a tree structure. It splits the data set into smaller and smaller subsets simultaneously leading to the development of an associated DT, incrementally. The outcome is a tree comprising various decision nodes and leaf nodes. A decision node consists of two or more branches, each depicting values for the attribute tested. Leaf node signifies a decision on the numerical target. The uppermost decision node in a tree corresponding to the best predictor is termed as *root node*. DTs are capable to analyze both categorical as well as numerical data.

### 13.2.2.5   Decision tree model building with R programming tools

The *rpart* (recursive partitioning and regression trees) package under R programming language was employed in this study for building decision regression trees, which is briefly delineated. As stated earlier, let the data consist of $n$ inputs and a response, for each of $P$ patterns, i.e., $\{x^{(p)}, t^{(p)}\}$ for $p = 1, 2, \ldots, P$ with $x^{(p)} = \{x^{(p1)}, x^{(p2)}, \ldots, x^{(pn)}\}$. This algorithm aims to automatically compute the optimum number of splitting variables and split points, as well as the best topology that the tree should possess (Hastie et al., 2009). Assume that the feature space is partitioned into $M$ regions $F_1, F_2, \ldots, F_M$ and it models the response as a constant $c_m$ in each region:

$$f(x) = \sum_{m=1}^{M} c_m I(x \in F_m) \tag{13.12}$$

Let the criterion be minimization of the SSE, $\sum \left[ t^{(p)} - f\left(x^{(p)}\right) \right]^2$. Then the best $\hat{c}_m$ is just the average of $t^{(p)}$ in region $F_m$:

$$\hat{c}_m = ave\left[ t^{(p)} \mid x^{(p)} \in F_m \right]$$

where $ave[\cdot]$ denotes the average. Now, determining the best binary partition in terms of minimum SSE, is generally computationally infeasible. Therefore, a top-down greedy-search algorithm is applied. Starting with all the data points, consider a splitting variable $j$ and split point $s$, and define the pair of half-planes:

$$F_1(j,s) = \{X \mid X_j \le s\} \text{ and } F_2(j,s) = \{X \mid X_j > s\} \tag{13.14}$$

Now, it is to seek the splitting variable $j$ and split point $s$ that solve:

$$\min_{j,s} \left[ \min_{c_1} \sum_{x^{(p)} \in F_1(j,s)} \left\{ t^{(p)} - c_1 \right\}^2 + \min_{c_2} \sum_{x^{(p)} \in F_2(j,s)} \left\{ t^{(p)} - c_2 \right\}^2 \right] \tag{13.15}$$

For any choice of $j$ and $s$, the inner minimization is solved by

$$\hat{c}_1 = ave\left[ t^{(p)} \mid x^{(p)} \in F_1(j,s) \right] \text{ and } \hat{c}_2 = ave\left[ t^{(p)} \mid x^{(p)} \in F_2(j,s) \right] \tag{13.16}$$

For each splitting variable, the split point $s$ can be determined quite promptly. Thus, determination of the best pair $(j, s)$ is feasible by browsing through all the inputs. Having found the best split, the data are partitioned into the two resulting regions, and repeat this splitting process on each of the two regions. Further, this process is employed on all the resulting regions. Now, the question is how large the tree should be grown. Naturally, a big tree could over-fit the data, whereas a short tree might not discover the underlying important structure, i.e., tree size is a tuning parameter leading to the model's complexity. Thus, the optimum tree size should be adaptively resolved from the data. One smart strategy would be to split tree nodes only if the decrease in SSE due to the split exceeds some threshold. Nevertheless, this tactic is too short-sighted as an imprecise split might produce a very good split underneath. The ideal scheme would be to grow a large tree $T_0$, stopping the splitting process only when a certain least node size is attained. Then this big tree is pruned using cost-complexity pruning technique that is delineated. Define a subtree $T \subset T_0$ to be any tree, which is attained as a result of pruning $T_0$, i.e., collapsing any number of its internal (nonterminal) nodes. Let the terminal nodes be indexed by $m$, with node $m$ representing region $F_m$. Consider $|T|$ to be the number of terminal nodes in $T$. Let the following quantities be stated in the present context:

$$\left.\begin{array}{c} N_m = \left\{x^{(p)} \in F_m\right\} \\[12pt] \hat{c}_m = \dfrac{1}{N_m} \sum_{x^{(p)} \in F_m} t^{(p)} \\[16pt] Q_m\left(T\right) = \dfrac{1}{N_m} \sum_{x^{(p)} \in F_m} \left\{t^{(p)} - \hat{c}_m\right\}^2 \end{array}\right\} \qquad (13.17)$$

The cost complexity criterion is defined as

$$C_\alpha(T) = \sum_{m=1}^{|T|} N_m Q_m(T) + \alpha\,|\,T\,| \qquad (13.18)$$

It is to find the subtree $T_\alpha \subseteq T_0$ to minimize $C_\alpha(T)$, for each $\alpha$. The tuning parameter $\alpha \geq 0$ leads the trade-off between tree size and its goodness-of-fit to the data. Large values of $\alpha$ produce smaller trees $T_\alpha$, and *vice versa*. It implies that with $\alpha = 0$, the outcome would be full tree $T_0$. Now, how to adaptively choose $\alpha$? For each $\alpha$, it can be shown that there is a unique smallest subtree $T_\alpha$ that minimizes $C_\alpha(T)$. To find $T_\alpha$, the weakest link pruning technique is used, which successively collapses the internal node that affects the least per-node increase in $\sum_m N_m Q_m(T)$, and continues until it attains the single-node (root) tree.

This leads to a finite sequence of subtrees, and it can be shown that this sequence must contain $T_\alpha$. Further pedagogical details can be found in Breiman et al. (1984). The $\hat{\alpha}$ is computed by the cross-validation method. The value of $\hat{\alpha}$ is so chosen that it minimizes the cross-validated SSE. The ultimate tree is $T_{\hat{\alpha}}$.

### 13.2.2.6   Random forest models

RFs (or random decision forests) are a kind of ensemble learning method for regression and other tasks; they operate by constructing several DTs during training time and producing the mean prediction (regression) of the individual trees (Breiman, 2001; Daumé III, 2012; James et al., 2013). RFs resolve drawback of over-fitting to the training set, which is inherent in decision trees. Random forest regression begins with the creation of decision trees. Decision trees recursively divide data in the regression space until the quantum of variation in the subspace becomes small. Subsequently, a predictor for the subspace is created just by taking the average value of the input patterns corresponding to the target patterns in the subspace. A typical algorithm (Hastie et al., 2009) illustrating various steps involved in RF for regression models, is given as follows:

Random forest algorithm for regression:

1. For $p = 1$ to $P$:
   a. Get a bootstrap sample $Z^*$ of size $N$ from the training data.
   b. Construct a random-forest tree $T_p$ on the bootstrapped data through recursive iteration for each terminal node of $T_p$ until the minimum node size $n_{\min}$ is attained as follows:
      i. Randomly choose $m$ variables from the set of $l$ variables.

    ii.  Select optimum variable/split-point based on the variables selected in previ-
ous step (i).

   iii.  Bifurcate the node into daughter nodes.

2. Compute the ensemble of trees $[T_p]_1^P$.

3. Carry out the prediction at a new point $x$, i.e., regression with: $\hat{f}_{rf}^P(x) = \frac{1}{P}\Sigma_{p=1}^P T_p(x)$.

The course of recursive partitioning can be envisaged as a regression tree. Predictions for new data are determined by searching the predictor pertinent to the division containing the new input variable. This splitting-up step for RF is greedy and, thus, does not generally converge to the globally optimal tree. This problem is resolved through an ensemble of locally optimal trees wherein each tree is grown by uniformly drawing a random sample from the original subset. This process is termed as *bagging* or *bootstrap aggregation*. Moreover, once the re-sampled version of the data set is constructed, all but a small number of features are sampled. Completion of each sampling strategy leads to a unique trained-tree. The set of such trees is called a forest. Finally, predictions are conducted on the basis of collection of the unique predictions made by each of these trees. This procedure is known as *voting*. An aggregate prediction achieves higher accuracy than any of the constituent trees. Thus, RFs are referred to as ensemble learning methods, which are currently regarded as the best regression tools among data scientists. Although RF models, generally, exhibit superior prediction performance, they are tough to interpret. That is why these are known as black box models.

    The *randomForest* package supported by R language has been used for building the RF models in this chapter. A complete description about computational aspects of this algorithm can be found in Liaw and Wiener (2002).

### 13.2.2.7 Linear regression models

Linear models are conventional methods of statistics, which were initially developed in the pre-computer era. They are simple and often provide an adequate and interpretable description about relationship between the inputs and the output variables. For predictive application, they can sometimes outperform modern nonlinear models, especially in situations involving small numbers of training observations, low signal-to-noise ratio or sparse data. Thus, even nowadays, with the advent of advanced computational algorithms for regression like ML algorithms as described in foregoing sections, conventional multiple linear regression (MLR) models are still widely used for predictive analytics. They study linear and additive relationships between variables.

    Let $t$ denote the dependent variable whose values are to be predicted; and let $X^{(i)}, i = 1, 2, \ldots, n$ denote independent variables based upon which, the prediction has to be performed. The prediction equation for computing the predicted values of $t$ is

$$\hat{t} = b_0 + b_1 X^{(1)} + b_2 X^{(2)} + \cdots + b_n X^{(n)} \tag{13.19}$$

This equation possesses the property that prediction for $t$ is a straight-line function of the $X$ variables. The slopes of their individual straight-line relationships with $t$ are the constants $b_i, i = 1, 2, \ldots, n$, called coefficients of the variables indicating variation in predicted value of $b_i$ per unit of change in $X^{(i)}$, keeping other items equal. The additional constant $b_0$, called *intercept*, is prediction that the model would produce if all the $X$ values were zero (if possible). The coefficients and intercept are estimated by the least squares method, i.e., setting them equal to the unique values that minimize SSE within the sample of data to which the model is fitted. Also, the model's prediction errors are generally assumed to be independently and identically normally distributed.

The linear model function, *glm*() supported by R programming language has been employed for the MLR analysis in this chapter. The detailed pedagogic description can be found in Kabacoff (2015).

### 13.2.2.8 Model evaluation error metrics

The prediction performances of various machine learning models developed in this study *vis-à-vis* the conventional models were evaluated by means of several statistical metrics such as mean absolute error (MAE), root mean square error (RMSE), and Akaike's information criterion (AIC). The mathematical formulas for these methods are as follows:

$$\text{MAE} = \frac{\sum_{i=1}^{n} |\text{Actual} - \text{Predicted}|}{n} \tag{13.20}$$

$$\text{RMSE} = \sqrt{\frac{1}{n} \sum_{i=1}^{n} \left( \frac{\text{Actual} - \text{Predicted}}{\text{Actual}} \right)^2} \tag{13.21}$$

$$\left. \begin{aligned} \text{AIC} &= n \ln\left(\frac{\text{RSS}}{n}\right) + 2k, \\[2mm] \text{RSS} &= \sum_{i=1}^{n} (\text{Actual} - \text{Predicted})^2 \end{aligned} \right\} \tag{13.22}$$

where $n$ is the number of data points (observations in the test set); $k$ is the number of estimated parameters (including the variance); and RSS is the estimated residual of the fitted model.

## 13.3 Results and discussion

### 13.3.1 Neural Network Models

The NN model for modeling fertility in Murrah breeding bulls was developed using *nnet* and *neuralnet* packages with different settings, i.e., varying number of hidden layers; number of neurons in the hidden layer(s); transfer function on the hidden layer neurons; data partitioning scheme; learning rate, error goal, epochs, learning algorithms, etc. The NN model constructed with the *neuralnet* package using *rprop* learning algorithm is depicted in Figure 13.2.

The optimal configurations for the NN models created with both the packages were determined empirically through a "trial and error" approach, and the same are given in Table 13.2.

### 13.3.2 Support vector regression models

The *eps-regression* under *e1071*-package was employed through R programming tools to fit the SVR model for modeling fertility in Murrah bulls. Accuracy of the model was enhanced by fine-tuning SVR model parameters, i.e., cost and $\varepsilon$ values. This was done by using the

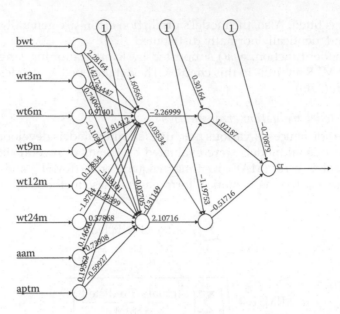

Error: 0.476917    Steps: 6746

*Figure 13.2* NN model constructed with *rprop* algorithm.

*Table 13.2* Neural network model's optimum configuration and predictive performance

| Type of neural network/R package | Learning algorithm | Number of neurons/transfer function in hidden layer(s) | Epochs/ steps | MAE | RMSE | AIC |
|---|---|---|---|---|---|---|
| Feed-forward (*nnet*) | BFGS | 6 (Log sigmoid) | 100 | 13.12 | 0.15 | 177.47 |
| Feed-forward (*neuralnet*) | Rprop | 2, 2 (Tangent sigmoid) | 6746 | 14.53 | 0.30 | 54.95 |

*tune* function in the same package. The tuning SVR, i.e., hyperparameter optimization or model selection was based on grid search method. A lot of models were trained for the different combinations of cost and $\varepsilon$, and the optimal one was selected (Table 13.3). The tune method was employed to train models with $\varepsilon = 0, 0.1, 0.2, \ldots, 1$ and cost $= 2^2, 2^3, 2^4, \ldots, 2^9$ (Figure 13.3).

In Figure 13.3, the darker the region is, the better the model is (i.e., RMSE is closer to zero in darker regions).

### 13.3.3 Decision tree regression model

The best decision tree model was developed using *rpart* package with *maxsurrogate* and *usersurrogate* parameters set to 0 values. The *maxsurrogate* parameter denotes the number of surrogate splits retained in the output. If this is set to 0, it signifies that half of the computational time is used in search for surrogate splits, thereby reducing computational time, while the feature *usesurrogate* specifies as to how to use surrogate in splitting process. The best decision tree model's configuration along with its predictive accuracy attained in terms of MAE, RMSE, and AIC is shown in Table 13.4.

*Table 13.3* Support vector regression machine's optimum configuration and predictive performance

| SVR type/R package | Epsilon | Cost c | Kernel | Data partitioning scheme | MAE | RMSE | AIC |
|---|---|---|---|---|---|---|---|
| *Eps-regression* (Grid search with tuning) | 1 (0–1) | 4 ($2^2$–$2^9$) | Gaussian RBF | 90:10 (10-fold cross validation) | 8.05 | 0.06 | 52.86 |

RBF: radial basis function.

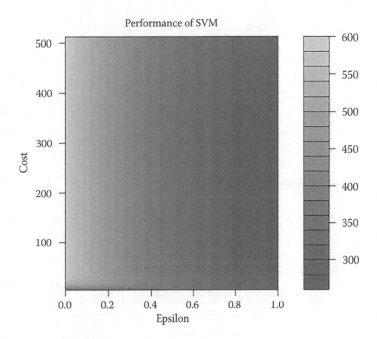

*Figure 13.3* SVR model's performance through grid search.

*Table 13.4* Decision tree regression model's optimum configuration and predictive performance

| Approach/R package | Maxsurrogate | Usesurrogate | Data partitioning scheme | MAE | RMSE | AIC |
|---|---|---|---|---|---|---|
| Top-down greedy search (*rpart* package) | 0 | 0 | 90:10 | 13.26 | 0.22 | 155.23 |

## 13.3.4   Random forest regression model

The RF models for prediction of fertility in Murrah bulls were developed using *random Forest* package with values of *ntree* and *mtry* empirically set at 500 and 2, respectively (Figure 13.4). The data partitioning scheme used was the same as in previous modeling experiments, i.e., 90:10.

The optimum configuration and predictive performance of the random forest model is given in Table 13.5.

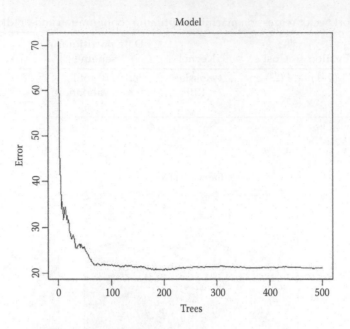

*Figure 13.4* Random forest regression model.

*Table 13.5* Random forest model's optimum configuration and predictive performance

| R package | Number of trees (*ntree*) | Number of variables per level (*mtry*) | Data partitioning scheme | MAE | RMSE | AIC |
|---|---|---|---|---|---|---|
| Random Forest | 500 | 2 | 90:10 | 9.58 | 0.03 | 127.94 |

### 13.3.5   Linear model for regression

The conventional linear model was fitted to the same data set using the same data partitioning scheme, i.e., 90:10, as ML models described above. The R function *glm*() has been employed for this purpose. The MAE, RMSE, and AIC attained with best LM were 6.44, 0.18, and 356.37, respectively.

### 13.3.6   Machine learning models vis-à-vis linear regression model

The predictive accuracies of ML models *vis-à-vis* the linear model to predict the fertility of Murrah bulls have been compared (Table 13.6).

*Table 13.6* Comparison of predictive accuracies of machine learning models *vis-à-vis* linear model to predict fertility of Murrah breeding bulls

| Accuracy metric | Models' prediction accuracies | | | | |
|---|---|---|---|---|---|
| | NN | SVR | DT | RF | LM |
| MAE | 13.12 | 8.05 | 13.26 | 9.58 | 6.44 |
| RMSE | 0.15 | 0.06 | 0.22 | 0.03 | 0.18 |
| AIC | 177.47 | 52.86 | 155.23 | 127.94 | 356.37 |

The experimental results (Table 13.6) that emerged from this study revealed that the ML models, i.e., RF, SVR, and NN models, outperformed the LM, whereas the DT model did not perform well due to its well-known inherent problem of over-fitting. Thus, the ML approach (especially the RF paradigm) is capable of efficiently predicting the fertility of Murrah bulls; which was, generally, found better than conventional linear models. Hence, ML algorithms can be employed as a plausible alternative to linear regression models in predicting the fertility of Murrah breeding bulls.

## 13.4   Conclusion

Various supervised ML algorithms, *viz.*, NN, SVR machine, DT, and RF have been investigated empirically in this chapter, for modeling breeding bulls' fertility in Murrah buffaloes. The performance of these intelligent models has been compared with that of the classical linear model for regression, also developed in this study. The results of this study revealed that the ML approach, generally, outperformed the classical linear models for regression. Hence, the ML models developed in this study are superior to precisely assess the conception rate in Murrah breeding bulls at organised dairy farm(s) like ICAR-NDRI, Karnal (India). These intelligent models will provide decision support to organized dairy farms for selecting good buffalo bulls.

## References

Anastasiadis, A.D., Magoulas, G.D. and Vrahatis, M.N. (2005). New globally convergent training scheme based on the resilient propagation algorithm. *Neurocomputing*, 64, 253–270.

Anonymous (2014). 19th Livestock Census-2012 All India Report. Department of Animal Husbandry, Dairying and Fisheries, Ministry of Agriculture, Govt. of India, New Delhi. www.dahd.nic.in/sites/default/files/Livestock5.pdf.

Anonymous (2017). Annual Report 2016–17. Department of Animal Husbandry, Dairying & Fisheries Ministry of Agriculture & Farmers Welfare Government of India.

Berglund, P. and Heeringa, S. (2014). *Multiple Imputation of Missing Data Using SAS*. SAS Institute Inc., Cary, NC.

Berry, D.P., Evans, R.D. and Mc Parland, S. (2011). Evaluation of bull fertility in dairy and beef cattle using cow field data. *Theriogenology*, 75, 172–181.

Borchers, M.R., Chang, Y.M., Proudfoot, K.L., Wadsworth, B.A., Stone, A.E. and Bewley, J.M. (2017). Machine-learning based calving prediction from activity, lying, and ruminating behaviors in dairy cattle. *Journal of Dairy Science*, 100, 5664–5674.

Breiman, L. (2001). Random forests. *Machine Learning*, 45, 5–32.

Breiman, L., Friedman, J., Olshen, R. and Stone, C. (1984). *Classification and Regression Trees*. Wadsworth, New York.

Caraviello, D.Z., Weigel, K.A., Craven, M., Gianola, D., Cook, N.B., Nordlund, K.V., Fricke, P.M. and Wiltbank, M.C. (2006). Analysis of reproductive performance of lactating cows on large dairy farms using machine learning algorithms. *Journal of Dairy Science*, 89, 4703–4722.

Cook, J.G. and Green, M.J. (2016). Use of early lactation milk recording data to predict the calving to conception interval in dairy herds. *Journal of Dairy Science*, 99, 4699–4706.

Daumé III, H. (2012). A course in machine learning. http://ciml.info.

De Haas, Y., Janss, L.L.G. and Kadarmideen, H.N. (2007). Genetic correlations between body condition scores and fertility in dairy cattle using bivariate random regression models. *Journal of Animal Breeding and Genetics*, 124, 277–285.

Du, K.-L. and Swamy, M.N.S. (2014). Fundamentals of machine learning. Chapter 2. In: *Neural Networks and Statistical Learning*. Springer, London. doi:10.1007/978-1-4471-5571-3_2.

Eriksson, S., Johansson, K., Axelsson, H.H. and Fikse, W.F. (2017). Genetic trends for fertility, udder health and protein yield in Swedish red cattle estimated with different models. *Journal of Animal Breeding and Genetics*, 134, 308–321.

Fenlon, C., O'Gradyy, L., Dunnion, J., Shallooz, L., Butlerz, S. and Doherty, M. (2016). A comparison of machine learning techniques for predicting insemination outcome in Irish dairy cows. In: *Proceedings of the 24th Irish Conference on Artificial Intelligence and Cognitive Science*, September 20–21, Dublin, Ireland, pp. 57–67. http://aics2016.ucd.ie/papers/full/AICS_2016_paper_30.pdf.

González-Recio, O., Rosa, G.J.M. and Gianola, D. (2014). Machine learning methods and predictive ability metrics for genome-wide prediction of complex traits. *Livestock Science*, 166, 217–231.

Gunn, S.R. (1998). Support vector machines for classification and regression. In: ISIS Technical Report, Image Speech & Intelligent Systems Group, University of Southampton, UK.

Hastie, T., Tibshirani, R. and Friedman, J. (2009). *Elements of Statistical Learning: Data Mining, Inference and Prediction*. Second Edition. Springer, New York.

Haykin, S. (2005). *Neural Networks: A Comprehensive Foundation*. Second Edition. Pearson Education (Singapore) Pte. Ltd., Delhi.

Hecht-Nielsen, R. (1990). *Neurocomputing*. Addison Wesley Longman Publishing Co., Inc. Boston, MA.

Hempstalk, K., McParland, S. and Berry, D.P. (2015). Machine learning algorithms for the prediction of conception success to a given insemination in lactating dairy cows. *Journal of Dairy Science*, 98, 5262–5273.

Hornik, K. (1991). Approximation capabilities of multilayer feedforward networks. *Neural Networks*, 4, 251–257.

Intrator, O. and Intrator, N. (1993). Using neural nets for interpretation of nonlinear models. In: *Proceedings of the Statistical Computing Section, American Statistical Society*, San Francisco, pp. 244–249.

James, G., Witten, D., Hastie, T. and Tibshirani, R. (2013). *An Introduction to Statistical Learning with Applications in* R. Springer, New York.

Kabacoff, R.I. (2015). *R in Action: Data Analysis and Graphics with* R. Manning Publications Co., New York.

Kominakis, A.P., Abas, Z., Maltaris, I. and Rogdakis, E. (2002). A preliminary study of the application of artificial neural networks to prediction of milk yield in dairy sheep. *Computers and Electronics in Agriculture*, 35, 35–48.

Kowalczyk, A. (2014). Support Vector Regression with R. www.svm-tutorial.com/2014/10/support-vector-regression-r/.

Kuhn, M. and Johnson, K. (2013). *Applied Predictive Modeling*. Springer, New York. doi:10.1007/978-1-4614-6849-3.

Li, F. (2008). Function approximation by neural networks. In: Sun, F., Zhang, J., Tan, Y., Cao, J. and Yu, W. (Eds.) *Advances in Neural Networks*. Lecture Notes in Computer Science, 5263, 384–390. Springer, Berlin, Germany.

Liaw, A. and Wiener, M. (2002). Classification and regression by *randomForest*. *R News*, 2/3, 18–22.

Mir, M.A., Chakravarty, A.K., Gupta, A.K., Naha, B.C., Jamuna, V., Patil, C.S. and Singh, A.P. (2015). Optimizing age of bull at first use in relation to fertility of Murrah breeding bulls. *Veterinary World*, 8, 518–522.

Murphy, M.D., O'Mahony, M.J., Shalloo, L., French, P. and Upton, J. (2014). Comparison of modeling techniques for milk-production forecasting. *Journal of Dairy Science*, 97, 3352–3363.

Panchal, I., Sawhney, I.K., Sharma, A.K. and Dang, A.K. (2016). Classification of healthy and mastitis Murrah buffaloes by application of neural network models using yield and milk quality parameters. *Computers and Electronics in Agriculture*, 127, 242–248.

Panchal, I., Sawhney, I.K., Sharma, A.K., Garg, M.K. and Dang, A.K. (2017). Mastitis detection in Murrah buffaloes with intelligent models based upon electro-chemical and quality parameters of milk. *Indian Journal of Animal Research*, 51, 922–926.

Patil, C.S., Chakravarty, A.K., Singh, A., Kumar, V., Jamuna, V. and Vohra, V. (2014). Development of a predictive model for daughter pregnancy rate and standardization of voluntary waiting period in Murrah buffalo. *Tropical Animal Health and Production*, 46, 279–284.

Piles, M., Díez, J., delCoz, J.J., Montañés, E., Quevedo, J.R., Ramon, J., Rafel, O., López-Béjar, M. and Tusell, L. (2013). Predicting fertility from seminal traits: Performance of several parametric and non-parametric procedures. *Livestock Science*, 155, 137–147.

Ramón, M., Martínez-Pastor, F., García-Álvarez, O., Maroto-Morales, A., Josefa-Soler, A., Jiménez-Rabadán, P., Fernández-Santos, M.R., Bernabéu, R. and Garde, J.J. (2012). Taking advantage of the use of supervised learning methods for characterization of sperm population structure related with freezability in the Iberian red deer. *Theriogenology*, 77, 1661–1672.

Riedmiller, M. and Braun, H. (1993). A direct adaptive method for faster backpropagation learning: The RPROP algorithm. In: *Proceedings of the IEEE International Conference on Neural Networks*, San Francisco, pp. 586–591.

Shahinfar, S., Mehrabani-Yeganeh, H., Lucas, C., Kalhor, A., Kazemian, M. and Weigel, K.A. (2012). Prediction of breeding values for dairy cattle using artificial neural networks and neuro-fuzzy systems. *Computational and Mathematical Methods in Medicine*. doi:10.1155/2012/127130.

Shahinfar, S., Page, D., Guenther, J., Cabrera, V., Fricke, P. and Weigel, K. (2014). Prediction of insemination outcomes in Holstein dairy cattle using alternative machine learning algorithms. *Journal of Dairy Science*, 97, 731–742.

Shalev-Shwartz, S. and Ben-David, S. (2014). *Understanding Machine Learning: From Theory to Algorithms*. Cambridge University Press, New York.

Smola, A.J. and Scholkopf, B. (2004). A tutorial on support vector regression. *Statistics and Computing*, 14, 199–222.

Smola, A. and Vishwanathan, S.V.N. (2008). *Introduction to Machine Learning*. Cambridge University Press, Cambridge.

Sharma, A. K., Jain, D.K., Chakravarty, A.K., Malhotra, R. and Ruhil, A.P. (2013). Predicting economic traits in Murrah buffaloes with connectionist models. *Journal of Indian Society of Agricultural Statistics*, 67, 1–11.

Sharma, A.K., Sharma, R.K. and Kasana, H.S. (2006). Empirical comparisons of feed-forward connectionist and conventional regression models for prediction of first lactation 305-day milk yield in Karan Fries dairy cows. *Neural Computing and Applications*, 15, 359–365.

Sharma, A.K., Sharma, R.K. and Kasana, H.S. (2007). Prediction of first lactation 305-day milk yield in Karan-Fries dairy cattle using ANN modelling. *Applied Soft Computing*, 7, 1112–1120.

Utt, M.D. (2016). Prediction of bull fertility. *Animal Reproduction Science*, 169, 37–44.

Vapnik, V. (1995). *The Nature of Statistical Learning Theory*. Springer, New York.

Venables, W.N. and Ripley, B.D. (2002). *Modern Applied Statistics with S*. Springer, New York.

Witten, I.H. and Frank, E. (2005). *Data Mining: Practical Machine Learning Tools and Techniques*. Second Edition. Morgan Kaufmann Publishers, San Francisco, CA.

# chapter fourteen

# Computational study of the Coanda flow for V/STOL

**Maharshi Subhash and Michele Trancossi**
*University of Modena and Reggio Emilia*

**Maharshi Subhash**
*Graphic Era University*

**Michele Trancossi**
*Sheffield Hallam University*

## Contents

## 14.1  Introduction

It is not only a dream but also a necessity of tomorrow to have vertical and short takeoffs and landings (V/STOL) in the civil aviation sector; because of rapid growth of aviation for humanitarian purposes. There are several methods [1–5], which can be useful to implement V/STOL for air vehicles. Out of these methods, the most adequate method is based on thrust vectoring. The project ACHEON (Aerial Coanda High Efficiency Orienting-jet Nozzle) encompasses a thrust vectoring propulsive nozzle called HOMER (High-speed Orienting Momentum with Enhanced Reversibility), which is supported in a patent developed at the University of Modena and Reggio-Emilia, Italy [6]. The idea encapsulated in the project is to use the Coanda surface for the thrust vectoring to achieve V/STOL. In the past, several works depicted the use of the Coanda surface for the flow control on the aircraft wing and other flow control devices [7–12]. However, this concept can also be used efficiently in other industrial applications like plasma spray gun (and for direct injection in combustion chamber to improve the combustion efficiency [13–15]).

The attachment of the jet over the adjacent curved surface was known about two centuries ago by Young [16] and patented around one century later by Henry Coanda, a Romanian engineer; therefore, this phenomenon is known as the "Conada" effect. The conditions of the stability of the flow over the curved surface has been described by Rayleigh [17] through the streamline curvature, although this flow feature did not attract much research attention until 1961. The mechanism of the flow over the convex surface causing Coanda flow can be found in the literature by Newman [18]. They found that the flow adheres the curved surface due to the momentum balance between centrifugal force and the pressure force [19]. Due to the interaction of the ambient fluid to the boundary layer on the flow over the curved surface, the static pressure increases gradually and when the pressure gradient becomes zero, that position on the curved surface is the verge of separation; beyond this position the pressure gradient becomes positive and causes the reverse flow. The detailed literature review in addition to its engineering applications, especially for generating lift on the curved surface has been performed by Trancossi [20].

Some past works were concentrated on the investigation of the mechanism of the flow over the curved surface. Wille and Fernholtz [21] have performed experiments on the flow over a convex surface and found that surface curvature has a significant influence on the jet deflection. They also studied the boundary layer phenomena and entrainment of ambient fluid, causing the jet to adhere to the curved surface. However, in counterpart the jet-spreading rate increases rapidly and causes the separation of the boundary layer. Therefore, in order to investigate the flow phenomena, one needs to go into the boundary layer, which has been attempted in previous work [22].

Experimental investigation on the flow over a circular cylinder by Fekete shows that the velocity profile on the curved surface is similar to the profile at the plane wall jet. He also investigated the surface pressure, position of separation, and wall shear stress. He has shown that the wall-shear stress is negligible as long as the ratio $b/R$ is not too small, stating experiments where $b/R < 0.0075$ may be prone to skin friction forces. He has discovered that $\theta_{sep}$ decreases with increased surface roughness; however, with large values of the Reynolds number the influence of surface roughness is negligible within the tested roughness limit [23].

Neuendorf and Wygnansky [24] investigated experimentally the flow over the curved surface; they found that the entrainment of the ambient fluid causes the jet to adhere to the curved surface, but, on the other side, also causes separation, because of the increment of the jet spreading rate. Therefore, boundary layer approximation fails. This indicates that the condition of the velocity gradient and pressure gradient for the separation needs to be reinvestigated in order to reveal the physics of the flow for the curved surface. Without knowing this flow behavior the design of such a nozzle would depend upon the trial methods, which would consume more time and cost. This part of work has already been invesigated [22]. In the present work, our main emphasis is on the flow and geometric parameters. Another work for identification of geometric parameters is by Patankar and Sridhar [25], which delineated the behavior of the Coanda flow, which is the function of the aspect ratio (ratio between jet orifice length to the jet orifice width), but the choice of aspect ratio depends upon the geometry of the flow. There are no unanimously defined parameters that influence the flow, due to the intricacies of the flow geometry. Here, an attempt has been made to define such parameters for this case and also to drive other researchers to validate these parameters. Therefore, for the nozzle flow over the Coanda surface, one can define the aspect ratio by diameter of the exit nozzle ($b$) to the radius of curvature ($R$) of the Coanda surface. In the present chapter this has been defined for design purposes.

There are more parameters which have been discussed in Section 14.4. The present work delineates relevant work, which is focused on the identification of the geometric and flow parameters for the design of such flow. To use this flow phenomenon for industrial applications, there is a necessity to identify the flow and geometric parameters for better optimized design.

## 14.2  Governing equations

The mass and momentum conservation equations are given in the Reynolds averaging form

$$\frac{\partial U_i}{\partial x_i} = 0 \tag{14.1}$$

$$\frac{\partial U_i}{\partial t} + U_j \frac{\partial U_i}{\partial x_j} = -\frac{1}{\rho}\frac{\partial P}{\partial x_i} + \nu \frac{\partial^2 U_i}{\partial x_j \partial x_j} - \frac{\partial}{\partial x_j}\overline{u_i' u_j'} \tag{14.2}$$

### 14.2.1  Spalart–Allmaras model

The Spalart–Allmaras equations are as follows:

$$\frac{\partial \hat{v}}{\partial t} + u \frac{\partial \hat{v}}{\partial x_j} = c_{b1}(1 - f_{t2})\hat{S}\hat{v} - \left[c_{w1}f_w - \frac{c_{b1}}{\kappa^2}f_{t2}\right]\left(\frac{\hat{v}}{d}\right)^2 + \frac{1}{\sigma}\left[\frac{\partial}{\partial x_j}\left\{(\nu + \hat{v})\frac{\partial \hat{v}}{\partial x_j}\right\} + c_{b2}\frac{\partial \hat{v}}{\partial x_i}\frac{\partial \hat{v}}{\partial x_i}\right] \tag{14.3}$$

and the turbulent eddy viscosity [26] is computed from

$$\mu_t = \rho \hat{v} f_{v1} \tag{14.4}$$

where

$$f_{v1} = \frac{\chi^3}{\chi^3 + c_{v1}^3} \tag{14.5}$$

$$\chi = \frac{\hat{v}}{\nu} \tag{14.6}$$

and $\rho$ is the density, $\nu = \mu/\rho$ is the molecular kinematic viscosity, and $\mu$ is the molecular dynamic viscosity. Additional definitions are given by the following equations:

$$\hat{S} = \Omega + \frac{\nu}{\kappa^2 d^2}f_{v2} \tag{14.7}$$

where $\Omega = \sqrt{2W_{ij}W_{ij}}$ is the magnitude of the vorticity, $d$ is the distance from the field point to the nearest wall, and

$$f_{v2} = 1 - \frac{\chi}{1 + \chi f_{v1}} \tag{14.8}$$

$$f_w = g\left[\frac{1+c_{w3}^6}{g^6+c_{w3}^6}\right] \tag{14.9}$$

$$g = r + c_{w2}\left(r^6 - r\right) \tag{14.10}$$

$$r = \min\left[\frac{\nu}{S\kappa^2 d^2}, 10\right] \tag{14.11}$$

$$f_{t2} = c_{t3}\exp\left(-c_{t4}\chi^2\right) \tag{14.12}$$

$$W_{ij} = \frac{1}{2}\left(\frac{\partial u_i}{\partial x_j} - \frac{\partial u_j}{\partial x_i}\right) \tag{14.13}$$

The constants are

$$c_{b1} = 0.1335,\ \sigma = 2/3,\ c_{b2} = 0.622,\ \kappa = 0.41,\ c_{w2} = 0.3,\ c_{w3} = 2,\ c_{v1} = 7.1,\ c_{t3} = 1.2$$

$$c_{t4} = 0.5,\ c_{w1} = \frac{c_{b1}}{\kappa^2} + \frac{1+c_{b2}}{\sigma}$$

### 14.2.2   *k–ε Model*

The equations are as follows [27]:

$$\rho\frac{\partial k}{\partial t} + \rho u_j\nabla k = \nabla\left[\left(\mu + \frac{\mu_T}{\sigma_k}\right)\nabla k\right] + \mu_T P - \rho\varepsilon \tag{14.14}$$

$$\rho\frac{\partial \varepsilon}{\partial t} + \rho u_j\nabla \varepsilon = \nabla\left[\left(\mu + \frac{\mu_T}{\sigma_\varepsilon}\right)\nabla \varepsilon\right] + \frac{\varepsilon}{k}\left(C_{\varepsilon 1}\mu_T P - \rho C_{\varepsilon 2}\varepsilon\right) - \rho\varepsilon \tag{14.15}$$

where

$$P = \nabla u \times \left[\nabla u + (\nabla u)^T\right] \tag{14.16}$$

$$\mu_T = \rho C_\mu \frac{k^2}{\varepsilon} \tag{14.17}$$

The model constants are

$$C_\mu = 0.09,\ C_{\varepsilon 1} = 1.44,\ C_{\varepsilon 2} = 1.92,\ \sigma_k = 1.0,\ \sigma_\varepsilon = 1.3$$

### 14.2.3   *SST k–ω model*

The kinematic eddy viscosity [28] is given by

$$\nu_T = \frac{a_1 \cdot k}{\max(a_1\ \omega, SF_2)} \tag{14.18}$$

Turbulent kinetic energy:

$$\rho\frac{\partial k}{\partial t}+U_j\frac{\partial k}{\partial x_j}=P_k-\beta^*k\omega+\left[(v+\sigma_k v_T)\frac{\partial k}{\partial x_j}\right] \tag{14.19}$$

Specific dissipation rate:

$$\rho\frac{\partial \omega}{\partial t}+U_j\frac{\partial \omega}{\partial x_j}=\alpha S^2-\beta\omega^2+\left[(v+\sigma_\omega v_T)\frac{\partial \omega}{\partial x_j}\right]+2(1-F_1)\sigma_{\omega 2}\frac{1}{\omega}\frac{\partial k}{\partial x_i}\frac{\partial \omega}{\partial x_i} \tag{14.20}$$

Closure coefficient and auxiliary equations are given by

$$F_2=\tanh\left[\left[\max\left(\frac{2\sqrt{k}}{\beta^*\omega y},\frac{500v}{y^2\omega}\right)\right]^2\right] \tag{14.21}$$

$$F_1=\tanh\left[\left[\min\left(\max\left(\frac{2\sqrt{k}}{\beta^*\omega y},\frac{500v}{y^2\ \omega}\right),\frac{4\sigma_{\omega 2}\ k}{CD_{k\omega}\ y^2}\right)\right]^4\right] \tag{14.22}$$

$$S^2=\left|\frac{1}{2}(\partial_j u_i+\partial_i u_j)\right|^2 \tag{14.23}$$

$$P_k=\min\left(G,10\beta^*k\omega\right) \tag{14.24}$$

$$G=v_T\frac{\partial U_i}{\partial x_j}\left(\frac{\partial U_i}{\partial x_j}+\frac{\partial U_j}{\partial x_i}\right) \tag{14.25}$$

$$CD_{k\omega}=\max\left(2\rho\sigma_{\omega 2}\frac{1}{\omega}\frac{\partial k}{\partial x_j}\frac{\partial \omega}{\partial x_j},10^{-10}\right) \tag{14.26}$$

$$\phi=\phi_1 F_1+\phi_2(1-F_1) \tag{14.27}$$

The values of model constants are given as

$\alpha_1=5/9,\ \alpha_2=0.44,\ \beta_1=3/40,\ \beta_2=0.0828,\ \beta^*=0.09,\ \sigma_{k1}=0.85,\sigma_{k2}=1,\ \sigma_{\omega 1}=0.5,\sigma_{\omega 2}=0.856$

## 14.2.4   *k–ε–ζ–f Model*

In order to improve numerical stability of the original $\overline{v^2}-f$ model by solving a transport equation for the velocity scale $\zeta=\frac{\overline{v^2}}{k}$ instead of velocity scale $\overline{v^2}$. The variable $\zeta$ represents a scalar whose near-wall behaviour resembles that of the normal-to-wall Reynolds stress component.

Incorporating the Durbin's [29] elliptic relaxation concept, a new eddy-viscosity turbulence model comprising four equations denoted as $k$–$\varepsilon$–$\zeta$–$f$ was developed by Hanjalic et al. [30].

The eddy-viscosity is obtained in the following form:

$$v_t = C_\mu \zeta \frac{k^2}{\varepsilon} \tag{14.28}$$

moreover, the rest of the variables are from the following set of model equations; thus,

$$\frac{\partial(\alpha_k \rho_k k_k)}{\partial t} + \frac{\partial}{\partial x_j}(\alpha_k \rho_k v_k k_k) = \frac{\partial}{\partial x_j}\left[\alpha_k \left(\mu_k + \frac{\mu_k^t}{\sigma_k}\right)\frac{\partial k_k}{\partial x_j}\right] + \alpha_k \rho_k (P_k - \varepsilon_k) \tag{14.29}$$

$$\frac{\partial(\alpha_k \rho_k \varepsilon_k)}{\partial t} + \frac{\partial}{\partial x_j}(\alpha_k \rho_k v_k \varepsilon_k) = \frac{(C_{\varepsilon 1}^* P_k \alpha_k \varepsilon_k - C_{\varepsilon 2}\alpha_k \varepsilon_k^2)}{k_k} + \frac{\partial}{\partial x_j}\left[\alpha_k \left(\mu_k + \frac{\mu_k^t}{\sigma_\varepsilon}\right)\frac{\partial \varepsilon_k}{\partial x_j}\right] \tag{14.30}$$

$$\frac{\partial(\alpha_k \rho_k \zeta_k)}{\partial t} + \frac{\partial}{\partial x_j}(\alpha_k \rho_k v_k \zeta_k) = \rho f - \rho \frac{\zeta}{k} P_k + \frac{\partial}{\partial x_j}\left[\alpha_k \left(\mu_k + \frac{\mu_k^t}{\sigma_\zeta}\right)\frac{\partial \zeta_k}{\partial x_j}\right] \tag{14.31}$$

where the following form of the $f$ equation is adopted as

$$f - L^2 \frac{\partial^2 f}{\partial x_j \partial x_j} = \left(C_1 + C_2 \frac{P_k}{\zeta}\right)\frac{\left(\frac{2}{3} - \zeta\right)}{T} \tag{14.32}$$

The turbulent time scale $T$ and length scale $L$ are given by

$$T = \max\left(\min\left(\frac{k}{\varepsilon}, \frac{\alpha k}{v^2 C_\mu \sqrt{6S^2}}\right), C_T \left(\frac{v}{\varepsilon}\right)^{1/2}\right) \tag{14.33}$$

$$L = C_L \max\left(\min\left(\frac{k^{3/2}}{\varepsilon}, \frac{k^{3/2}}{v^2 C_\mu \sqrt{6S^2}}\right), C_\eta \left(\frac{v^3}{\varepsilon}\right)^{1/2}\right) \tag{14.34}$$

Additional modifications to the $\varepsilon$-equation are that the constant $C_{\varepsilon 1}$ is dampened close to the wall; thus,

$$C_{\varepsilon 1}^* = C_{\varepsilon 1}\left(1 + 0.045\sqrt{1/\zeta}\right) \tag{14.35}$$

This is computationally more robust than the original model $\overline{v^2} - f$.

## 14.3 *Grid independence test and solution methodology*

The main geometric parameters such as the ratio of the nozzle throat diameter ($b$) to the exit curvature radius ($R$) have been considered for the present study. The grid has been generated in the Gambit [31] and the boundary layer resolved appropriately as shown in Figures 14.1 and 14.2 for two different $b/R$ ratios. Computations have been performed on

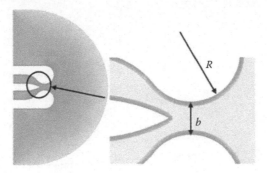

*Figure 14.1* Grid generation of the first flow geometry.

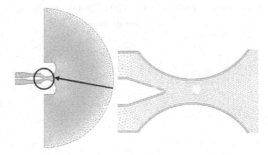

*Figure 14.2* Grid generation of the second flow geometry.

commercial software AVL Fire [32] for $k$–$\zeta$–$f$ turbulence model and the rest of the model on Fluent 6.3 (2016) for Ref. [33].

The grid independent check has been performed according to the ERCOFTAC [30] guidelines and as depicted in the following papers [34,35]. The optimum number of grid (numerically stable grid) has been determined through the numerical computation of the grid at different refinement levels of the grid at the curved surface (first grid from the wall at 80, 40, and 20 μm). It has been found that, when the grid resolved the viscous sublayer until $y+$ value less than two (first grid from the wall at 20 μm), then one can get the jet deflection angle independent of the grid. In addition, the $x+$ value is less than 60 for the stable solution of the flow along the downstream. We have employed the four turbulence models such as the Spalart–Allmaras (SA) model [26], SST $K$–$\omega$ [28] model, $k$–$\varepsilon$ with enhanced wall treatment [27] model, and $k$–$\zeta$–$f$ model [30].

In this chapter, the descritization error has been minimized using the second-order upwind scheme for the momentum equation and for modified turbulent viscosity in the Spalart–Allmaras (SA) model (Equation 14.3). The pressure and velocity have been coupled through the SIMPLE (Semi-Implicit Method for Pressure Linked Equation) algorithm [36].

The first-order implicit method has been used to discretize the unsteady term. The advantage of the fully implicit scheme is that it is unconditionally stable with respect to time step size. However, the time step has been taken $\Delta t = 1 \times 10^{-3}$ s. Figure 14.3 (a–c) shows the residual RMS plot of the three models has been given; it has been found that for the SA model the error is in the range of $10^{-9}$–$10^{-10}$, that is lowest in comparison to the other turbulence model.

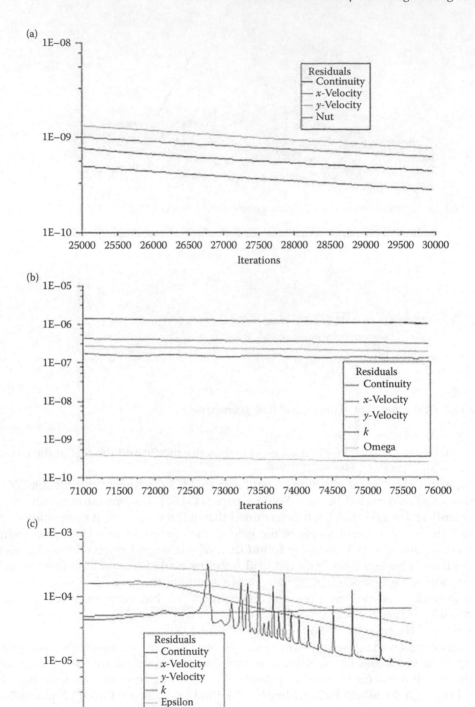

*Figure 14.3* Residual plot for different turbulence models: (a) SA, (b) SST $k$–$\omega$, (c) $k$–$\varepsilon$ model.

Moreover, we have found that the $k$–$\zeta$–$f$ model has more numerical stability than other models; therefore, in this computational study this model has been used.

## 14.4 Results and discussion

Before going into the details of the computational study, it is customary to define the road map of the design parameters. There are several parameters for the design of such a nozzle. But, the key parameters on which the flow and geometry have vital importance are as follows:

Nozzle exit throat diameter: $b$

Nozzle exit curvature radius: $R$

Nozzle exit Reynolds number: $\mathrm{Re}_{exit}$

Flow Reynolds number over the curved surface: $\mathrm{Re}_{flow}$

Exit jet deflection angle: $\theta$

Mach number of the flow (calculated at the exit of the nozzle for compressible flow): $M$

Velcity ratio of two jets or mass flux ratio of two jets: $V_1/V_2$ or $M_1/M_2$

Thrust of the jet in the $x$-direction (along the flow) and the $y$-direction (normal to the flow): $F_x$ and $F_y$

It has been easily realized that, for such application in V/STOL the flow velocity will be in a compressible range. However, in the present work, the study has been started from the incompressible flow to at the verge of the compressible flow ($M = 0.3$).

For the geometry (Figure 14.1) the radius of exit curvature ($R$) is 101.566 mm, the exit throat diameter ($b$) is 40.163 mm and inlet diameter ($d_1 = d_2$) is 56 mm and the ratio of $b/R$ is 0.395. Assume the average velocity and velocity ratio are constant for all the computations within the range of average velocity 20–40 m/s. Initial observation on the computational results shows that the jet deflection angle is the function of velocity ratio. The flow visualization has been given through velocity contours from Figures 14.4 to 14.6 for some selected cases. Now, these contours reveal the fact that the highest velocity occurs at the exit of the nozzle attached to the upper curvature of the nozzle, as the upper nozzle has been designated as $V_1$ velocity and lower nozzle as $V_2$. The ratio is always greater than one; therefore, the attachment of the flow is near the upper exit curvature and the maximum velocity occurs there. Consequently, the boundary layer thickness is very small in the range of the micrometer.

In another configuration as shown in Figure 14.2, the throat diameter ($b$) is 46 mm, the radius of curvature ($R$) is 179.543 mm, the inlet diameter ($d_1 = d_2$) is 77 mm, and the $b/R$ ratio is 0.256. The velocity contours have been shown in Figure 14.7. The complete attachement of the flow on the Coanda surface has been seen for the velocity ratio greater than 1.3. The reason for the larger attachment angle can be explained by the velocity vector (Figure 14.8) and the velocity plot (Figures 14.9 and 14.10).

Figure 14.8 depicts the velocity profile at the exit of the nozzle. The subsequent plots in Figures 14.9 and 14.10 show the velocity profile near the exit and far from the exit, respectively. Near the exit of the flow, the effect of pulling the low velocity jet toward the high velocity jet is low, and as the flow progresses the effect is more pronounced. We can see that far from the exit, the velocity profile attained the higher gradient than the low velocity jet; in effect the low velocity jet attracted toward the high velocity jet.

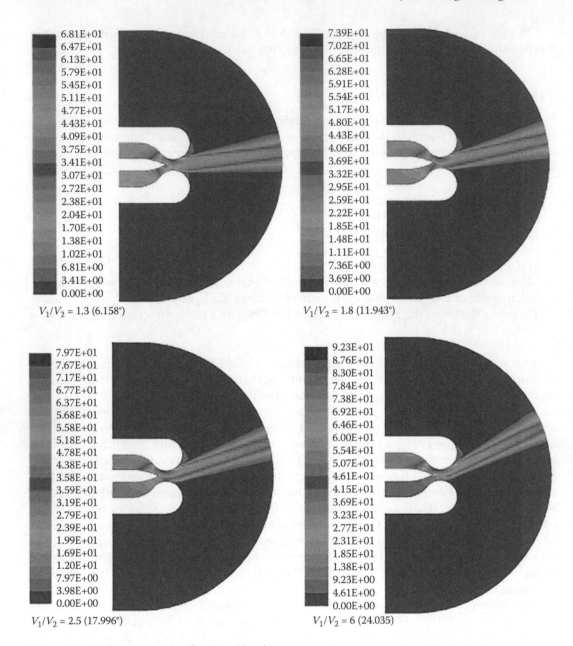

| | | | |
|---|---|---|---|
| 6.81E+01 | | 7.39E+01 | |
| 6.47E+01 | | 7.02E+01 | |
| 6.13E+01 | | 6.65E+01 | |
| 5.79E+01 | | 6.28E+01 | |
| 5.45E+01 | | 5.91E+01 | |
| 5.11E+01 | | 5.54E+01 | |
| 4.77E+01 | | 5.17E+01 | |
| 4.43E+01 | | 4.80E+01 | |
| 4.09E+01 | | 4.43E+01 | |
| 3.75E+01 | | 4.06E+01 | |
| 3.41E+01 | | 3.69E+01 | |
| 3.07E+01 | | 3.32E+01 | |
| 2.72E+01 | | 2.95E+01 | |
| 2.38E+01 | | 2.59E+01 | |
| 2.04E+01 | | 2.22E+01 | |
| 1.70E+01 | | 1.85E+01 | |
| 1.38E+01 | | 1.48E+01 | |
| 1.02E+01 | | 1.11E+01 | |
| 6.81E+00 | | 7.36E+00 | |
| 3.41E+00 | | 3.69E+00 | |
| 0.00E+00 | | 0.00E+00 | |

$V_1/V_2 = 1.3 \ (6.158°)$    $V_1/V_2 = 1.8 \ (11.943°)$

| | | | |
|---|---|---|---|
| 7.97E+01 | | 9.23E+01 | |
| 7.67E+01 | | 8.76E+01 | |
| 7.17E+01 | | 8.30E+01 | |
| 6.77E+01 | | 7.84E+01 | |
| 6.37E+01 | | 7.38E+01 | |
| 5.68E+01 | | 6.92E+01 | |
| 5.58E+01 | | 6.46E+01 | |
| 5.18E+01 | | 6.00E+01 | |
| 4.78E+01 | | 5.54E+01 | |
| 4.38E+01 | | 5.07E+01 | |
| 3.58E+01 | | 4.61E+01 | |
| 3.59E+01 | | 4.15E+01 | |
| 3.19E+01 | | 3.69E+01 | |
| 2.79E+01 | | 3.23E+01 | |
| 2.39E+01 | | 2.77E+01 | |
| 1.99E+01 | | 2.31E+01 | |
| 1.69E+01 | | 1.85E+01 | |
| 1.20E+01 | | 1.38E+01 | |
| 7.97E+00 | | 9.23E+00 | |
| 3.98E+00 | | 4.61E+00 | |
| 0.00E+00 | | 0.00E+00 | |

$V_1/V_2 = 2.5 \ (17.996°)$    $V_1/V_2 = 6 \ (24.035)$

*Figure 14.4* Velocity contours for $V_{av} = 20 \, \text{m/s}$.

Therefore, it can be said that due to the decrement of the $b/R$ ratio from 0.395 to 0.256 a large attachment angle has been found. In this way, the jet adhesion angle is a strong function of the $b/R$ ratio, which is a geometric parameter. The $b/R$ ratio is the main controlling parameter for the jet adhesion angle.

After computation for the above-mentioned velocity ranges, the relation between the velocity ratio ($V_1/V_2$) and the jet deflection angle ($\theta$) is plotted in Figure 14.11. For the average velocities 20, 25, and 30 m/s, the deflection angle has nearly the same value until the large velocity ratio. For the $V_{av} = 35 \, \text{m/s}$, there is little difference in the jet

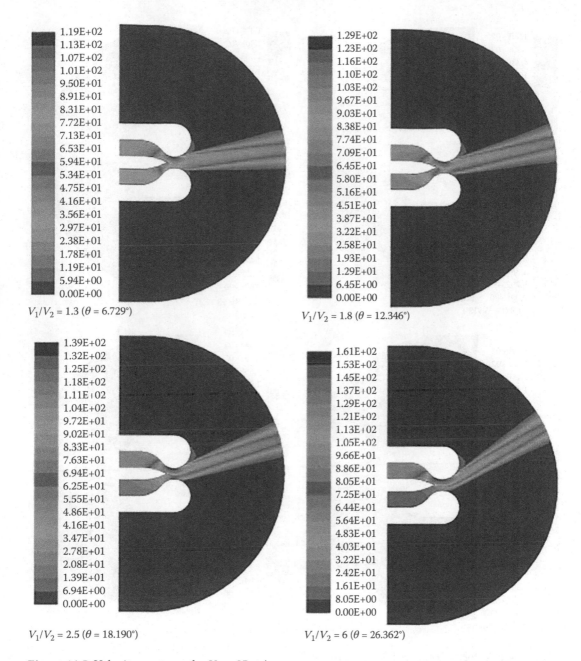

| 1.19E+02 | | 1.29E+02 |
| 1.13E+02 | | 1.23E+02 |
| 1.07E+02 | | 1.16E+02 |
| 1.01E+02 | | 1.10E+02 |
| 9.50E+01 | | 1.03E+02 |
| 8.91E+01 | | 9.67E+01 |
| 8.31E+01 | | 9.03E+01 |
| 7.72E+01 | | 8.38E+01 |
| 7.13E+01 | | 7.74E+01 |
| 6.53E+01 | | 7.09E+01 |
| 5.94E+01 | | 6.45E+01 |
| 5.34E+01 | | 5.80E+01 |
| 4.75E+01 | | 5.16E+01 |
| 4.16E+01 | | 4.51E+01 |
| 3.56E+01 | | 3.87E+01 |
| 2.97E+01 | | 3.22E+01 |
| 2.38E+01 | | 2.58E+01 |
| 1.78E+01 | | 1.93E+01 |
| 1.19E+01 | | 1.29E+01 |
| 5.94E+00 | | 6.45E+00 |
| 0.00E+00 | | 0.00E+00 |

$V_1/V_2 = 1.3$ ($\theta = 6.729°$)        $V_1/V_2 = 1.8$ ($\theta = 12.346°$)

| 1.39E+02 | | 1.61E+02 |
| 1.32E+02 | | 1.53E+02 |
| 1.25E+02 | | 1.45E+02 |
| 1.18E+02 | | 1.37E+02 |
| 1.11E+02 | | 1.29E+02 |
| 1.04E+02 | | 1.21E+02 |
| 9.72E+01 | | 1.13E+02 |
| 9.02E+01 | | 1.05E+02 |
| 8.33E+01 | | 9.66E+01 |
| 7.63E+01 | | 8.86E+01 |
| 6.94E+01 | | 8.05E+01 |
| 6.25E+01 | | 7.25E+01 |
| 5.55E+01 | | 6.44E+01 |
| 4.86E+01 | | 5.64E+01 |
| 4.16E+01 | | 4.83E+01 |
| 3.47E+01 | | 4.03E+01 |
| 2.78E+01 | | 3.22E+01 |
| 2.08E+01 | | 2.42E+01 |
| 1.39E+01 | | 1.61E+01 |
| 6.94E+00 | | 8.05E+00 |
| 0.00E+00 | | 0.00E+00 |

$V_1/V_2 = 2.5$ ($\theta = 18.190°$)        $V_1/V_2 = 6$ ($\theta = 26.362°$)

*Figure 14.5*  Velocity contours for $V_{av} = 35$ m/s.

deflection angle from $V_1/V_2 = 4$; however, for the lower velocity ratio it is nearly the same. In other words, it can be said that the rate of increment of the deflection angle for average velocity 35 m/s is higher than the lower velocity range (20, 25, and 30 m/s) and for the 40 m/s. For the $V_{av} = 40$ m/s, the deflection angle is larger than the other average velocity. The reason may lie in the fact that, at this inlet velocity, the exit velocity of jet is quite higher near to the Mach number 0.3. Nevertheless, the slope is the same as for 20, 25, and 30 m/s for all velocity ratios. Now, the different behaviors of the slope for

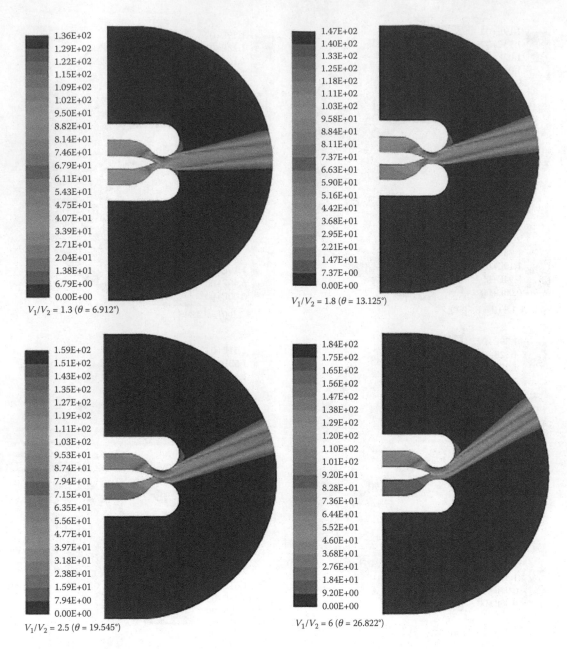

*Figure 14.6* Velocity contours for $V_{av} = 40$ m/s.

average velocity 35 m/s at larger velocity ratios have been investigated in detail. An investigation of the reason for aberration leads us to see the flow phenomena in detail; therefore, the calculation of the $Re_{exit}$ and $Re_{flow}$ performed for all average velocity range as shown in Table 14.1.

     An interesting phenomenon has been observed, for the average velocity 20, 25, 30, and 40 m/s, i.e., the exit Reynolds number is always higher than the flow Reynolds number over the exit curvature. This behavior pointed out the laminarization of the flow; however, for most cases the flow Reynolds numbers are in the order of $10^5$ (we are using the word

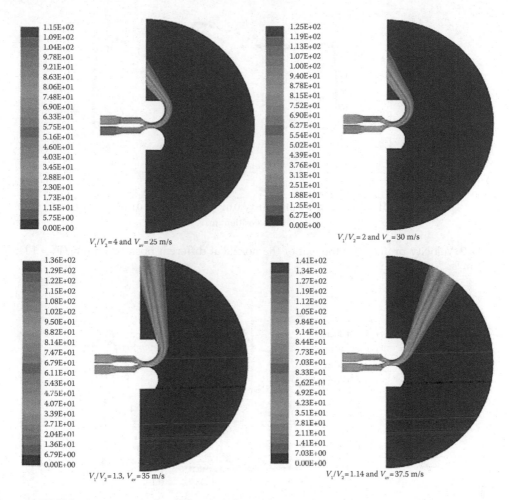

$V_1/V_2 = 4$ and $V_{av} = 25$ m/s

$V_1/V_2 = 2$ and $V_{av} = 30$ m/s

$V_1/V_2 = 1.3$, $V_{av} = 35$ m/s

$V_1/V_2 = 1.14$ and $V_{av} = 37.5$ m/s

*Figure 14.7* Velocity contours for different velocity ratio and various average velocities.

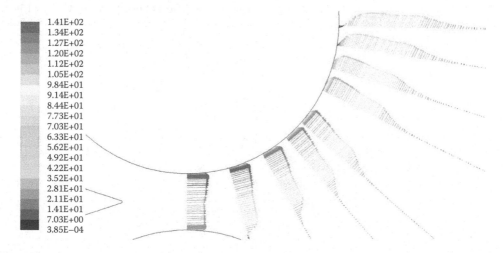

*Figure 14.8* Velocity vector at different positions of the outer wall of the Coanda surface for $V_1/V_2 = 1.14$ and $V_{av} = 37.5$ m/s.

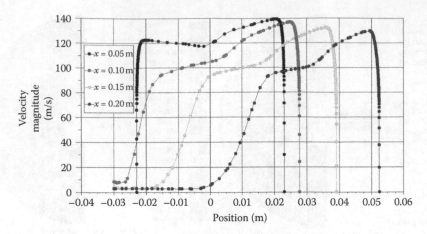

**Figure 14.9** Velocity profile near the exit of the nozzle at different positions for $V_1/V_2 = 1.14$ and $V_{av} = 37.5\,\text{m/s}$.

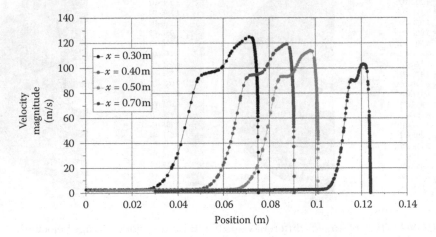

**Figure 14.10** Velocity profile far from the exit of the nozzle at different positions for $V_1/V_2 = 1.14$ and $V_{av} = 37.5\,\text{m/s}$.

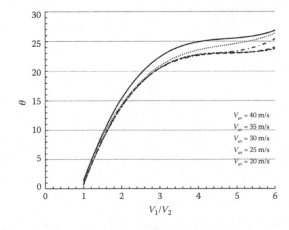

**Figure 14.11** Plots between velocity ratio and the deflection angle.

*Table 14.1* The calculation of the exit Reynolds number and the flow Reynolds number over the exit curvature

| $V_1/V_2$ | $Re_{exit}$ | $Re_{flow}$ |
|---|---|---|
| | $V_{av} = 40\,m/s$ | |
| 1.3 | 3.080E+05 | 9.395E+04 |
| 1.8 | 3.080E+05 | 1.784E+05 |
| 2.5 | 3.080E+05 | 2.657E+05 |
| 6.0 | 1.269E+05 | 1.502E+05 |
| | $V_{av} = 35\,m/s$ | |
| 1.3 | 2.695E+05 | 3.135E+05 |
| 1.8 | 2.695E+05 | 3.135E+05 |
| 2.5 | 2.302E+05 | 2.679E+05 |
| 6.0 | 1.110E+05 | 1.291E+05 |
| | $V_{av} = 30\,m/s$ | |
| 1.3 | 1.977E+05 | 1.572E+05 |
| 1.8 | 2.310E+05 | 1.836E+05 |
| 2.5 | 9.467E+04 | 7.525E+04 |
| 6.0 | 7.989E+04 | 6.350E+04 |
| | $V_{av} = 25\,m/s$ | |
| 1.3 | 7.870E+04 | 2.116E+04 |
| 1.8 | 1.646E+05 | 8.599E+04 |
| 2.5 | 7.888E+04 | 6.227E+04 |
| 6.0 | 6.658E+04 | 7.012E+04 |
| | $V_{av} = 20\,m/s$ | |
| 1.3 | 5.28E+04 | 1.435E+04 |
| 1.8 | 1.32E+05 | 6.942E+04 |
| 2.5 | 1.32E+05 | 1.004E+05 |
| 6.0 | 1.31E+05 | 1.392E+05 |

"laminarization" because the flow retards on the curved surface with respect to the flow at the exit of the nozzle). Very few cases go below this range, but for the average velocity 35 m/s, the flow accelerates over the curved surface, and consequently, the flow Reynolds number is higher than the exit Reynolds number. Now, we reach at this point that the flow over the curved surface can exhibit two phenomena: one is laminarization and the other is acceleration. These two distinct behaviors have different characteristics of the jet deflection angle with respect to the higher velocity ratio. To envisage this behavior from another point, the Mach number has been calculated at the exit flow velocity and found around 0.3. This is near the compressible flow behavior, and for higher average velocity the exit Mach number is a little bit above this value. Until, now, we have these two behaviors of the flow. This should be further investigated experimentally.

Our next attempt is to envisage the effect of laminarization and the acceleration of flow upon the thrust. Here, the thrust means the force exerted by the flow in along the flow and normal to the flow. The thrust in *x*-direction would contribute in cruising and the normal thrust in $y$-direction would contribute in the lift force of the aircraft. This has been calculated by multiplying the mass flux with the exit velocity as shown in Table 14.2. The force normal to the flow direction can contribute to the lift force. Therefore, it has been observed from Table 14.2 that the normal force is maximum for the range of velocity ratio from 1.8 to 2.5. This is a strong function of the flow phenomena over the curved surface depicted above. Therefore, it can be said that the force that contributes the lift can be in the range of the velocity ratio of 1.8–2.5. Through this computational study, the range of the maximum force has been investigated, which is vital for the design of such a nozzle for maximum lift force.

Until now, we have discussed the influence of given parameters on the flow. For the given velocity ratio, for which we can have maximum thrust, the jet deflection angle is a function of $b/R$ ratio.

For the velocity ratios 1.3, 2, and 4, we can see the complete attachments of the flow over the exit curvature for the lower value of the $b/R$ ratio. This parameter has a strong

**Table 14.2** Force calculation along the flow ($x$-direction) direction and normal to the flow ($y$-axis)

| $V_1/V_2$ | $F_x$ (N) | $F_y$ (N) |
|---|---|---|
| | $V_{av} = 40\,\text{m/s}$ | |
| 1.3 | 614.711 | 42.863 |
| 1.8 | 614.704 | 92.796 |
| 2.5 | 614.704 | 124.714 |
| 6.0 | 614.711 | 42.863 |
| | $V_{av} = 35\,\text{m/s}$ | |
| 1.3 | 470.639 | 32.272 |
| 1.8 | 470.634 | 56.322 |
| 2.5 | 402.063 | 73.827 |
| 6.0 | 193.852 | 57.003 |
| | $V_{av} = 30\,\text{m/s}$ | |
| 1.3 | 295.984 | 23.385 |
| 1.8 | 345.773 | 43.333 |
| 2.5 | 141.717 | 27.383 |
| 6.0 | 119.593 | 35.587 |
| | $V_{av} = 25\,\text{m/s}$ | |
| 1.3 | 98.175 | 6.280 |
| 1.8 | 205.371 | 28.274 |
| 2.5 | 98.407 | 18.219 |
| 6.0 | 83.056 | 23.674 |
| | $V_{av} = 20\,\text{m/s}$ | |
| 1.3 | 52.668 | 3.587 |
| 1.8 | 131.428 | 18.13 |
| 2.5 | 131.229 | 24.462 |
| 6.0 | 130.925 | 31.676 |

influence on the attachment of the flow and can be controlled for the desirable thrust for the propulsion.

## 14.5  Conclusions

The flow behaviors have been studied in detail. The flow and geometric parameters for the design of such a nozzle have been recognized through the computational fluid dynamics analysis. It has been realized that the influence of the $b/R$ ratio is higher than other parameters. Therefore, this parameter can be one of the controlling elements for the design of the nozzle in order to have maximum lift and cruising velocity. The range of $V_1/V_2$ has been defined for the maximum thrust. The correlation will be developed after the experimental database for calculation of thrust and deflection angle.

## Acknowledgments

Some of the computational works of the present chapter were performed as part of Project ACHEON (Aerial Coanda High Efficiency Orienting-jet Nozzle) with ref. 309041, supported by the European Union through the 7th Framework Programme during the stay of the first author at UNIMORE Italy. Some of the computational studies were performed recently using AVL-Fire. The first author acknowledges AVL List GmbH, Hans-List-Platz 1, A-8020, Graz, Austria for providing AVL AST software for research and development purposes under the University Partnership Program.

## Nomenclature

$b$ = Exit throat diameter of the nozzle (m)
$d_1$ = Diameter of the upper nozzle (m)
$d_2$ = Diameter of the lower nozzle (m)
$F_x$ = X component of the resultant thrust at the exit of the nozzle (N)
$F_y$ = Y component of the resultant thrust at the exit of the nozzle (N)
$P$ = Mean pressure (N/m²)
$R$ = Radius of the Coanda surface attached with the nozzle exit (m)
$Re_{exit}$ = Reynolds number of the flow at the exit of the nozzle, $V_{exit,av} \times b/\nu$ (–)
$Re_{flow}$ = Reynolds number of the flow at the curved surface of the nozzle, $V_{flow,av} \times R\theta/\nu$ (–)
$U_i$ = Reynolds averaged velocity tensor (m/s)
$u'_i$ = Fluctuating velocity tensor (m/s)
$V_1$ = Velocity of the flow in upper nozzle (m/s)
$V_2$ = Velocity of the flow in lower nozzle (m/s)

## References

1. Yoshitani, N., Hashimoto, S.-I., Kimura, T., Motohashi, K., and Ueno, S., "Flight Control Simulators for Unmanned Fixed-Wing and VTOL Aircraft," *ICROS-SICE International Joint Conference 2009*, August 18–21, 2009, Fukuoka International Congress Center, Japan.
2. Thomason, T., "Bell-Boeing JVX Tilt Rotor Program – Flight Test Program," *American Institute of Aeronautics and Astronautics*, 1983, AIAA Paper No. 83-2726.
3. Saeed, B., Gratton, G., and Mares, C., "A Feasibility Assessment of Annular Winged VTOL Flight Vehicles," *Aeronautical Journal*, Vol. 115, 2011, pp. 683–692.
4. Kim, H., Rajesh, G., Setoguchi, T., and Matsuo, S., "Optimization Study of a Coanda Ejector," *Journal of Thermal Science*, Vol. 15, No. 4, 2006, pp. 331–336.

5. Alvi, F., Strykowski, P., Krothapalli, A., and Forliti, D., "Vectoring Thrust in Multiaxis Using Confined Shear Layers," *Journal of Fluids Engineering*, Vol. 122, No. 1, 2000, pp. 3–13.

6. Trancossi, M., Dumas, A. Giuliani, I., and Baffigi, I., "Ugello capace di deviare in modo dinamico e con-trollabile un getto sintetico senza parti meccaniche in movimento e suo sistema di controllo," Patent No.RE2011A000049, Italy, 2011.

7. Freund J. B. and Mungal, M. G., "Drag and Wake Modification of Axisymmetric Bluff Bodies Using Coanda Blowing," *Journal of Aircraft*, Vol. 31, No. 3, May–June 1994, pp. 572–578.

8. Chng, T. L., Rachman, A., Tsai, H. M., and Zha, Ge-C., "Flow Control of an Airfoil via Injection and Suction," *Journal of Aircraft*, Vol. 46, No. 1, 2009, pp. 291–300.

9. Lee, D.-W., Hwang, J.-G., Kwon, Y.-D., Kwon, S.-B., Kim, G-Y., and Lee, D.-E., "A Study on the Air Knife Flow with Coanda Effect," *Journal of Mechanical Science and Technology*, Vol. 21, 2007, pp. 2214–2220.

10. Lalli, F., Bruschi, A., Lama, R., Liberti, L., Mandrone, S., and Pesarino, V., "Coanda Effects in Coastal Flows," *Coastal Engineering*, Vol. 57, 2010, pp. 278–289.

11. Collis, S.S., Joslin, R.D., Seifert, A., and Theofilis, V., "Issues in Active Flow Control: Theory, Control, Simulation, and Experiment," *Progress in Aerospace Science*, Vol. 40, 2004, pp. 237–289.

12. Florin, F., Alexandru, D., Octavian, P., and Horia, D., "Control of Two-Dimensional Turbulent Wall Jet on a Coanda Surface," *Proceedings in Applied. Mathematics and Mechanics*, Vol. 11, 2011, pp. 651–652.

13. Mabey, K., Smith, B., Whichard, G., and McKechnie, T., "Coanda-Assisted Spray Manipulation Collar for a Commercial Plasma Spray Gun," *Journal of Thermal Spray Technology*, Vol. 20, No. 4, 2011, pp. 782–790.

14. Kim, H., Rajesh, G., Setoguchi, T., and Matsuo, S., "Optimization Study of a Coanda Ejector," *Journal of Thermal Science*, Vol. 15, No. 4, 2006, pp. 331–336.

15. Vanierschot, M., Persoons, T., and Van den Bulck, E., "A New Method for Annular Jet Control Based on Cross-Flow Injection," *Physics of Fluids*, Vol. 21, 2009, pp. 025103-1–025103-9.

16. Young, T., "Outlines of Experiments and Inquires Respecting Sound and Light," *Philosophical Transactions of Royal Society of London*, Vol. 90, 1 January 1800, pp. 106–150.

17. Rayleigh, L., "On the Dynamics of Revolving Fluid," *Proceedings of Royal Society of London*, Series A, Vol. 93, No. 648, 1 March, 1917, pp. 148–154.

18. Newman, B. G., The Deflexion of Plane Jets by Adjacent Boundaries, in Coanda Effect, In *Boundary Layer and Flow Control*, edited by G. V. Lachmann, Vol. 1, Pergamon Press, Oxford, 1961, pp. 232–264.

19. Carpenter, P. W., and Green, P. N., "The Aeroacoustics and Aerodynamics of High-Speed Coanda Devices, Part 1: Conventional Arrangement of Exit Nozzle and Surface," *Journal of Sound and Vibration*, Vol. 208, No. 5, 1997, pp. 777–801.

20. Trancossi, M., "An Overview of Scientific and Technical Literature on Coanda Effect Applied to Nozzles," *SAE Technical Papers* No. 2011-01-2591, Issn 0148-7191, 2011.

21. Wille, R., and Fernholtz, H., "Report on the First European Mechanics Colloquium, on the Coanda Effect," *Journal of Fluid Mechanics*, Vol. 23, No. 4, 1965, pp. 801–819.

22. Subhash, M., and Dumas, A., "Computational Study of Coanda Adhesion over Curved Surface," *SAE International Journal of Aerospace*, 2013, paper number: 13ATC-0018/2013-01-2302 (Accepted for Publication).

23. Fekete, G. I., "Coanda Flow of a Two-Dimensional Wall Jet on the Outside of a Circular Cylinder," Mechanical Engineering Research Laboratories, Rept. 63-11, McGill University, 1963.

24. Neuendorf, R., and Wygnansky, I., "On a Turbulent Wall Jet Flowing over a Circular Cylinder," *Journal of Fluid Mechanics*, Vol. 381, 1999, pp. 1–25.

25. Patankar, U., and Sridhar, K., "Three-Dimensional Curved Wall Jets," *Journal of Basic Engineering*, Vol. 94, No. 2, 1972, pp. 339–344.

26. Spalart, P. R., and Allmaras, S. R., "A One-Equation Turbulence Model for Aerodynamic Flows," *AIAA* Paper No. 92-0439, 1992.

27. Launder, B. E., and Sharma, B. I., "Application of the Energy Dissipation Model of Turbulence to the Calculation of Flow Near a Spinning Disc," *Letters in Heat and Mass Transfer*, Vol. 1, No. 2, 1974, pp. 131–138.

28. Menter, F. R., "Two-Equation Eddy-Viscosity Turbulence Models for Engineering Applications," *AIAA Journal*, Vol. 32, No. 8, August 1994, pp. 1598–1605.
29. Durbin, P. A., "Separated Flow Computations with the k-$\varepsilon$-v$^2$ Model," *AIAA Journal*, Vol. 33, 1995, pp. 659–664.
30. Hanjalic, K., Popovac, M., and Hadziabdic, M., "A Robust Near-Wall Elliptic-Relaxation Eddy-Viscosity Turbulence Model for CFD," *International Journal of Heat Fluid Flow*, Vol. 25, No. 6, 2004, pp. 1047–1051.
31. ANSYS Fluent User manual, 2016.
32. AVL-Fire User manual, 2014.
33. Casey, M., and Wintergerste, T., "ERCOFTAC Special Interest Group on 'Quality and Trust in Industrial CFD' Best Practice Guidelines," Version 1.0, January 2000.
34. Rizzi, A., and Vos, J., "Towards Establishing Credibility in Computational Fluid Dynamics," *AIAA Journal*, Vol. 36, No. 5, 1998, pp. 668–675.
35. Celik, I., Li, J., Hu, G., and Shaffer, C., "Limitations of Richardson Extrapolation and Some Possible Remedies," *Journal of Fluids Engineering*, Vol. 127, July 2005, pp. 795–805.
36. Patankar, S. V., and Spalding, D. B., "A Calculation Procedure for Heat, Mass and Momentum Transfer in Three-Dimensional Parabolic Flows," *International Journal of Heat and Mass Transfer*, Vol. 15, 1972, pp. 1787–1806.

## chapter fifteen

# Introduction to collocation method with application of B-spline basis functions to solve differential equations

**Geeta Arora**
*Lovely Professional University*

## Contents

## 15.1  Introduction

Due to the wide existence and applicability of ordinary and partial differential equations in various branches of science and engineering, a variety of nonlinear systems of initial and boundary value problems have been extensively studied in the literature. Many of the mathematical models of engineering problems can be expressed in terms of partial differential equations such as in describing the physics of various phenomena in science, in the study of the physical laws of fluid flow diffusion in transport problems, electromagnetic waves, neural networks, tissue engineering, quantum phenomena, etc. These are some of the application areas where existing phenomena or processes can be easily described in the form of initial and boundary value problems. Since it is not always feasible to calculate the analytical solutions of obtained modeled equations, there emerges the need for and role of advanced numerical methods.

A variety of numerical methods are available to obtain the numerical as well as analytical solutions of partial differential equations. Two of the most popular techniques for solving partial differential equations include the finite difference method and the finite element method. In the finite difference method, the solution is derived at a finite number of points by approximating the derivatives at each of the selected points. The accuracy of this method is based on the refinement of the grid points where the solution is being evaluated [1]. In the finite element method the focus is on dividing the domain into a finite number of elements with allocated nodes at predefined locations around the boundary elements [2]. The elements as well as the nodes result in a mesh that can be refined to minimize the error.

In the last few years, the collocation method, which is a type of finite element method, has been an emerging popular technique to solve various ordinary and partial differential equations. This method has been developed from the finite element method using the concepts of the finite difference method. This method has been applied to solve a variety of mathematical problems with different types of basis functions with the aim to obtain the best possible numerical solutions of various linear and nonlinear mathematical problems. It involves satisfying a differential equation to some tolerance at a selected finite number of points, called *collocation points*.

In this chapter, the collocation method will be discussed using B-spline basis functions in standard as well as in trigonometric form. A numerical problem of an advection diffusion equation is solved to describe the application of the method with details. The obtained results are presented in the form of tables depicting the absolute and maximum absolute errors.

## 15.2 Collocation method

In the collocation method, the numerical solution of a differential equation is obtained as a linear combination of basis functions with unknown coefficients to be determined. In this approach, a given function is approximated by a polynomial at collocation points chosen by some predefined way that can be either uniform or nonuniform.

Let us discuss the application of this approach to a general differential equation represented in the following form:

$$f(x, u, u_x) = 0 \tag{15.1}$$

to be solved by the collocation method in domain $[x_L, x_R]$ with the known values of given boundary conditions defined as

$$u(x_L) = \varphi_0, \quad u(x_R) = \varphi_1 \tag{15.2}$$

To initialize, the method requires a proper choice of basis functions $\{\varphi_1, \varphi_2, \ldots, \varphi_N\}$ and a set of points given by $x_L = x_1 < x_2 < \cdots < x_N = x_R$.

The numerical solution can be approximated as

$$U = \sum_{i=1}^{N} c_i \varphi_i(x) \tag{15.3}$$

Here, $c_i's$ are the unknowns to be calculated, and $N$ is the total number of domain partitions. The domain partition also affects the performance of the method, with more domain partitions, the closer the approximate solution approaches to the exact solution. To apply the approach of the collocation method, the approximated solution value at the boundary is taken from the boundary conditions, and the solution is obtained at internal node points.

## 15.3 B-spline

The theory of B-spline function is well known in obtaining the approximate numerical solution of boundary value problems, either ordinary or partial differential equations due to their distinct properties.

Schoenberg [3] in 1946 was the first researcher to refer to the word *B-spline* ("B" refers to basis) in his research work related to the field of mathematics. He described B-spline as a short form of basis spline that represents a smooth, piecewise polynomial. The concept of B-spline is an extended form of splines with some additional properties. A B-spline basis function is a spline function described upon the knot sequence $x_i$ having minimal support with respect to a given degree, smoothness, and domain partition. Following are some related definitions related to nodes:

- A set of real numbers $x_0 \le x_1 \le x_2 \le \cdots \le x_{N-1} \le x_N$, which is the uniform partition of domain $[x_L, x_R]$, are called the *node points* or *knots*.
- The *i*th *knot span* is the open-half interval given by $[x_i, x_{i+1}]$.
- Node points are said to be *uniform* if they are equally spaced, i.e., $x_{i+1} - x_i = h$, $0 \le i \le N - 1$, else they are *nonuniform* nodes.

The first definition of the B-spline basis functions was given by Schumaker [4] using the idea of divided differences. After this, a recurrence relation was independently obtained by Cox [5] and de Boor [6] in the early 1970s to compute B-spline of various orders and degrees. The recursive formula is used to calculate the *m*th B-spline basis function of the *l*th degree in a recursive manner by implementation of the Leibniz' theorem, which can be stated as follows:

$$B_{m,l}(x) = V_{m,l}B_{m,l-1}(x) + (1 - V_{m+1,l})B_{m+1,l-1}(x) \tag{15.4}$$

$$\text{where, } V_{m,l} = \left( \frac{x - x_m}{x_{m+l} - x_m} \right).$$

This is the well-known Cox–de Boor recursion formula to calculate a particular degree B-spline basis function as a linear combination of basis functions of smaller degree. Here $B_{m,l}(x)$ is an *m*th B-spline basis function of degree *l*, and *x* is a parameter variable.

The above-defined recurrence relation (15.4) can be used for $l = 1$ as the initial value to generate the first-degree B-splines. It then results in construction of the higher-order basis functions. The basis function $B_{m,l}(x)$ for degree $l \ge 1$ can hence be written as a linear combination of two $(l - 1)^{th}$ degree basis functions.

### 15.3.1 B-spline of degree zero

On substituting $l = 0$ in (15.4), the basis function of zero-degree B-spline is obtained, which is a step function, an elementary B-spline basis function defined by

$$B_{m,0} = \begin{cases} 1, & x \in [x_m, x_{m+1}) \\ 0, & \text{otherwise} \end{cases} \tag{15.5}$$

From the definition, a zero-degree B-spline can be described as a function that is nonzero and has value one, on the half open interval $[x_m, x_{m+1})$ while at all other points it is zero. The appearance of zero-degree B-spline is as presented in Figure 15.1.

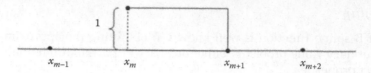

*Figure 15.1* Zero-degree B-spline.

### 15.3.2   First-degree (linear) B-spline

Linear B-spline is the first-degree B-spline calculated by using the Cox–de Boor recursive formula (15.4) by substituting $l = 1$ and implementing the value of zero-degree B-spline (15.5). The linear B-spline basis function is depicted as a tent or hat function having a non-zero value in two knot spans $[x_m, x_{m+1})$ and $[x_{m+1}, x_{m+2})$ as follows:

$$B_{m,1} = \begin{cases} \dfrac{x - x_m}{x_{m+1} - x_m}, & x \in [x_m, x_{m+1}) \\[2ex] \dfrac{x_{m+2} - x}{x_{m+2} - x_{m+1}}, & x \in [x_{m+1}, x_{m+2}) \\[2ex] 0, & \text{otherwise} \end{cases} \tag{15.6}$$

and can be represented as in Figure 15.2.

### 15.3.3   Second-degree (quadratic) B-spline

The quadratic B-spline is a second-degree B-spline that is obtained from the de Boor recursion formula (15.4) for $l = 2$ and linear B-spline (15.6) as

$$B_{m,2} = \begin{cases} \dfrac{(x - x_m)^2}{(x_{m+2} - x_m)(x_{m+1} - x_m)} & x \in [x_m, x_{m+1}) \\[3ex] \dfrac{(x - x_m)(x_{m+2} - x)}{(x_{m+2} - x_m)(x_{m+2} - x_{m+1})} + \dfrac{(x_{m+3} - x)(x - x_{m+1})}{(x_{m+3} - x_{m+1})(x_{m+2} - x_{m+1})} & x \in [x_{m+1}, x_{m+2}) \\[3ex] \dfrac{(x_{m+3} - x)^2}{(x_{m+3} - x_{m+1})(x_{m+3} - x_{m+2})} & x \in [x_{m+2}, x_{m+3}) \\[3ex] 0, & \text{otherwise} \end{cases} \tag{15.7}$$

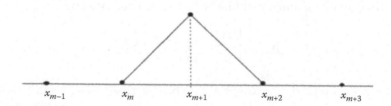

*Figure 15.2* First-degree B-spline.

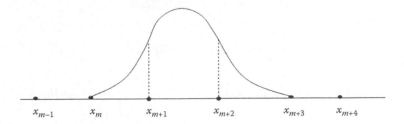

*Figure 15.3* Second-degree B-spline.

From the definition it can be concluded that this second-degree B-spline basis function is nonzero for three consecutive knot spans and has the presentation of a curve as depicted in Figure 15.3.

Using a similar approach, the formula for higher-degree B-splines can be obtained.

## 15.4   Characteristics of B-spline basis functions

B-spline basis functions have some distinguished characteristics described as follows:

1. $B_{m,l}(x)$ is a nonzero polynomial on $[x_m, x_{m+l+1})$ for degree $l \geq 0$.
2. On any span $[x_m, x_{m+1})$ at most $l + 1$ basis functions of degree $l$ are nonzero: $B_{m-l,l}(x), B_{m-l+1,l}(x), B_{m-l+2,l}(x), \ldots, B_{m,l}(x)$.
3. Nonnegativity:
   For all $m$, $l$, and $x$, $B_{m,l}(x)$ is nonnegative in the interval $[x_m, x_{m+l+1})$. The closed interval is called the support of $B_{m,l}(x)$.
4. Local knots:
   The $m$th B-spline $B_{m,l}(x)$ depends only on the knots $x_m, x_{m+1}, x_{m+2}, \ldots, x_{m+l+1}$.
5. Local support:
   If $x$ is outside the interval $[x_m, x_{m+l+1})$, then $B_{m,l}(x) = 0$:
   The local support property indicates that each segment of a B-spline curve is influenced by only $l$ control points or each control point affects only $l$ curve segments.

## 15.5   Types of B-spline

B-spline basis functions are also in trigonometric and exponential forms. The most commonly used basis functions are the cubic B-spline (B-spline of degree three) because of the property that they are symmetric with respect to the origin. Following are the trigonometric and exponential B-spline basis functions of degree three.

### 15.5.1   Trigonometric B-spline basis functions

The trigonometric B-spline basis functions of degree three are defined as

$$TB_m(x) = \frac{1}{w} \begin{cases} p^3(x_m) & x \in [x_m, x_{m+1}) \\ p(x_m)(p(x_m)q(x_{m+2}) + q(x_{m+3})p(x_{m+1})) + q(x_{m+4})p^2(x_{m+2}) & x \in [x_{m+1}, x_{m+2}) \\ q(x_{m+4})(p(x_{m+1})q(x_{m+3}) + q(x_{m+4})p(x_{m+2})) + p(x_m)q^2(x_{m+3}) & x \in [x_{m+2}, x_{m+3}) \\ q^3(x_{m+4}) & x \in [x_{m+2}, x_{m+3}) \end{cases} \tag{15.8}$$

where

$$h = \frac{b-a}{n}, \text{ is the step size for domain } x \in [a,b]$$

$$p(x_m) = \sin\left(\frac{x-x_m}{h}\right), \quad q(x_m) = \sin\left(\frac{x_m - x}{2}\right), \quad w = \sin\left(\frac{h}{2}\right)\sin(h)\sin\left(\frac{3h}{2}\right)$$

This is a polynomial cubic trigonometric function with some geometric properties like $C^\infty$ continuity, nonnegativity, and partition of unity.

### 15.5.2  Exponential B-spline basis functions

The exponential B-spline basis functions of degree three are defined as

$$EB_m(x) = \begin{cases} b_2\left((x_{m-2}-x)-\dfrac{1}{p}\left(\sinh\left(p(x_{m-2}-x)\right)\right)\right) & x \in [x_{m-2}, x_{m-1}) \\[2mm] a_1 + b_1(x_m - x) + c_1\exp\left(p(x_m - x)\right) + d_1\exp\left(-p(x_m - x)\right) & x \in [x_{m-1}, x_m) \\[2mm] a_1 + b_1(x - x_m) + c_1\exp\left(p(x - x_m)\right) + d_1\exp\left(-p(x - x_m)\right) & x \in [x_m, x_{m+1}) \\[2mm] b_2\left((x - x_{m+2})-\dfrac{1}{p}\left(\sinh\left(p(x - x_{m+2})\right)\right)\right) & x \in [x_{m+1}, x_{m+2}) \end{cases} \tag{15.9}$$

$$\text{where, } a_1 = \frac{phc}{phc - s}, b_1 = \frac{p}{2}\left(\frac{c(c-1)+s^2}{(phc - s)(1-c)}\right), b_2 = \frac{p}{2(phc - s)}$$

$$c_1 = \frac{1}{4}\left(\frac{e^{-ph}(1-c)+s(e^{-ph}-1)}{(phc - s)(1-c)}\right), d_1 = \frac{1}{4}\left(\frac{e^{ph}(c-1)+s(e^{ph}-1)}{(phc - s)(1-c)}\right)$$

$$c = \cosh(ph), s = \sinh(ph), \text{ and } h = \frac{b-a}{n}, \text{ is the step size for domain } x \in [a,b]$$

and $p$ is a parameter to be chosen from $p = \max\limits_{0 \le i \le N} p_i$.

## 15.6  Methodology: Collocation method using B-spline basis function

In the collocation method, using B-spline approximation leads to a technique that requires only the unknown parameters at certain node points to generate the solution. Collocation points and degree of B-spline are the key factors that play an important role in the implementation of the method and also affect the results to be obtained up to a desired level of accuracy. Consider a mesh $a = x_0 < x_1, \ldots, x_{N-1} < x_N = b$ as a uniform partition of the solution domain $a \le x \le b$ by the knots $x_m$ with step size $h = x_{m+1} - x_m$, where $m = 0, \ldots, N-1$.

Let us discuss the method to solve a partial differential equation by assuming the solution approximated by $U(x, t)$ that can be written in linear combination form of B-spline basis functions as follows:

$$U(x,t) = \sum_{j=m-l+2}^{m+l-2} c_j B_j(x) \tag{15.10}$$

Here, $l$ defines the degree of the B-spline basis functions, $m$ defines the number of collocation points, and $c_m$ are the constants to be calculated from the generated matrix system to be solved using any numerical method.

The formula for the cubic B-spline basis function using the definition and second-order B-spline basis function was first given by Prenter [7] to solve a partial differential equation given by

$$B_{m,3}(x) = \frac{1}{h^3} \begin{cases} (x - x_{m-2})^3 & x \in [x_{m-2}, x_{m-1}) \\ (x - x_{m-2})^3 - 4(x - x_{m-1})^3 & x \in [x_{m-1}, x_m) \\ (x_{m+2} - x)^3 - 4(x_{m+1} - x)^3 & x \in [x_m, x_{m+1}) \\ (x_{m+2} - x)^3 & x \in [x_{m+1}, x_{m+2}) \\ 0 & \text{otherwise} \end{cases} \tag{15.11}$$

The cubic B-spline basis function as defined above will be nonzero at four knot spans and is presented as in Figure 15.4.

From this definition, the values of $B_m(x)$ at the node points with its first and second derivatives can be tabulated as in Table 15.1.

By substituting $l = 3$ in (15.10), the solution can be approximated as

$$U(x,t) = \sum_{j=m-1}^{m+1} c_j(t)B_j(x) \tag{15.12}$$

Hence, it can be simplified in the form

$$U(x_m, t) = c_{m-1}B_{m-1}(x_m) + c_m B_m(x_m) + c_{m+1}B_{m+1}(x_m)$$

$$\text{or } U(x_m, t) = c_{m-1}B_m(x_{m+1}) + c_m B_m(x_m) + c_{m+1}B_m(x_{m-1}) \tag{15.13}$$

It is evident that the nonzero part of $B_m$ is localized to a small neighborhood of $x_m$, namely, in the interval $x_{m-2} < x_m < x_{m+2}$. Because of this, only $B_{m-1}, B_m, B_{m+1}$ contribute to the value of $U$ at $x_m$. Using the values of basis functions at the node points from Table 15.1 in Equation (15.13), the approximate solution and its derivatives up to second order can be determined in terms of parameters $c'_m s$ that can be written as

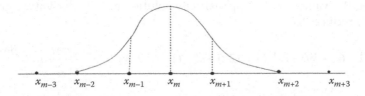

*Figure 15.4* Third-degree B-spline.

**Table 15.1** Value of $B_m(x)$ for cubic B-spline and its derivatives at the nodal points

|            | $x_{m-2}$ | $x_{m-1}$ | $x_m$      | $x_{m+1}$ | $x_{m+2}$ |
|------------|-----------|-----------|------------|-----------|-----------|
| $B_m(x)$   | 0         | 1         | 4          | 1         | 0         |
| $B_m'(x)$  | 0         | $3/h$     | 0          | $-3/h$    | 0         |
| $B_m''(x)$ | 0         | $6/h^2$   | $-12/h^2$  | $6/h^2$   | 0         |

$$U(x_m,t) = c_{m-1} + 4c_m + c_{m+1}$$

$$hU'(x_m,t) = 3(c_{m+1} - c_{m-1}) \tag{15.14}$$

$$h^2U''(x_m,t) = 6(c_{m-1} - 2c_m + c_{m+1})$$

## 15.7 Numerical solution of advection diffusion equation using collocation method

To gain insight into the application of the collocation method to solve a partial differential equation, let us apply the above described method to find the solution of a well-known advection diffusion equation in one dimension, given by

$$u_t + \alpha u_x - \beta u_{xx} = 0, \quad x \in [0,1] \tag{15.15}$$

with boundary conditions $u(0,t) = \varphi_0$, $u(1,t) = \varphi_1$, and initial condition $u(x,0) = \phi(x)$.

Here, $u$ is the concentration of fluid with uniform velocity, $\alpha$ and $\beta$ are constant representing the coefficient of diffusion, and $l$ is the length of the channel.

The numerical solution for the equation is obtained on finite domain $[0, l]$. The domain is discretized with uniform partition as $x_0 < x_1,...,x_{N-1} < x_N$ with common difference $h = x_{m+1} - x_m$, where $m = 0,..., N-1$.

To implement the collocation method, the time derivative is discretized using the finite difference approach, and the Crank–Nicolson scheme is applied on spatial variables in Equation (15.15) to get

$$\left[\frac{u^{n+1} - u^n}{\Delta t}\right] = \beta\left[\frac{u_{xx}^{n+1} + u_{xx}^n}{2}\right] - \alpha\left[\frac{(u_x)^{n+1} + (u_x)^n}{2}\right] \tag{15.16}$$

Here, $\Delta t$ represents the time step.

On separating the values of solution $u(x, t)$ at $n$th and $(n + 1)$th levels and hence substituting the values of $u(x, t)$ and its derivative in terms of the approximate function given by (15.4), equation can be written in terms of unknown time parameters, $c_m's$.

### 15.7.1 Using B-spline basis functions

On substituting the values of the approximate solution and its derivative given by (15.14), Equation (15.16) reduces to

$$c_{m-1}^{n+1}(1 - 6z - 3y) + c_m^{n+1}(4 + 12z) + c_{m+1}^{n+1}(1 - 6z + 3y) = u^n - \frac{\alpha\Delta t}{2}u_x^n + \frac{\beta\Delta t}{2}u_{xx}^n \tag{15.17}$$

Here, $y = \dfrac{\alpha\Delta t}{2h}$, $z = \dfrac{\beta\Delta t}{2h^2}$ and $m = 0,..., N$.

This system consists of $(N+1)$ linear equations in $(N+3)$ unknowns $\{c_{-1}, c_0, c_1, \ldots, c_N, c_{N+1}\}$.

To obtain a unique solution to this system, two additional constraints $c_{-1}$ and $c_{N+1}$ are required. These additional constraints can be obtained by substituting the approximation for $u(x, t)$ in the boundary conditions.

At $m = 0$, using boundary condition, $u(x_0, t) = \varphi_0$, gives

$$c_0^{n+1}(36z + 12y) + c_1^{n+1}(6y) = \left(u^n - \frac{\alpha \Delta t}{2} u_x^n + \frac{\beta \Delta t}{2} u_{xx}^n\right) - \varphi_0(1 - 6z - 3y) \tag{15.18}$$

Similarly at $m = N$, using boundary condition $u(x_N, t) = \varphi_1$, we obtain

$$c_{N-1}^{n+1}(-6y) + c_N^{n+1}(36z - 12y) = \left(u^n - \frac{\alpha \Delta t}{2} u_x^n + \frac{\beta \Delta t}{2} u_{xx}^n\right) - \varphi_1(1 - 6z + 3y) \tag{15.19}$$

This results in a $(N + 1) \times (N + 1)$ matrix system, given by $AC = B$ where $C$ is the unknown coefficients, $\{c_0, c_1, \ldots, c_{N-1}, c_N\}$.

The coefficient matrix $A$ is given by

$$\begin{bmatrix} 36z + 12y & 6y & 0 & 0 & 0 & 0 \\ 1 - 6z - 3y & 4 + 12z & 1 - 6z + 3y & \cdot & \cdot & 0 \\ 0 & 1 - 6z - 3y & 4 + 12z & 1 - 6z + 3y & 0 & 0 \\ 0 & 0 & \ddots & \ddots & \ddots & 0 \\ 0 & \cdot & \cdot & 1 - 6z - 3y & 4 + 12z & 1 - 6z + 3y \\ 0 & 0 & 0 & 0 & -6y & 36z - 12y \end{bmatrix}$$

with right-hand side

$$\begin{bmatrix} \left(u^n - \left(\frac{\alpha \Delta t}{2}\right) u_x^n + \left(\frac{\alpha \Delta t}{2}\right) u_{xx}^n\right) - \varphi_0(1 - 6z - 3y) \\ u^n - \left(\frac{\alpha \Delta t}{2}\right) u_x^n + \left(\frac{\beta \Delta t}{2}\right) u_{xx}^n \\ \vdots \\ u^n - \left(\frac{\alpha \Delta t}{2}\right) u_x^n + \left(\frac{\beta \Delta t}{2}\right) u_{xx}^n \\ \left(u^n - \left(\frac{\alpha \Delta t}{2}\right) u_x^n + \left(\frac{\alpha \Delta t}{2}\right) u_{xx}^n\right) - \varphi_1(1 - 6z + 3y) \end{bmatrix}$$

### 15.7.2    *Using trigonometric B-spline basis functions*

Using the definition of the trigonometric B-spline basis function, the values of basis functions and the first and second derivatives can be tabulated as in Table 15.2, where $a_i's$ are defined as

$$a_1 = \frac{\sin^2\left(\dfrac{h}{2}\right)}{\sin(h)\sin\left(\dfrac{3h}{2}\right)}, \quad a_2 = \frac{2}{1 + 2\cos(h)},$$

$$a_3 = \frac{-3}{4\sin\left(\dfrac{3h}{2}\right)}, \quad a_4 = \frac{3}{4\sin\left(\dfrac{3h}{2}\right)},$$

$$a_5 = \frac{3(1 + \cos(h))}{16\sin^2\left(\dfrac{h}{2}\right)\left(2\cos\left(\dfrac{h}{2}\right) + \cos\left(\dfrac{3h}{2}\right)\right)}, \quad a_6 = -\frac{3\cos^2\left(\dfrac{h}{2}\right)}{\sin^2\left(\dfrac{h}{2}\right)(2 + 4\cos(h))}$$

Using the linear combination formula to write the approximate solution with trigonometric B-spline basis functions up to the second-order derivative, the approximate solution can be determined in terms of the time parameters $c'_m s$ as

$$U(x_m, t) = a_1 c_{m-1} + a_2 c_m + a_1 c_{m+1}$$
$$U'(x_m, t) = a_4 c_{m-1} + a_3 c_{m+1}$$
$$U''(x_m, t) = a_5 c_{m-1} + a_6 c_m + a_5 c_{m+1}$$

On substituting the values of basis functions in Equation (15.16), the system can be written as

$$c_{m-1}^{n+1}(a_1 - \gamma a_5 + \eta a_4) + c_m^{n+1}(a_2 - \gamma a_6) + c_{m+1}^{n+1}(a_1 - \gamma a_5 + \eta a_3) = u^n - \eta u_x^n + \gamma u_{xx}^n \quad (15.20)$$

Here, $\gamma = \dfrac{\beta \Delta t}{2}, \eta = \dfrac{\alpha \Delta t}{2}$, and $m = 0,\ldots, N$.

This system consists of $(N + 1)$ linear equations in $(N + 3)$ unknowns $\{c_{-1}, c_0, c_1, \ldots, c_N, c_{N+1}\}$. To obtain a unique solution to this system, two additional constraints $c_{-1}$ and $c_{N+1}$ are required. These additional constraints can be obtained by substituting the approximation for $u(x, t)$ in the boundary conditions. At $m = 0$, using boundary condition, $u(0, t) = \varphi_0$, gives

$$c_0^{n+1}\left(\gamma\left(\frac{a_2 a_5}{a_1} - a_6\right) - \eta\left(\frac{a_2 a_4}{a_1}\right)\right) + c_1^{n+1}\eta(a_3 - a_4) = \left(u^n - \frac{\alpha \Delta t}{2}u_x^n + \frac{\beta \Delta t}{2}u_{xx}^n\right)$$

$$- \frac{\varphi_0}{a_1}(a_1 - \gamma a_5 + \eta a_4) \quad (15.21)$$

**Table 15.2** Value of $B_m(x)$ for trigonometric cubic B-spline and its derivatives at the nodal points

| | $x_{m-2}$ | $x_{m-1}$ | $x_m$ | $x_{m+1}$ | $x_{m+2}$ |
|---|---|---|---|---|---|
| $TB_m(x)$ | 0 | $a_1$ | $a_2$ | $a_1$ | 0 |
| $TB'_m(x)$ | 0 | $a_3$ | 0 | $a_4$ | 0 |
| $TB''_m(x)$ | 0 | $a_5$ | $a_6$ | $a_5$ | 0 |

Similarly at $m = N$, using boundary condition $u(1, t) = \varphi_1$, we obtain

$$c_{N-1}^{n+1} \eta (a_4 - a_3) - c_N^{n+1} \left( \gamma \left( \frac{a_2 a_5}{a_1} - a_6 \right) - \eta \left( \frac{a_2 a_3}{a_1} \right) \right) = \left( u^n - \frac{\alpha \Delta t}{2} u_x^n + \frac{\beta \Delta t}{2} u_{xx}^n \right)$$

$$- \frac{\varphi_1}{a_1} (a_1 - \gamma a_5 + \eta a_3) \tag{15.22}$$

This results in a $(N + 1) \times (N + 1)$ matrix system, given by $AC = B$, where $C$ is the unknown coefficients, $\{c_0, c_1, \ldots, c_{N-1}, c_N\}$.

Application of the collocation method with both of the above discussed B-spline basis functions results in a tridiagonal matrix system that can be solved by using the Thomas algorithm [8]. On solving this tridiagonal matrix, the values of unknown $c_m's$ are obtained. By using these values of $c_m's$, the approximate solution at a particular time level is obtained. But to find the solution at the first time level, values of $c_m's$ at the initial time level are required, and then the system can be solved recursively. The solution at the zeroth time level can be computed from the initial condition.

## 15.8   Numerical example

The B-spline basis functions are widely used to solve various linear and nonlinear ordinary and partial differential equations, see Refs. [9–16]. This chapter is an effort to discuss the basics of basis functions and their implementation to solve differential equations.

To get insight into the method, consider a problem with $\alpha = 0$, $\beta = 1$ in Equation (15.15) that reduces to the heat equation given as

$$u_t - u_{xx} = 0, \quad x \in [0, 1] \tag{15.23}$$

with boundary conditions $u(0, t) = 0$, $u(1, t) = 0$ and initial condition $u(x, 0) = \sin(\pi x)$.

The exact solution of the equation is given by $u(x, t) = \exp(-\pi^2 t) \sin(\pi x)$.

Numerical solution of the concerned equation is obtained by collocation method using standard B-spline and trigonometric B-spline basis functions at $t = 1$. To discuss the accuracy of the obtained solutions, errors are calculated using both types of basis functions and are depicted in Table 15.3. The max absolute errors are depicted in Table 15.4 for two different values of time step and domain partition. First, values are calculated at time step 0.01 and 40 domain partitions for $t = 1$ to $t = 3$ and then the time step is considered 0.001 with 160 domain partitions for same time levels.

It can be concluded from Tables 15.3 and 15.4 that the solution is comparable with the exact solution from both forms of the B-spline basis functions. In case of standard B-spline

*Table 15.3*  Absolute errors using B-spline and trigonometric B-spline

| Node point | B-spline | Trigonometric B-spline |
|---|---|---|
| 0.1 | 8.15E-05 | 5.60E-06 |
| 0.2 | 3.67E-07 | 1.45E-05 |
| 0.4 | 7.87E-07 | 2.32E-05 |
| 0.5 | 8.27E-07 | 2.42E-05 |
| 0.6 | 7.87E-07 | 2.26E-05 |
| 0.8 | 6.71E-07 | 1.87E-05 |
| 0.9 | 8.15E-05 | 9.83E-06 |

*Table 15.4* Maximum absolute errors using B-spline and trigonometric B-spline

| Time | B-spline | Trigonometric B-spline |
|------|----------|------------------------|
| | At $N = 40$ and $\Delta t = 0.01$ | |
| 1 | 2.0112E-04 | 2.4165E-05 |
| 2 | 1.2937E-05 | 3.2392E-07 |
| 3 | 1.0714E-06 | 2.7468E-08 |
| | At $N = 160$ and $\Delta t = 0.001$ | |
| 1 | 2.4242E-08 | 6.5252E-06 |
| 2 | 2.8568E-11 | 6.1191E-10 |
| 3 | 1.5320E-15 | 4.4115E-14 |

basis functions, the solution is very much improved on increasing the domain partitions with small values of time steps, while in the case of the trigonometric B-spline, the solution is also improved but at a normal pace.

## References

1. W. Zahra, *Numerical Treatment of Boundary Value Problems Using Spline Functions*, LAP Lambert Academic Publishing, GmbH & Co.KG and licensors, Germany 2010.
2. K. S. Surana, J. N. Reddy, *The Finite Element Method for Boundary Value Problems*, CRC Press, Taylor & Francis Group, Boca Raton, FL, 2016.
3. I. J. Schoenberg, Contribution to the problem of approximation of equidistant data by analytical functions, *Quarterly Applied Mathematics*, 4, 1946, 45–99.
4. L. L. Schumaker, *Spline Functions, Basic Theory*, Wiley, Cambridge University Press, New York, 1981.
5. M. G. Cox, The numerical evaluation of B-splines, *Journal of the Institute of Mathematical Applications*, 10, 1972, 134–149.
6. C. de Boor, *A Practical Guide to Splines*, Springer Verlag, New York, 1978.
7. P. M. Prenter, *Splines and Variational Methods*, John Wiley & Sons, New York, 1975.
8. D. U. Von Rosenberg, *Methods for Solution of Partial Differential Equations*, Vol. 113, American Elsevier Publishing Inc., New York, 1969.
9. M. K. Kadalbajoo, L. P. Tripathi, A. Kumar, A cubic B-spline collocation method for a numerical solution of the generalized Black–Scholes equation, *Mathematical and Computer Modelling*, 55 (3–4), 2012, 1483–1505.
10. A. K. Khalifa, K. R. Raslan, H. M. Alzubaidi, A collocation method with cubic B-splines for solving the MRLW equation, *Journal of Computational and Applied Mathematics*, 212 (2), 2008, 406–418.
11. R. Pourgholi, Applications of cubic B-splines collocation method for solving nonlinear inverse parabolic partial differential equations, *Numerical Methods for Partial Differential Equations*, 33 (1), 2017, 88–104.
12. M. Gholamian, J. Saberi-Nadjafi, Cubic B-splines collocation method for a class of partial integro-differential equation, *Alexandria Engineering Journal*, 2017. doi:10.1016/j.aej.2017.06.004.
13. G. Arora, V. Joshi, A computational approach using modified trigonometric cubic B-spline for numerical solution of Burgers' equation in one and two dimensions, *Alexandria Engineering Journal*, 2017. doi:10.1016/j.aej.2017.02.017.
14. G. Arora, V. Joshi, A computational approach for solution of one dimensional parabolic partial differential equation with application in biological processes, *Ain Shams Engineering Journal*, 2016. doi:10.1016/j.asej.2016.06.013.
15. M. Abbas, A. A. Majid, A. I. Md. Ismail, A. Rashid, Numerical method using cubic trigonometric B-spline technique for nonclassical diffusion problems, *Abstract and Applied Analysis*, 2014, Article ID 849682, 11 pages.
16. O. Ersoy, I. Dag, The exponential cubic B-spline collocation method for the Kuramoto–Sivashinsky equation, *Filomat*, 30 (3), 2016, 853–861, DOI 10.2298/FIL1603853E.

# chapter sixteen

# Rayleigh's approximation method on reflection/refraction phenomena of plane SH-wave in a corrugated anisotropic structure

**Neelima Bhengra**
*Indian Institute of Technology (ISM)*

## Contents

## 16.1 Introduction

In recent years, the phenomena of elastic wave scattering due to different obstacles present in the media have drawn the considerable attention of many distinct researchers across the globe. This is because this investigation enables us to unravel deep subsurface structures that have immense operational usage in oil exploration, earthquake engineering, and much more. Various types of materials provide distinct propagative behavior to the waves underneath the earth. During SH-wave propagation, the boundaries present between the layers distribute the individual waves to reflect or transmit through the interface depending on the angle of incidence. This defines the distributive characteristics of the interface. Furthermore, the boundaries are majorly irregular or corrugated, which further increases the complexity of the investigation. The variation of the wave propagation also depends largely on the physical characteristics of the medium. Therefore, propagation through such layers can also enlighten us with some important facts about faults and anticlinal structures beneath the earth. The phenomena of reflection and transmission have been the principal concept behind the subjects of geophysics and seismology. Explorations of oil and gas companies have been using this concept for years to detect the accumulation of hydrocarbons beneath the earth. Literature is already present on the reflection and transmission of SH-waves, such as Ewing et al. (1957), Keith et al. (1977), Aki and Richards (2002), etc. Fokkema (1980) investigated these phenomena using the time-harmonic waves. The study of stress free boundary between two incompressible materials using the reflection and transmission phenomena was done by Pal and Chattopadhyay (1984).

Nowadays, the reflection and refraction through porous media have become one of the core subjects of investigation due to the dynamic behavior. A separate field of study has emerged concerning the propagation through the porous media. A typical porous media is the one that has some pores in it, which is usually filled with fluid. Such materials are often characterized by their porosity values. Porosity is defined as the ratio of the volume of void space to the total volume. Porosity values range from 0 to 1. The connected pore space enables the filtration of pore fluid through the porous media. Pumice, sandstone, and soil are some of the naturally occurring porous materials found in the earth. The dynamic nature of such porous materials is the major aspect of rock study, which is effectively used in seismic exploration for detailed investigation of subsurface structures to explore sedimentary basins for hydrocarbon production. Deresiewicz (1961) first studied the boundary effects on the wave propagation in a liquid-filled porous media. Wu et al. (1990) investigated reflection and refraction of elastic waves from a fluid-saturated porous solid boundary. Sharma and Gogna (1992) used the plane harmonic waves to investigate the reflection and refraction phenomena through an interface between an elastic solid and a liquid-saturated porous media by making purposeful use of the asymptotic approximation of dissipation function. Tajuddin and Haussaini (2005) analyzed the reflection phenomena of plane waves at the boundaries of a liquid-filled poroelastic half-space. Tomar and Arora (2007) studied the reflection and refraction phenomena of elastic waves through an elastic/porous solid filled with immiscible fluids.

Wave propagation through an anisotropic media is fundamentally very different to an isotropic media. In seismology, if there are variations in phase velocity that depend largely on factors such as wave propagation direction, particle motion direction, the orientation of the material and the stress and strain of the propagating media, then it is said that there is an anisotropy in the propagating medium. The anisotropic properties of the material have a significant contribution on the reflection and refraction coefficients. Information of such coefficients can help us to understand the mechanical properties of the medium. Anisotropy also occurs due to the presence of thin laminates arranged in a particular order. Other factors such as micro-fracturing and orientation of the mineral can also result in a general anisotropy. Normally, it is difficult to derive a general anisotropy from a specific anisotropy; therefore, it is necessary that during a wave propagation problem, the anisotropy should be of the general type. These general problems have motivated the present study. Crampin (1977) was the first researcher who differentiated anisotropy with isotropy. He established that the variation in velocity due to anisotropy is one of the many anomalies that can occur in the media. The concepts of reflection and transmission phenomenon in the anisotropic half-space have been the base of geological study to explore continental margins for mineral exploration. Daley and Hron (1979) investigated the ellipsoidal anisotropic media to derive reflection and transmission coefficients for seismic waves. Rokhlin et al. (1986) studied this wave scattering phenomena of elastic waves on a plane interface lying between two generally anisotropic media. Then, Thomsen (1988) published a paper on reflection seismology over azimuthally anisotropic media.

Rayleigh (1907) made the first attempt to find the solution to the reflection problem of light or sound when incident perpendicularly on an uneven boundary surface. Then, Sato (1955) applied Rayleigh's concept on the elastic waves, which was later extended by Asano (1960, 1961, 1966). Abubakar (1962a–c) attempted to study the problem of scattering of elastic waves incident on a corrugated interface by utilizing the perturbation technique. Saini and Singh (1977) studied the effect of anisotropy on the reflection of SH-waves at an interface. In general terms, the Rayleigh's method approximates the exponential term associated with the corrugated interface. For the solution of first-order approximation of corrugation, the linear terms are retained, and since the amplitude and slope of the corrugated boundary are assumed to be very small, the higher orders are neglected. Several other kinds of literature have also been published on Rayleigh's method implemented on elastic wave scattering in the corrugated interface, such as Tomar and Saini (1997), Tomar et al. (2002), Tomar and Kaur (2003), Tomar and Singh (2007), etc. Tomar and Kaur (2007) then investigated the behavior of the SH-wave at a corrugated interface that lies in between a dry sandy half-space and an anisotropic elastic half-space.

In the present chapter, utilizing Rayleigh's approximation method, an attempt has been made to study the reflection and refraction pattern in a corrugated interface sandwiched between an initially stressed fluid-saturated poroelastic half-space and a highly anisotropic half-space. Here the highly anisotropic half-space is considered as triclinic. Closed form formulae for the reflection and refraction coefficients have been derived. Rayleigh's method has been effectively used to derive first- and second-order approximations of the coefficients. Some special cases have also been deduced. The energy ratios of the reflected and refracted waves are also presented. Various two-dimensional plots have been drawn to show the effects of some affecting parameters such as initial stress parameter, corrugation amplitude, wavelength and frequency factor.

## 16.2   Problem formulation and its solution

Let us assume $z = \varsigma(x)$ as the equation of the corrugated interface that separates two media, namely, an initially stressed fluid-saturated poroelastic half-space and a highly anisotropic half-space. In the above equation, $\zeta$ is considered as a periodic function of $x$, which is independent of $y$ whose mean value is zero. The x-axis is taken on the horizontal plane, while z-axis is taken vertically downward. Let $F_2$ be the upper half-space occupying the region $-\infty < z \leq \varsigma(x)$ and $F_1$ be the lower half-space occupying the region $\varsigma(x) \geq z > \infty$. The geometry of the problem is presented in Figure 16.1.

The Fourier series representation of the function can be taken as

$$\zeta = \sum_{n=1}^{\infty} \left[ \zeta_n e^{in\lambda x} + \zeta_{-n} e^{-in\lambda x} \right] \tag{16.1}$$

Here, $\zeta_n$ and $\zeta_{-n}$ are Fourier expansion coefficients, $\lambda$ is the wave number and $n$ is series expansion order and the wavelength of corrugation is $2\pi/\lambda$.

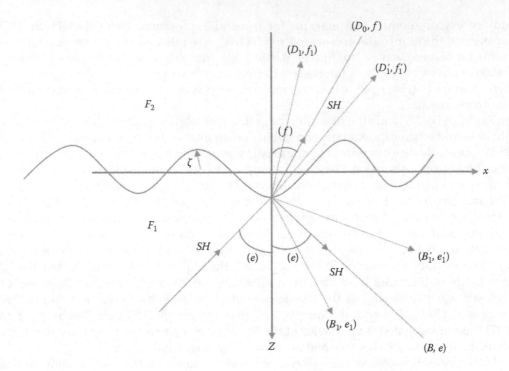

*Figure 16.1* Geometry of the problem.

Introducing the constant terms $d$, $c_n$, and $S_n$ such that

$$\zeta_1 = \zeta_{-1} = \frac{d}{2}, \quad \zeta_{\pm n} = \frac{c_n \mp s_n}{2}, \quad n = 2,3,\ldots \tag{16.2}$$

Using Equation (16.2), the series in Equation (16.1) can be written as

$$\zeta = d\cos\lambda x + \sum_{n=2}^{\infty}\left[c_n\cos n\lambda x + s_n\sin n\lambda x\right], \tag{16.3}$$

If the shape of the corrugated interface is represented by only one cosine term, i.e., $\zeta = d\cos\lambda x$; then $2\pi/\lambda$ and $d$ is the wavelength and amplitude of corrugation, respectively. Let $u_i, v_i$ and $w_i(I = 1, 2)$ be the displacement components along $x$, $y$, and $z$ directions, respectively.

For the propagation of SH-wave, it is assumed that

$$u_i = 0, w_i = 0, v_i = v_i(x, z, t), \frac{\partial}{\partial y} \equiv 0 \tag{16.4}$$

Indices 1 and 2 stand for the highly anisotropic half-space and the fluid-saturated poroelastic half-space. The first and second partial derivatives with respect to time are represented as $\partial_t$ and $\partial_{tt}$, respectively. Moreover, $\partial_z$ and $\partial_{zz}$ stand for $\frac{d}{dz}$ and $\frac{d}{dz^2}$, respectively.

## 16.2.1    Solution for the lower highly anisotropic half-space

Consider a homogeneous highly anisotropic elastic medium that has 21 elastic constants. By using the Hooke's law, the stress–strain relations of the highly anisotropic media are given by

$$T_{11}^t = F_{11}e_{11} + F_{12}e_{22} + F_{13}e_{33} + F_{14}e_{23} + F_{15}e_{13} + F_{16}e_{12},$$

$$T_{22}^t = F_{12}e_{11} + F_{22}e_{22} + F_{23}e_{33} + F_{24}e_{23} + F_{25}e_{13} + F_{26}e_{12},$$

$$T_{33}^t = F_{13}e_{11} + F_{23}e_{22} + F_{33}e_{33} + F_{34}e_{23} + F_{35}e_{13} + F_{36}e_{12},$$

$$T_{23}^t = F_{14}e_{11} + F_{24}e_{22} + F_{34}e_{33} + F_{44}e_{23} + F_{45}e_{13} + F_{46}e_{12}, \tag{16.5}$$

$$T_{13}^t = F_{15}e_{11} + F_{25}e_{22} + F_{35}e_{33} + F_{45}e_{23} + F_{55}e_{13} + F_{56}e_{12},$$

$$T_{12}^t = F_{16}e_{11} + F_{26}e_{22} + F_{36}e_{33} + F_{46}e_{23} + F_{56}e_{13} + F_{66}e_{12}.$$

Here, $T_{ij}^t$, $F_{ij}$, and $e_{ij}$ are the components of the stress tensor, stiffness coefficients, and components of the strain tensor, respectively.

The equations of motion in the absence of body forces are given by Biot (1965),

$$\partial_x T_{11}^t + \partial_y T_{12}^t + \partial_z T_{13}^t - S_{11}^t \left( \partial_y \omega_z - \partial_z \omega_y \right) = \rho_1 \partial_{tt} u_1,$$

$$\partial_x T_{21}^t + \partial_y T_{22}^t + \partial_z T_{23}^t - S_{11}^t \partial_x \omega_z = \rho_1 \partial_{tt} v_1, \tag{16.6}$$

$$\partial_x T_{31}^t + \partial_y T_{32}^t + \partial_z T_{33}^t + S_{11}^t \partial_x \omega_y = \rho_1 \partial_{tt} w_1.$$

where $\rho_1$ denotes the mass density, and $\omega_x$, $\omega_y$, and $\omega_z$ are the rotational components given by

$$\omega_x = \tfrac{1}{2}(\partial_y w - \partial_z v), \ \ \omega_y = \tfrac{1}{2}(\partial_z u - \partial_x w), \ \text{and} \ \omega_z = \tfrac{1}{2}(\partial_x v - \partial_y u).$$

Now, using Equations (16.4)–(16.6), we get the governing equation of motion and the stress–strain components

$$\partial_x T_{21} + \partial_z T_{23} - \frac{S_{11}^t}{2} \partial_x v_1 = \rho_1 \partial_{tt} v_1 \tag{16.7}$$

and

$$T_{21} = F_{66} \partial_x v_1 + F_{64} \partial_z v_1, \ T_{23} = F_{46} \partial_x v_1 + F_{44} \partial_z v_1 \tag{16.8}$$

Using Equations (16.7) and (16.8), we obtain

$$F_{66} \partial_{xx} v_1 + 2F_{46} \partial_{xz} v_1 + F_{44} \partial_{zz} v_1 - \frac{S_{11}^t}{2} \partial_{xx} v_1 = \rho_1 \partial_{tt} v_1 \tag{16.9}$$

Assuming the solution as $v(x,z,t) = V(z)e^{i(\omega t - k_1 x)}$, where $\omega$ is the angular frequency, $k_1 = \dfrac{\omega \sin e}{\beta_1}$ is the horizontal component of the wave number, $e$ being the angle of incidence and substituting these in Equation (16.9), we have

$$\partial_{zz}V - 2ik_1\mu_1\,\partial_z V + k_1^2\mu_2\left[\frac{\omega^2}{k_1^2\beta_1^{\,2}} - 1 + \xi\right]V = 0 \tag{16.10}$$

where

$$\mu_1 = \frac{F_{46}}{F_{44}}\,,\mu_2 = \frac{F_{66}}{F_{44}},\,\xi = \frac{S_{11}^t}{F_{66}},\text{ and }\beta_1^2 = \frac{F_{66}}{\rho_1}$$

The solution of Equation (16.10) is given as

$$v_1(x,z,t) = \left(A_0 e^{i\Omega_0 z} + B_0 e^{-i\Omega z}\right) \tag{16.11}$$

where

$$\Omega_0 = k_1\left(\mu_1 + \sqrt{\mu_1^2 + \mu_2\left(\cot^2 e + \xi\right)}\,\right) \text{ and } \Omega = k_1\left(-\mu_1 + \sqrt{\mu_1^2 + \mu_2\left(\cot^2 e + \xi\right)}\,\right)$$

Hence, the displacement for the lower half-space is given by

$$v_1(x,z,t) = \left(B_0 e^{i\Omega_0 z} + B e^{-i\Omega z}\right)e^{(\omega t - k_1 x)} \tag{16.12}$$

### 16.2.2   Solution for the upper fluid-saturated poroelastic half-space

Let us consider the upper half-space as a fluid-saturated transversely isotropic poro-elastic infinite medium. At first, we deduce the equation governing the propagation of SH-wave in a poroelastic medium. If $(u_2, v_2, w_2)$ and $(U_2, V_2, W_2)$ denote the components of solid-phase displacements and fluid-phase displacements of the poroelastic medium, respectively, then for SH-wave propagating in $x, z$ direction and causing displacement in $y$ direction only, we have

$$u_2 = 0, w_2 = 0, v_2 = v_2(x,z,t) \text{ and } U_2 = 0, W_2 = 0, V_2 = V_2(x,z,t) \tag{16.13}$$

For the upper half-space, the stress–strain relations are

$$T_{11}^p = (2N + A)e_{11} + Ae_{22} + Fe_{33} + ME,$$

$$T_{22}^p = Ae_{11} + (2N + A)e_{22} + Fe_{33} + ME,$$

$$T_{33}^p = Fe_{11} + Fe_{22} + 2Ce_{33} + KE,$$

$$T_{12}^p = 2Ne_{12}, \tag{16.14}$$

$$T_{23}^p = 2Ge_{23},$$

$$T_{31}^p = 2Ge_{31},$$

$$\sigma = Me_{11} + Me_{22} + Ke_{33} + DE,$$

where, $A,C,D,F,G,K,M,N$ are the material constants; $T_{ij}^p$ are the components of the stress tensors acting on the solid phase of the poroelastic material; $E = div\vec{U}_i$ is the fluid volumetric strain; $\sigma = -fp$ is the stress acting on the fluid phase of poroelastic material in which $P$ is the pressure in the fluid and $f$ is the porosity of the poroelastic material.

With the help of Equations (16.13) and (16.14), the equation of motion for SH-wave propagation in fluid-saturated initially stressed poroelastic medium in the absence of body forces and the viscoelasticity of the fluid based on Biot (1956a,b, 1962, 1965) can be written as

$$\partial_x T_{21}^p + \partial_z T_{23}^p - S_{11}^p \partial_z \omega_{13} = \partial_{tt} \left[ \rho_{11} v_2 + \rho_{12} V_2 \right],$$

$$\partial_{tt} \left[ \rho_{11} v_2 + \rho_{12} V_2 \right] = 0,$$

(16.15)

where $\omega_{ij} = \dfrac{1}{2}(u_{i,j} - u_{j,i})$; $S_{11}^p$ are horizontal initial stress; $\rho_{11}, \rho_{22},$ and $\rho_{12}$ take into account as the inertial effects of the moving fluids and are associated with the densities of the solid part $\rho_s$, fluid part $\rho_f$, and the aggregate medium $\rho_2$ by the relations such that the mass density of the aggregate is $\rho_2 = \rho_{11} + 2\rho_{12} + \rho_{22} = \rho_s + f(\rho_f - \rho_s)$.

Moreover, the following inequalities also hold for the dynamic coefficients $\rho_{11} > 0, \rho_{12} \le 0, \rho_{22} > 0, \rho_{11}\rho_{22} - \rho_{12}^2 > 0$

On further simplification, Equation (16.15) results in

$$\left( N + \frac{S_{11}^p}{2} \right) \partial_{xx} v_2 + G \partial_{zz} v_2 = d' \partial_{tt} v_2$$

(16.16)

where $d' = \left( \rho_{11} - \dfrac{\rho_{12}^2}{\rho_{22}} \right)$.

To solve Equation (16.16), we assume $v_2(x, z, t) = V(z)e^{i(\omega t - k_2 x)}$ and after applying this in Equation (16.16), we get

$$\partial_{zz} V + k_2^2 \eta^2 V = 0,$$

(16.17)

where

$$\eta = k_2 \left[ \mu_1' \left( -1 + \mu_2' \mathrm{cosec}^2 f \right) \right], \mu_1' = \frac{N}{G}(1 + \xi_1), \mu_2' = d^p (1 + \xi_1)^{-1}, \xi_1 = \frac{S_{11}^p}{2N}, \text{and } d^p = \left( \gamma_{11} - \frac{\gamma_{12}^2}{\gamma_{22}} \right)$$

The solution of Equation (16.17) is given by

$$V(x, z, t) = \left( C_0 e^{-i\eta z} + D_0 e^{i\eta z} \right)$$

(16.18)

The displacement for upper initially stressed poroelastic half-space, i.e., $F_2$, is given as

$$v_2(x, z, t) = \left( C_0 e^{-i\eta z} + D_0 e^{i\eta z} \right) e^{i(\omega t - k_2 x)}$$

(16.19)

where $C_0$ and $D_0$ are constants, and $k_2$ is the wave number defined by the law of refraction

$$k_2 : k_1 = \sin e : \sin f$$

$f$ is the refraction angle, and $\rho_2$ is the density of the upper half-space.

Let us assume that a ray of plane SH-wave of unit amplitude is propagating in the lower half-space ($F_1$) and is incident at the corrugated interface $z = \zeta$, making an angle $e$ with the z-axis. Due to the corrugation at the interface, the reflection and refraction phenomena will be affected, and the incident SH-wave will give rise to (1) a regularly reflected and a regularly transmitted wave at angles $e$ and $f$ with the z-axis, in the lower ($F_1$) and upper half-space ($F_2$), respectively; (2) a spectrum of $n$th order of irregularly reflected and irregularly refracted waves at angles $e_n$ and $f_n$ in the left side of regularly reflected and regularly refracted waves, respectively; and (3) a similar spectrum of irregularly reflected and irregularly refracted waves at angles $e'_n$ and $f'_n$ in the right side of regularly reflected and regularly refracted waves, respectively, at the corrugated interface.

The angle of refraction $f$ is related to the angle of incidence $e$ through Snell's law

$$\frac{\sin e}{\beta_1} = \frac{\sin f}{\beta_2} \tag{16.20}$$

The angles $e_n, e'_n, f_n$, and $f'_n$, are given by the following spectrum theorem by Abubakar (1962a–c):

$$\sin e_n - \sin e = \frac{n\lambda\beta_1}{\omega}, \sin e'_n - \sin e = -\frac{n\lambda\beta_1}{\omega},$$

$$\sin f_n - \sin f = \frac{n\lambda\beta_2}{\omega}, \sin f'_n - \sin f = -\frac{n\lambda\beta_2}{\omega} \tag{16.21}$$

The total displacement in the lower highly anisotropic half-space ($F_1$) is then given by the sum of the incident, regularly reflected, and irregularly refracted waves

$$v(x,z,t) = \left[ B_0 e^{i\Omega_0 z} + B e^{-i\Omega z} + \sum_{n=1}^{\infty} B_n e^{-i\Omega_n z} e^{-in\lambda x} + \sum_{n=1}^{\infty} B'_n e^{-i\Omega'_n z} e^{in\lambda x} \right] e^{i\omega\left(t - \frac{x\sin e}{\beta_1}\right)} \tag{16.22}$$

where

$$\Omega_n = \frac{\omega \sin e_n}{\beta_1}\left(-\mu_1 + \sqrt{\mu_1^2 + \mu_2\left(\cot^2 e_n + \xi\right)}\right) \text{ and } \Omega'_n = \frac{\omega \sin e_n}{\beta_1}\left(-\mu_1 + \sqrt{\mu_1^2 + \mu_2\left(\cot^2 e'_n + \xi\right)}\right)$$

Similarly, the total displacement in upper initially stressed fluid-saturated poroelastic half-space ($F_2$) is the sum of regularly refracted and irregularly refracted waves:

$$v_1(x,z,t) = \left[ D_0 e^{i\eta z} + \sum_{n=1}^{\infty} D_n e^{i\eta_n z} e^{-in\lambda x} + \sum_{n=1}^{\infty} D'_n e^{i\eta'_n z} e^{in\lambda x} \right] e^{i\omega\left(t - \frac{x\sin f}{\beta_2}\right)} \tag{16.23}$$

where

$$\eta_n = \frac{\omega \sin f_n}{\beta_2}\left[\mu'_1\left(-1 + \mu'_2 \cosec^2 f_n\right)\right] \text{ and } \eta'_n = \frac{\omega \sin f_n}{\beta_2}\left[\mu'_1\left(-1 + \mu'_2 \cosec^2 f'_n\right)\right]$$

The constants $B_0$ and $B$ are the reflection and refraction coefficients at plane interface, respectively, and the constants $B_n, B'_n$ and $D_n, D'_n$ are the reflection and refraction coefficients, respectively, for the first-order approximation of corrugation. All these constants are determined from the boundary conditions at the interface.

## 16.3 Boundary conditions

The boundary conditions at the corrugated interface $z = \zeta$ ensure the continuity of displacement and stress, i.e.,

$$v_1 = v_2 \tag{16.24}$$

$$\left(T_{23}^t - \zeta' T_{12}^t\right) = \left(T_{23}^p - \zeta' T_{12}^p\right) \tag{16.25}$$

where $\zeta'$ is the derivative of $\zeta$ with respect to $x$. Substituting Equations (16.12) and (16.19) in the above boundary conditions, we obtain

$$
\left[ B_0 e^{i\Omega_0 \zeta} + B e^{-i\Omega \zeta} + \sum_{n=1}^{\infty} B_n e^{-i\Omega_n \zeta} e^{-in\lambda x} + \sum_{n=1}^{\infty} B'_n e^{-i\Omega'_n \zeta} e^{in\lambda x} \right]
$$
$$
= \left[ D_0 e^{i\eta z} + \sum_{n=1}^{\infty} D_n e^{i\eta_n z} e^{-in\lambda x} + \sum_{n=1}^{\infty} D'_n e^{i\eta'_n z} e^{in\lambda x} \right] \tag{16.26}
$$

and

$$
B_0 \left\{ \Omega_0 \left(F_{44} - \zeta' F_{64}\right) - \frac{\omega \sin e}{\beta_1} \left(F_{46} - \zeta' F_{66}\right) \right\} e^{i\Omega_0 \zeta} + B\{ -\Omega \left(F_{44} - \zeta' F_{64}\right) - \frac{\omega \sin e}{\beta_1}
$$

$$
\times \left(F_{46} - \zeta' F_{66}\right)\} e^{-i\Omega \zeta} + \sum_{n=1}^{\infty} B_n e^{-i\Omega_n \zeta} e^{-in\lambda x} \left\{ -\left(F_{44} - \zeta' F_{64}\right)\Omega_n - \left(\frac{\omega \sin e}{\beta_1} + n\lambda\right)\left(F_{46} - \zeta' F_{66}\right) \right\}
$$

$$
+ \sum_{n=1}^{\infty} B_n' e^{-i\Omega'_n \zeta} e^{in\lambda x} \left\{ -\left(F_{44} - \zeta' F_{64}\right)\Omega'_n + \left(-\frac{\omega \sin e}{\beta_1} + n\lambda\right)\left(F_{46} - \zeta' F_{66}\right) \right\}
$$

$$
= D_0 \left\{ G\eta + \zeta' N \frac{\omega \sin f}{\beta_2} \right\} e^{i\eta \zeta} + \sum_{n=1}^{\infty} D_n e^{-i\eta_n \zeta} e^{-in\lambda x} \left\{ \eta_n G + n\lambda \zeta' N + \zeta' N \frac{\omega \sin f}{\beta_2} \right\}
$$

$$
+ \sum_{n=1}^{\infty} D'_n e^{i\eta'_n \zeta} e^{in\lambda x} \left\{ \eta'_n G - n\lambda \zeta' N + \zeta' N \frac{\omega \sin f}{\beta_2} \right\} \tag{16.27}
$$

From Equations (16.26) and (16.27), the reflection and refraction coefficients of $n$th order of approximation of the corrugated interface can be determined.

## 16.4 Solution of the first-order approximation of the corrugation

As discussed earlier, the amplitude and slope of corrugation are very small, so the higher powers of $\zeta$ can be neglected. The exponential function involving $\zeta$ can then be approximated as (Rayleigh's approximation of first order)

$$e^{\pm i\alpha \zeta} = 1 \pm i\alpha \zeta. \tag{16.28}$$

In view of Equation (16.28), using Equations (16.26) and (16.27) by collecting the terms independent of $x$ and $\zeta$ to both sides,

$$B_0 + B = D_0 \tag{16.29}$$

$$\left( F_{44}\Omega_0 - \frac{F_{46}\omega \sin e}{\beta_1} \right) B_0 - \left( F_{44}\Omega + \frac{F_{46}\omega \sin e}{2\beta_1} \right) B = D_0 G \eta \tag{16.30}$$

These equations provide the values of reflection and refraction coefficients of the regularly reflected and refracted SH-wave at a plane interface.

Solving Equations (16.29) and (16.30), we have

$$\frac{B}{B_0} = \frac{\left[ -G\eta + \left( F_{44}\Omega_0 - \dfrac{F_{46}\omega \sin e}{\beta_1} \right) \right]}{\left[ G\eta + \left( F_{44}\Omega + \dfrac{F_{46}\omega \sin f}{\beta_1} \right) \right]}, \tag{16.31}$$

$$\frac{D_0}{B_0} = \frac{\left[ F_{44}\Omega_0 + F_{44}\Omega \right]}{\left[ G\eta + \left( F_{44}\Omega + \dfrac{F_{46}\omega \sin f}{\beta_1} \right) \right]} \tag{16.32}$$

Equations (16.31) and (16.32) give the reflection and refraction coefficients of SH-waves at a plane interface between initially stressed fluid-saturated poroelastic half-space and highly anisotropic half-space.

In order to find the solutions of the first-order approximation for the coefficients $B_n$ and $D_n$, we arrange the coefficients of $e^{-in\lambda x}$ on both sides of Equations (16.26) and (16.27), and then we have

$$B_n - D_n = i\zeta_{-n}\left( -B_0\Omega_0 + B\Omega + \eta D_0 \right) \tag{16.33}$$

$$b_n B_n - d_n D_n d_n = -i\zeta_{-n}\left( t_3 D_0 - t_2 B - t_1 B_0 \right) \tag{16.34}$$

$$b_n = \left\{ -F_{44}\Omega_n - F_{46}\left( \frac{\omega \sin e}{\beta_1} + n\lambda \right) \right\}, d_n = \eta_n G,$$

$$t_1 = \left\{ n\lambda \left( -F_{46}\Omega_0 + F_{66}\frac{\omega \sin e}{\beta_1} \right) + \Omega_0 \left( -F_{44}\Omega_0 + F_{46}\frac{\omega \sin e}{\beta_1} \right) \right\},$$

where

$$t_2 = \left\{ n\lambda \left( F_{46}\Omega + F_{66}\frac{\omega \sin e}{\beta_1} \right) + \Omega \left( -F_{44}\Omega - F_{46}\frac{\omega \sin e}{\beta_1} \right) \right\},$$

$$t_3 = \left\{ -G\eta^2 + \frac{n\lambda N\omega \sin f}{\beta_2} \right\}$$

Equating the coefficients of $e^{in\lambda x}$, we obtain the first-order approximation for coefficients $B'_n$ and $D'_n$,

$$B'_n - D'_n = i\zeta_n \left( -B_0 \Omega_0 + B\Omega + \eta D_0 \right) \tag{16.35}$$

$$B'_n b'_n - D'_n d'_n = -i\zeta_n \left( t_6 D_0 - t_5 B - t_4 B_0 \right) \tag{16.36}$$

where

$$b'_n = \left\{ -F_{44}\Omega'_n + F_{46}\left( -\frac{\omega \sin e}{\beta_1} + n\lambda \right) \right\}, d'_n = \eta'_n G,$$

$$t_4 = \left\{ n\lambda \left( F_{46}\Omega_0 - F_{66}\frac{\omega \sin e}{\beta_1} \right) + \Omega_0 \left( -F_{44}\Omega_0 + F_{46}\frac{\omega \sin e}{\beta_1} \right) \right\},$$

$$t_5 = \left\{ n\lambda \left( -F_{46}\Omega - F_{66}\frac{\omega \sin e}{\beta_1} \right) + \Omega \left( -F_{44}\Omega - F_{46}\frac{\omega \sin e}{\beta_1} \right) \right\},$$

$$t_6 = \left\{ -G\eta^2 - \frac{n\lambda N\omega \sin f}{\beta_2} \right\}$$

From Equations (16.33) to (16.36), we obtain the reflection and refraction coefficients of irregularly reflected and refracted waves for the first-order approximation:

$$\frac{B_n}{B_0} = \frac{\Pi^+_{B_n}}{\Pi^+_n}, \frac{D_n}{B_0} = \frac{\Pi^+_{D_n}}{\Pi^+_n}, \frac{B'_n}{B_0} = \frac{\Pi^-_{B'_n}}{\Pi^-_n}, \frac{D'_n}{B_0} = \frac{\Pi^-_{D'_n}}{\Pi^-_n} \tag{16.37}$$

where

$$\Pi^+_{B_n} = i\zeta_{-n} \left\{ (d_n - b_n)\left( -\Omega_0 + \Omega\frac{B}{B_0} + \eta\frac{D_0}{B_0} \right) + (t_3 - \eta b_n)\frac{D_0}{B_0} - (t_2 + b_n\Omega)\frac{B}{B_0} + (-t_1 + b_n\Omega_0) \right\},$$

$$\Pi^+_{D_n} = i\zeta_{-n} \left\{ (t_3 - \eta b_n)\frac{D_0}{B_0} - (t_2 + b_n\Omega)\frac{B}{B_0} + (-t_1 + b_n\Omega_0) \right\},$$

$$\Pi^-_{B'_n} = i\zeta_n \left\{ (d'_n - b'_n)\left( -\Omega_0 + \Omega\frac{B}{B_0} + \eta\frac{D_0}{B_0} \right) + (t_6 + \eta b'_n)\frac{D_0}{B_0} - (t_5 + b'_n\Omega)\frac{B}{B_0} + (-t_4 - b'_n\Omega_0) \right\},$$

$$\Pi^-_{D'_n} = i\zeta_n \left\{ (t_6 + \eta b'_n)\frac{D_0}{B_0} - (t_5 + b'_n\Omega)\frac{B}{B_0} + (-t_4 - b'_n\Omega_0) \right\},$$

$$\Pi^+_n = (b_n - a_n), \Pi^-_n = (d'_n - b'_n)$$

## 16.5   Solution for second-order approximation of the corrugation

For the solution of second-order approximation, we disregard the terms involving the third and higher powers of $\zeta$ so that

$$\exp(\pm i\alpha\zeta) = 1 \pm i\alpha\zeta - (\alpha\zeta)^2 / 2 \tag{16.38}$$

Using Equations (16.1) and (16.38) into Equations (16.26) and (16.27) and comparing the terms independent of $x$, the coefficients of $e^{-in\lambda x}$ and those of $e^{in\lambda x}$, separately on both sides of the equations thus obtained, we get the following system of six equations which on solving will give reflection and refraction coefficients for the second-order approximation:

$$B_0\left(1-\Omega_0^2\zeta_{-n}\zeta_n\right)+B\left(1-\Omega^2\zeta_{-n}\zeta_n\right)-i\Omega_n\zeta_nB_n-i\Omega_n'\zeta_{-n}B_n'=D_0\left(1-\eta^2\zeta_{-n}\zeta_n\right)+i\eta_n\zeta_nD_n+i\eta_n'\zeta_{-n}D_n',$$

$$i\Omega_0\zeta_{-n}B_0-i\Omega\zeta_{-n}B+\left(1-\Omega_n^2\zeta_{-n}\zeta_n\right)B_n-\frac{\Omega_n'^2\zeta_{-n}^2B_n'}{2}=i\eta\zeta_nD_0+\left(1-\eta_n^2\zeta_{-n}\zeta_n\right)D_n-\frac{\eta_n'^2\zeta_{-n}^2D_n'}{2},$$

$$i\Omega_0\zeta_nB_0-i\Omega\zeta_nB-\frac{\Omega_n^2\zeta_n^2B_n}{2}+\left(1-\Omega_n'^2\zeta_{-n}\zeta_n\right)B_n'=i\eta\zeta_nD_0-\frac{\eta_n^2\zeta_n^2D_n}{2}+\left(1-\eta_n'^2\zeta_{-n}\zeta_n\right)D_n',$$

$$B_0\left\{iF_{44}\Omega_0\left(1-\Omega_0^2\zeta_{-n}\zeta_n\right)+\frac{iF_{46}\omega\sin e}{\beta_1}\left(-1+\Omega_0^2\zeta_{-n}\zeta_n\right)\right\}$$

$$+B\left\{iF_{44}\Omega\left(-1+\Omega^2\zeta_{-n}\zeta_n\right)+\frac{iF_{46}\omega\sin e}{\beta_1}\left(-1+\Omega_0^2\zeta_{-n}\zeta_n\right)\right\}$$

$$+B_n\left\{-F_{44}\Omega_n^2\zeta_n+\lambda n\left(-1+\frac{\Omega_n^2\zeta_{-n}\zeta_n}{2}\right)\left\{\zeta_n\Omega_nF_{64}+\lambda n\zeta_nF_{66}+\frac{\zeta_nF_{66}\omega\sin e}{\beta_1}\right.\right.$$

$$\left.+F_{46}\Omega_n\zeta_n\left\{-\lambda n+\frac{F_{46}\omega\sin e}{\beta_1}\right\}\right\}+B_n'\left\{-F_{44}\Omega_n'^2\zeta_{-n}+\lambda n\left(1-\frac{\Omega_n'^2\zeta_{-n}\zeta_n}{2}\right)\right.$$

$$\left.\times\left\{\zeta_{-n}\Omega_n'F_{64}-\lambda n\zeta_{-n}F_{66}+\frac{\zeta_{-n}F_{66}\omega\sin e}{\beta_1}\right\}+F_{46}\Omega_n'\zeta_{-n}\left\{\lambda n-\frac{F_{46}\omega\sin e}{\beta_1}\right\}\right\}$$

$$=D_0iG\eta\left(1-\eta_n^2\zeta_{-n}\zeta_n\right)$$

$$+D_n\left[-G\eta_n^2\zeta_n-\lambda nN\left\{\lambda n\zeta_n\left(-1+\frac{\eta_n^2\zeta_{-n}\zeta_n}{2}\right)+\frac{\zeta_n\omega\sin f}{\beta_2}\left(-1+\frac{\eta_n^2\zeta_{-n}\zeta_n}{2}\right)\right\}\right]$$

$$+D_n'\left[-G\eta_n'^2\zeta_{-n}-\lambda nN\left\{\lambda n\zeta_{-n}\left(1+\frac{\eta_n'^2\zeta_{-n}\zeta_n}{2}\right)+\frac{\zeta_{-n}\omega\sin f}{\beta_2}\left(-1+\frac{\eta_n'^2\zeta_{-n}\zeta_n}{2}\right)\right\}\right],$$

$$B_0\left\{F_{64}\lambda n\Omega_0\zeta_{-n}\left(1-\frac{\Omega_0^2\zeta_{-n}\zeta_n}{2}\right)-F_{44}\Omega_0^2\zeta_{-n}+\frac{\lambda nF_{66}\omega\sin e\zeta_{-n}}{\beta_1}\left(1+\frac{\Omega_0^2\zeta_{-n}\zeta_n}{2}\right)+\frac{\Omega_0F_{46}\omega\sin e\zeta_{-n}}{\beta_1}\right\}$$

$$+B\left\{F_{64}\lambda n\Omega\zeta_{-n}\left(1-\frac{\Omega^2\zeta_{-n}\zeta_n}{2}\right)-F_{44}\Omega^2\zeta_{-n}-\frac{\lambda nF_{66}\omega\sin e\zeta_{-n}}{\beta_1}\left(1+\frac{\Omega^2\zeta_{-n}\zeta_n}{2}\right)-\frac{\Omega F_{46}\omega\sin e\zeta_{-n}}{\beta_1}\right\}$$

$$+B_n\left\{F_{64}\lambda n\Omega_n\zeta_{-n}\left(-1+\Omega_n^2\zeta_{-n}\zeta_n\right)+iF_{64}\lambda n\left(-1+\Omega_n^2\zeta_{-n}\zeta_n\right)-\frac{F_{46}\omega\sin e}{\beta_1}\left(1-\Omega_n^2\zeta_{-n}\zeta_n\right)\right\}$$

$$+ B_n' \left\{ i\Omega_n'^2 \zeta_{-n}^2 \left( -\lambda n F_{64} + \frac{F_{44}\Omega_n'}{2} \right) + iF_{66}\lambda n \Omega_n' \zeta_{-n}^2 \left( \lambda n - \frac{\omega \sin e}{\beta_1} \right)\left( F_{66} - \frac{F_{46}\Omega_n'}{2} \right) \right\}$$

$$= D_0 \left[ -G\eta^2 \zeta_{-n} - \frac{\lambda n N \zeta_{-n} \omega \sin f}{\beta_2} \left\{ -1 + \frac{\eta^2 \zeta_{-n} \zeta_n}{2} \right\} \right]$$

$$+ D_n iG\eta_n \left[ 1 - \eta_n^2 \zeta_{-n} \zeta_n \right] + D_n' \left[ G\eta_n \zeta_{-n}^2 \left( \frac{-i\eta_n^2}{2} + \lambda^2 n^2 \right) - \frac{i\lambda n N}{\beta_2} \left\{ \eta_n' \frac{\zeta_{-n}^2 \omega \sin f}{\beta_2} \right\} \right],$$

$$B_0 \left\{ F_{64}\lambda n \Omega_0 \zeta_n \left( 1 - \frac{\Omega_0^2 \zeta_{-n} \zeta_n}{2} \right) - F_{44}\Omega_0^2 \zeta_{-n} + \frac{\lambda n F_{66} \omega \sin e \zeta_n}{\beta_1} \left( -1 + \frac{\Omega_0^2 \zeta_{-n} \zeta_n}{2} \right) + \frac{\Omega_0 F_{46} \omega \sin e \zeta_n}{\beta_1} \right\}$$

$$+ B \left\{ F_{64}\lambda n \Omega \zeta_{-n} \left( -1 + \frac{\Omega^2 \zeta_{-n} \zeta_n}{2} \right) - F_{44}\Omega^2 \zeta_n - \frac{\lambda n F_{66} \omega \sin e \zeta_n}{\beta_1} \left( -1 + \frac{\Omega^2 \zeta_{-n} \zeta_n}{2} \right) - \frac{\Omega F_{46} \omega \sin e \zeta_n}{\beta_1} \right\}$$

$$+ B_n \left\{ i\lambda n \Omega_n^2 \zeta_n^2 \left( F_{64} + \frac{\Omega_n^2 F_{44}}{2} \right) + iF_{66}\eta_n \lambda n \zeta_n^2 \left( \lambda n + \frac{\omega \sin e}{\beta_1} \right) + \frac{iF_{46}\eta_n^2 \zeta_n^2}{2} \left( \lambda n + \frac{\omega \sin e}{\beta_1} \right) \right\}$$

$$+ B_n' \left\{ F_{44}\eta_n' \left( -1 + \Omega_n'^2 \zeta_{-n} \zeta_n \right) + iF_{46}\left( \lambda n - \frac{\omega \sin e}{\beta_1} - \lambda n \Omega_n'^2 \zeta_{-n} \zeta_n + \frac{\zeta_{-n} \zeta_n \omega \sin e \eta_n'^2}{\beta_1} \right) \right\}$$

$$= D_0 \left[ -G\eta^2 \zeta_n + \frac{\lambda n N \zeta_n \omega \sin f}{\beta_2} \left\{ 1 - \frac{\eta^2 \zeta_{-n} \zeta_n \omega \sin f}{2} \right\} \right]$$

$$+ D_n' iG\eta_n' \left[ 1 - \eta_n'^2 \zeta_{-n} \zeta_n \right] + D_n \left[ G\eta_n \zeta_{-n}^2 \left( \frac{-i\eta_n^2}{2} \right) + \frac{i\lambda n N}{\beta_2} \left\{ \lambda n \beta_2 \zeta_n^2 + \omega \sin f \eta_n \zeta_n^2 \right\} \right]$$

## 16.6   Special case of a simple harmonic interface

We now obtain the reflection and refraction coefficients of incident plane SH-wave at an interface that is given by $\zeta = d\cos\lambda x$. Thus, the equation for the interface can be obtained by setting

$$\zeta_n = \zeta_{-n} = \begin{cases} \dfrac{d}{2}; & \text{when } n = 1 \\[2mm] 0; & \text{when } n = 2, 3, \ldots \end{cases}$$

In this case, $2\pi/\lambda$ is the wavelength and $d$ is the amplitude of the corrugation. Thus, the reflection and refraction coefficients for the first-order approximation of the corrugation can be obtained by setting $n = 1$ in Equation (16.37), and we obtain

$$\frac{B_1}{B_0} = \frac{\Pi_{B_1}^+}{\Pi_1^+}, \frac{D_1}{B_0} = \frac{\Pi_{D_1}^+}{\Pi_1^+}, \frac{B_1'}{B_0} = \frac{\Pi_{B_1'}^-}{\Pi_1^-}, \frac{D_1'}{B_0} = \frac{\Pi_{D_1'}^-}{\Pi_1^-} \qquad (16.39)$$

where

$$\Pi_{B_1}^+ = i\frac{d}{2}\left\{ (d_1 - b_1)\left( -\Omega_0 + \Omega\frac{B_1}{B_0} + \eta\frac{D_0}{B_0} \right) + (t_3' - \eta b_1)\frac{D_0}{B_0} - (t_2' + b_1\Omega)\frac{B_1}{B_0} + (-t_1' + b_1\Omega_0) \right\},$$

$$\Pi_{D_1}^+ = i\frac{d}{2}\left\{ (t_3' - \eta b_1)\frac{D_0}{B_0} - (t_2' + b_1\Omega)\frac{B_1}{B_0} + (-t_1' + b_1\Omega_0) \right\},$$

$$\Pi_{B_1'}^- = i\frac{d}{2}\left\{ (d_1' - b_1')\left( -\Omega_0 + \Omega\frac{B_1}{B_0} + \eta\frac{D_0}{B_0} \right) + (t_6' + \eta b_1')\frac{D_0}{B_0} - (t_5' + b_1'\Omega)\frac{B_1}{B_0} + (-t_4' - b_1'\Omega_0) \right\},$$

$$\Pi_{D_1'}^- = i\frac{d}{2}\left\{ (t_6' + \eta b_1')\frac{D_0}{B_0} - (t_5' + b_1'\Omega)\frac{B_1}{B_0} - (t_4' + b_1'\Omega_0) \right\},$$

$$\Pi_1^+ = (b_1 - a_1), \Pi_1^- = (d_1' - b_1')$$

where

$$b_1 = \left\{ -F_{44}\Omega_1 - F_{46}\left( \frac{\omega\sin e}{\beta_1} + \lambda \right) \right\}, d_1 = \eta_1 G, t_1' = \left\{ \lambda\left( -F_{46}\Omega_0 + F_{66}\frac{\omega\sin e}{\beta_1} \right) + \Omega_0\left( -F_{44}\Omega_0 + F_{46}\frac{\omega\sin e}{\beta_1} \right) \right\},$$

$$t_2' = \left\{ \lambda\left( F_{46}\Omega + F_{66}\frac{\omega\sin e}{\beta_1} \right) + \Omega\left( -F_{44}\Omega - F_{46}\frac{\omega\sin e}{\beta_1} \right) \right\}, t_3' = \left\{ -G\eta^2 + \frac{\lambda N\omega\sin f}{\beta_2} \right\},$$

$$b_1' = \left\{ -F_{44}\Omega_1' + F_{46}\left( -\frac{\omega\sin e}{\beta_1} + \lambda \right) \right\}, d_1' = \eta_1' G, t_4' = \left\{ \lambda\left( F_{46}\Omega_0 - F_{66}\frac{\omega\sin e}{\beta_1} \right) + \Omega_0\left( -F_{44}\Omega_0 + F_{46}\frac{\omega\sin e}{\beta_1} \right) \right\},$$

$$t_5' = \left\{ \lambda\left( -F_{46}\Omega - F_{66}\frac{\omega\sin e}{\beta_1} \right) + \Omega\left( -F_{44}\Omega - F_{46}\frac{\omega\sin e}{\beta_1} \right) \right\}, t_6' = \left\{ -G\eta^2 - \frac{\lambda N\omega\sin f}{\beta_2} \right\},$$

$$\Omega_1 = \frac{\omega\sin e_1}{\beta_1}\left( -\mu_1 + \sqrt{\mu_1^2 + \mu_2\left(\cot^2 e_1 + \xi\right)} \right), \Omega_1' = \frac{\omega\sin e_1}{\beta_1}\left( -\mu_1 + \sqrt{\mu_1^2 + \mu_2\left(\cot^2 e_1' + \xi\right)} \right),$$

$$\eta_1 = \frac{\omega\sin f_1}{\beta_2}\left[ \mu_1'\left(-1 + \mu_2'\cos ec^2 f_1\right) \right] \text{ and } \eta_1' = \frac{\omega\sin f_1}{\beta_2}\left[ \mu_1'\left(-1 + \mu_2'\mathrm{cosec}^2 f_1'\right) \right]$$

## 16.7 Particular cases for special case

*Case I:* When lower half-space is considered as isotropic medium without initial stress, i.e., when $F_{11} = F_{22} = F_{33} = \lambda_1 + 2\mu$, $F_{12} = F_{13} = F_{23} = \mu$, $F_{44} = F_{66} = \mu$, $F_{46} = 0$ and $S_{11}^i = 0$ and upper half-space in initially stressed fluid-saturated poroelastic half-space, then Equation (16.39) becomes

$$\frac{B_1}{B_0} = \frac{\Pi_{B_1}^+}{\Pi_1^+}, \frac{D_1}{B_0} = \frac{\Pi_{D_1}^+}{\Pi_1^+}, \frac{B_1'}{B_0} = \frac{\Pi_{B_1'}^-}{\Pi_1^-}, \frac{D_1'}{B_0} = \frac{\Pi_{D_1'}^-}{\Pi_1^-} \qquad (16.40)$$

where

$$\Pi_{B_1}^+ = i\frac{d}{2}\left\{(d_1 - b_1)\left(-\Omega_0 + \Omega\frac{B_1}{B_0} + \eta\frac{D_0}{B_0}\right) + (t_3' - \eta b_1)\frac{D_0}{B_0} - (t_2' + b_1\Omega)\frac{B_1}{B_0} + (-t_1' + b_1\Omega_0)\right\},$$

$$\Pi_{D_1}^+ = i\frac{d}{2}\left\{(t_3' - \eta b_1)\frac{D_0}{B_0} - (t_2' + b_1\Omega)\frac{B_1}{B_0} + (-t_1' + b_1\Omega_0)\right\},$$

$$\Pi_{B_1}^- = i\frac{d}{2}\left\{(d_1' - b_1')\left(-\Omega_0 + \Omega\frac{B_1}{B_0} + \eta\frac{D_0}{B_0}\right) + (t_6' + \eta b_1')\frac{D_0}{B_0} - (t_5' + b_1'\Omega)\frac{B_1}{B_0} + (-t_4' - b_1'\Omega_0)\right\},$$

$$\Pi_{D_1}^- = i\frac{d}{2}\left\{(t_6' + \eta b_1')\frac{D_0}{B_0} - (t_5' + b_1'\Omega)\frac{B_1}{B_0} - (t_4' + b_1'\Omega_0)\right\},$$

$$\Pi_1^+ = (b_1 - a_1), \Pi_1^- = (d_1' - b_1')$$

where

$$b_1 = \{-\mu\Omega_1\}, d_1 = \eta_1 G, t_1' = \left\{\lambda\left(\frac{\mu\omega\sin e}{\beta_1}\right) + \Omega_0(-\mu\Omega_0)\right\}, t_2' = \left\{\lambda\left(\frac{\mu\omega\sin e}{\beta_1}\right) + \Omega(-\mu\Omega)\right\},$$

$$t_3' = \left\{-G\eta^2 + \frac{\lambda N\omega\sin f}{\beta_2}\right\}, b_1' = \{-\mu\Omega_1'\}, d_1' = \eta_1' G, t_4' = \left\{\lambda\left(-\frac{\mu\omega\sin e}{\beta_1}\right) + \Omega_0(-\mu\Omega_0)\right\},$$

$$t_5' = \left\{\lambda\left(-\frac{\mu\omega\sin e}{\beta_1}\right) + \Omega(-\mu\Omega)\right\}, t_6' = \left\{-G\eta^2 - \frac{\lambda N\omega\sin f}{\beta_2}\right\}, \Omega_1 = \frac{\omega\sin e_1}{\beta_1}\left(\sqrt{(\cot^2 e_1)}\right),$$

$$\Omega_1' = \frac{\omega\sin e_1}{\beta_1}\left(\sqrt{(\cot^2 e_1')}\right), \eta_1 = \frac{\omega\sin f_1}{\beta_2}\left[\mu_1'\left(-1 + \mu_2'\text{cosec}^2 f_1\right)\right], \eta_1' = \frac{\omega\sin f_1}{\beta_2}\left[\mu_1'\left(-1 + \mu_2'\cos ec^2 f_1'\right)\right],$$

$$\mu_1 = 0, \mu_2 = 1, \xi = 0, \beta_1^2 = \frac{\mu}{\rho_1}, \mu_1' = \frac{N}{G}(1 + \xi_1), \mu_2' = d(1 + \xi_1)^{-1}, \xi_1 = \frac{S_{11}^p}{2N}, d^p = \left(\gamma_{11} - \frac{\gamma_{12}^2}{\gamma_{22}}\right)$$

Equation (16.40) is deduced for the case when SH-wave is incident at a corrugated interface between initially stressed fluid-saturated poroelastic half-space and isotropic elastic half-space.

*Case II:* When the upper half-space becomes isotropic elastic medium without initial stress and without poro-elasticity, i.e., $S_{11}^p = 0$, $d^p \to 1$, $N = G = \mu^p$ and lower half-space is considered as highly anisotropic half-space, then Equation (16.39) reduces to

$$\frac{B_1}{B_0} = \frac{\Pi_{B_1}^+}{\Pi_1^+}, \frac{D_1}{B_0} = \frac{\Pi_{D_1}^+}{\Pi_1^+}, \frac{B_1'}{B_0} = \frac{\Pi_{B_1}^-}{\Pi_1^-}, \frac{D_1'}{B_0} = \frac{\Pi_{D_1}^-}{\Pi_1^-} \tag{16.41}$$

where

$$\Pi_{B_1}^+ = i\frac{d}{2}\left\{(d_1 - b_1)\left(-\Omega_0 + \Omega\frac{B_1}{B_0} + \eta\frac{D_0}{B_0}\right) + (t_3' - \eta b_1)\frac{D_0}{B_0} - (t_2' + b_1\Omega)\frac{B_1}{B_0} + (-t_1' + b_1\Omega_0)\right\},$$

$$\Pi_{D_1}^+ = i\frac{d}{2}\left\{(t_3' - \eta b_1)\frac{D_0}{B_0} - (t_2' + b_1\Omega)\frac{B_1}{B_0} + (-t_1' + b_1\Omega_0)\right\},$$

$$\Pi_{\bar{B}_1^-} = i\frac{d}{2}\left\{(d_1' - b_1')\left(-\Omega_0 + \Omega\frac{B_1}{B_0} + \eta\frac{D_0}{B_0}\right) + (t_6' + \eta b_1')\frac{D_0}{B_0} - (t_5' + b_1'\Omega)\frac{B_1}{B_0} + (-t_4' - b_1'\Omega_0)\right\},$$

$$\Pi_{\bar{D}_1^-} = i\frac{d}{2}\left\{(t_6' + \eta b_1')\frac{D_0}{B_0} - (t_5' + b_1'\Omega)\frac{B_1}{B_0} - (t_4' + b_1'\Omega_0)\right\},$$

$$\Pi_1^+ = (b_1 - a_1), \Pi_1^- = (d_1' - b_1')$$

where

$$b_1 = \left\{-F_{44}\Omega_1 - F_{46}\left(\frac{\omega \sin e}{\beta_1} + \lambda\right)\right\}, d_1 = \eta_1\mu^p, t_1' = \left\{\lambda\left(-F_{46}\Omega_0 + F_{66}\frac{\omega \sin e}{\beta_1}\right) + \Omega_0\left(-F_{44}\Omega_0 + F_{46}\frac{\omega \sin e}{\beta_1}\right)\right\},$$

$$t_2' = \left\{\lambda\left(F_{46}\Omega + F_{66}\frac{\omega \sin e}{\beta_1}\right) + \Omega\left(-F_{44}\Omega - F_{46}\frac{\omega \sin e}{\beta_1}\right)\right\}, t_3' = \left\{-\mu^p\eta^2 + \frac{\lambda\mu^p\omega \sin f}{\beta_2}\right\},$$

$$b_1' = \left\{-F_{44}\Omega_1' + F_{46}\left(-\frac{\omega \sin e}{\beta_1} + \lambda\right)\right\}, d_1' = \eta_1'\mu^p, t_4' = \left\{\lambda\left(F_{46}\Omega_0 - F_{66}\frac{\omega \sin e}{\beta_1}\right) + \Omega_0\left(-F_{44}\Omega_0 + F_{46}\frac{\omega \sin e}{\beta_1}\right)\right\},$$

$$t_5' = \left\{\lambda\left(-F_{46}\Omega - F_{66}\frac{\omega \sin e}{\beta_1}\right) + \Omega\left(-F_{44}\Omega - F_{46}\frac{\omega \sin e}{\beta_1}\right)\right\}, t_6' = \left\{-\mu^p\eta^2 - \frac{\lambda\mu^p\omega \sin f}{\beta_2}\right\},$$

$$\Omega_1 = \frac{\omega \sin e_1}{\beta_1}\left(-\mu_1 + \sqrt{\mu_1^2 + \mu_2\left(\cot^2 e_1 + \xi\right)}\right), \Omega_1' = \frac{\omega \sin e_1}{\beta_1}\left(-\mu_1 + \sqrt{\mu_1^2 + \mu_2\left(\cot^2 e_1' + \xi\right)}\right),$$

$$\eta_1 = \frac{\omega \sin f_1}{\beta_2}\left[\left(-1 + \csc^2 f_1\right)\right], \eta_1' = \frac{\omega \sin f_1}{\beta_2}\left[\left(-1 + \csc^2 f_1'\right)\right],$$

$$\mu_1 = \frac{F_{46}}{F_{44}}, \mu_2 = \frac{F_{66}}{F_{44}}, \xi = \frac{S_{11}^t}{F_{66}}, \beta_1^2 = \frac{F_{66}}{\rho_1}, \eta = k_2\left[\left(-1 + \csc^2 f\right)\right],$$

$$\mu_1' = 1, \mu_2' = 1, \xi_1 = 0, \text{ and } d^p = \left(\gamma_{11} - \frac{\gamma_{12}^2}{\gamma_{22}}\right)$$

Equation (16.41) is deduced for the case when SH-wave is incident at a corrugated interface between isotropic elastic half-space and highly anisotropic half-space.

## 16.8    *Energy distribution*

It is apparent that when a plane SH-wave is incident on any surface, the energy of the incident wave is distributed among the reflected and refracted waves. The energy flux for the incident and each of the individually reflected and refracted waves can be obtained by multiplying total energy per unit volume with the wave velocity and the area of the wave front. In our case, the total energy per unit volume is twice the mean kinetic energy density. Also, the wave front area is proportional to the cosine of the angle of wave intersected against normal. Therefore, by Snell's law and the spectrum theorem, the energy equation for each of the individual waves, i.e., the incident, regularly reflected and refracted, and irregularly reflected and refracted SH-wave for the *nth*-order approximation of the corrugation can be written as (Abubakar 1962b, Tomar and Kaur 2007)

$$1 = \left|\frac{B}{B_0}\right|^2 + \sum_{n=1}^{\infty} \frac{\cos e_n}{\cos e}\left|\frac{B_n}{B_0}\right|^2 + \sum_{n=1}^{\infty} \frac{\cos e_n'}{\cos e}\left|\frac{B_n'}{B_0}\right|^2 + \frac{\rho_2\beta_2 \cos f}{\rho_1\beta_1 \cos e}\left|\frac{D_0}{B_0}\right|^2 + \sum_{n=1}^{\infty} \frac{\rho_2\beta_2 \cos f_n}{\rho_1\beta_1 \cos e}\left|\frac{D_n}{B_0}\right|^2$$

(16.42)

The energy distribution at the interface between two different types of half-spaces can be deduced using Equation (16.42) by equating the coefficients of $B_n, D_n, B_n'$, and $D_n'$ to 0, as they are positively dependent on corrugation amplitude,

$$1 = \left|\frac{B}{B_0}\right|^2 + \frac{\rho_2\beta_2^2 \tan e}{\rho_1\beta_1^2 \tan f}\left|\frac{D_0}{B_0}\right|^2$$

From Equation (16.42), when $n = 1$ it becomes

$$\sum_{i=1}^{6} E_i \approx 1$$

Here, $E_1$ and $E_2$ are the energy ratios of the regularly reflected and regularly refracted waves. Energy ratio is particularly defined as the ratio of the energy of reflection/refraction wave and energy of the incident wave. Similarly, $E_3$, $E_5$ and $E_4$, $E_6$ can be defined as the energy ratios of the irregularly reflected waves and irregularly refracted waves, respectively, for the first-order approximation of corrugation. Thus, the energy ratios are given as

$$E_1 = \left|\frac{B}{B_0}\right|^2, E_2 = \frac{\rho_2\beta_2 \cos f}{\rho_1\beta_1 \cos e}\left|\frac{D_0}{B_0}\right|^2, E_3 = \frac{\cos e_1}{\cos e}\left|\frac{B_1}{B_0}\right|^2,$$

$$E_4 = \frac{\rho_2\beta_2 \cos f_1}{\rho_1\beta_1 \cos e}\left|\frac{D_1}{B_0}\right|^2, E_5 = \frac{\cos e_1'}{\cos e}\left|\frac{B_1'}{B_0}\right|^2, E_6 = \frac{\rho_2\beta_2 \cos f_1'}{\rho_1\beta_1 \cos e}\left|\frac{D_1'}{B_0}\right|^2$$

## 16.9   Numerical discussion and results

A thorough numerical analysis has been performed to study the influence of various parameters such as corrugation amplitude, wavelength, frequency factor, and initial stress parameter associated with both the half-spaces, on the reflection and refraction coefficients against the angle of incidence. The following relevant elastic parameters have been used in the calculation and the results thus obtained are illustrated graphically. For medium $F_1$, the data taken are as follows (Tiersten 1969):

$$F_{11} = 86.74 \text{ Gpa}, \quad F_{22} = 129.77 \text{ Gpa}, \quad F_{33} = 102.83 \text{ Gpa},$$
$$F_{12} = -8.25 \text{ Gpa}, \quad F_{13} = 27.15 \text{ Gpa}, \quad F_{14} = -3.66 \text{ Gpa},$$
$$F_{23} = -7.42 \text{ Gpa}, \quad F_{24} = 5.7 \text{ Gpa}, \quad F_{34} = 9.92 \text{ Gpa},$$
$$F_{44} = 38.61 \text{ Gpa}, \quad F_{46} = 0.9 \text{ Gpa}, \quad F_{55} = 68.81 \text{ Gpa},$$
$$F_{66} = 29.01 \text{ Gpa}, \quad \rho_1 = 2649 \text{ kg/m}^3$$

For medium $F_2$, the data are as follows:

$$G = 0.1387 \times 10^{10} \text{ N/m}^2, N = 0.2774 \times 10^{10} \text{ N/m}^2, \rho_{11} = 1.926137 \times 10^3 \text{ kg/m}^3,$$

$$\rho_{12} = -0.002137 \times 10^3 \text{ kg/m}^3, \rho_{22} = 0.215337 \times 10^3 \text{ kg/m}^3$$

### 16.9.1 Effect of corrugation amplitude

Figures 16.2 and 16.3 have been drawn to demonstrate the variation of amplitude ratios represented as $(B_1/B_0)$ and $(D_1/B_0)$, respectively, against the incident angles for different values of corrugation amplitude ($d$). In Figure 16.2 it is seen that initially, the amplitude ratio is decreased with incidence angle, but on increasing the angle further, $(B_1/B_0)$ tends to increase until it reaches its individual maxima. After the maxima, the amplitude ratio

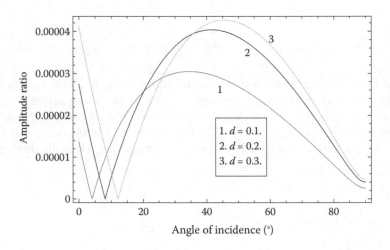

**Figure 16.2** Variation of modulus of amplitude ratio $(B_1/B_0)$ with respect to angle of incidence for different values of amplitude of corrugation ($d$).

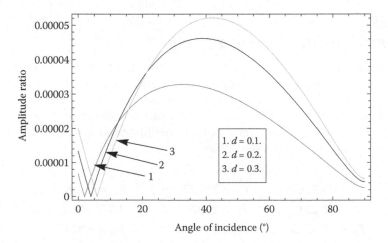

**Figure 16.3** Variation of modulus of amplitude ratio $(D_1/B_0)$ with respect to angle of incidence for different values of amplitude of corrugation ($d$).

gets decreased with increasing incident angles. When corrugation amplitude ($d$) is concerned, the amplitude ratios experience a positive effect for most parts of the incident angle. Furthermore, the ratio ($D_1/B_0$) in Figure 16.3 has a similar type behavior as seen in Figure 16.2 with respect to incidence angle and corrugation amplitude. Here, the effect of corrugation amplitude is less pronounced, which is identified by less spacing in between the curves.

Figures 16.4 and 16.5 have been plotted to discuss the variation of $(B_1'/B_0)$ and $(D_1'/B_0)$ against the incident angle while considering corrugation amplitude ($d$) as the effecting parameter. In Figure 16.4, the amplitude ratio $(B_1'/B_0)$ has a gradual increase throughout the incident angle. However, with the increasing values of ($d$), the ratio $(B_1'/B_0)$ gets decreased. Amplitude ratio $(D_1'/B_0)$ has the same characteristics as $(B_1'/B_0)$, which is clearly visible in Figure 16.5.

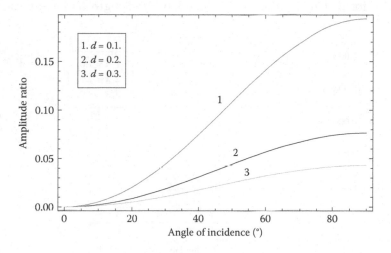

***Figure 16.4*** Variation of modulus of amplitude ratio $(B_1'/B_0)$ with respect to angle of incidence for different values of amplitude of corrugation ($d$).

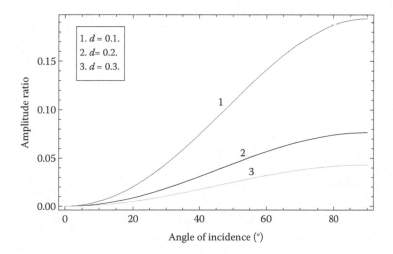

***Figure 16.5*** Variation of modulus of amplitude ratio $(D_1'/B_0)$ with respect to angle of incidence for different values of amplitude of corrugation ($d$).

## 16.9.2   *Effect of corrugation wavelength*

The purpose of Figures 16.6 and 16.7 is to depict the variation of amplitude ratios $(B_1/B_0)$ and $(D_1/B_0)$, respectively, against the incident angles for different values of wavelength of corrugation $(\lambda)$. In Figure 16.6, it is seen that the amplitude ratios decrease with increasing incidence angle and reach 0. On further increasing the incident angle, the ratios begin to increase for the rest of some parts. After that, the ratios have a decreasing behavior. However, the ratios are increased due to $\lambda$ during smaller incident angles. But after attaining minima, $(B_1/B_0)$ is decreased. In Figure 16.7, all the ratios have similar behavioral characteristics with respect to incident angle. However, there is a slight increase in $(D_1/B_0)$ for smaller angles and a slight decrease for higher incident angles due to corrugation wavelength.

Two-dimensional plots between amplitude ratios $(B_1'/B_0)$ and $(D_1'/B_0)$ with respect to incident angles for different values of corrugation wavelength $(\lambda)$ have been sketched in

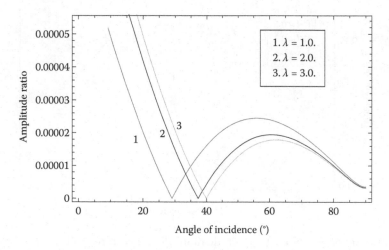

**Figure 16.6** Variation of modulus of amplitude ratio $(B_1/B_0)$ with respect to angle of incidence for different values of wavelength of corrugation $(\lambda)$.

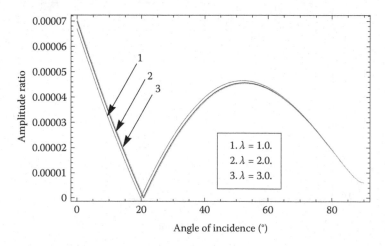

**Figure 16.7** Variation of modulus of amplitude ratio $(D_1/B_0)$ with respect to angle of incidence for different values of wavelength of corrugation $(\lambda)$.

Figures 16.8 and 16.9. In Figure 16.8, $(B_1'/B_0)$ has a decreasing characteristic for the smaller incident angles until it attains its minima. On further increase in incident angle, the values of amplitude ratio tend to increase and reach its maxima. Then on further increase in angle, the ratio gets decreased. However, it is observed that with increasing values of $\lambda$, the value of $(B_1'/B_0)$ is increased for most incident angles. A very different case is observed in Figure 16.9, where $(D_1'/B_0)$ starts increasing from 0° onward and then decreases afterward. The $\lambda$ puts a positive influence in $(D_1'/B_0)$ but the influence gets decreased in the higher incident angles.

## 16.9.3 Effect of frequency factor

The curves in Figures 16.10 and 16.11 have been traced out to demonstrate the behavior of $(B_1/B_0)$ and $(D_1/B_0)$ against incident angle for various values of frequency factor $(\omega d/\beta_1)$. In Figure 16.10, the curves have a similar characteristic as seen in the above cases. However,

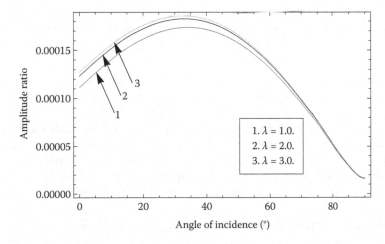

**Figure 16.8** Variation of modulus of amplitude ratio $(B_1'/B_0)$ with respect to angle of incidence for different values of wavelength of corrugation $(\lambda)$.

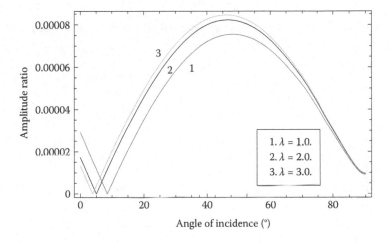

**Figure 16.9** Variation of modulus of amplitude ratio $(D_1'/B_0)$ with respect to angle of incidence for different values of wavelength of corrugation $(\lambda)$.

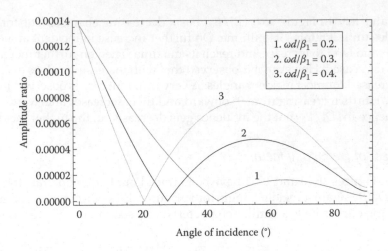

**Figure 16.10** Variation of modulus of amplitude ratio ($B_1/B_0$) with respect to angle of incidence for different values of frequency factor ($\omega d/\beta_1$).

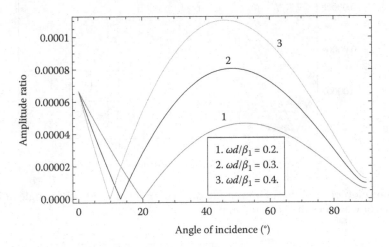

**Figure 16.11** Variation of modulus of amplitude ratio ($D_1/B_0$) with respect to angle of incidence for different values of frequency factor ($\omega d/\beta_1$).

the effect of ($\omega d/\beta_1$) is much more pronounced as the spaces between the curves are much higher. But, ($B_1/B_0$) is decreased in smaller incident angles and gets increased for higher angles. Similar behavior is observed in Figure 16.11 for ($D_1/B_0$).

In Figure 16.12, the amplitude ratio first decreases slightly with incident angle and then gets increased and finally is decreased. Frequency factor puts a favorable influence on ($B_1'/B_0$) for most part of the incidence angle. In Figure 16.13, the value of ($D_1'/B_0$) is increased initially and then is decreased with increasing incident angle values. Moreover, frequency factor has a positive influence on the amplitude ratios throughout the incident angle range.

### 16.9.4  *Influence of initial stress parameter on poroelastic half-space*

Figures 16.14–16.19 have been sketched to manifest the variation of different amplitude ratios against incidence angle related to the initially stressed fluid-saturated poroelastic

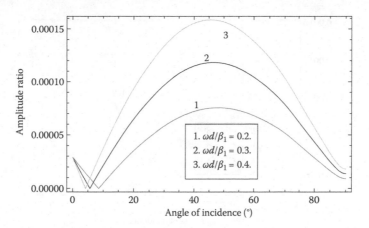

**Figure 16.12** Variation of modulus of amplitude ratio $\left(B_1'/B_0\right)$ with respect to angle of incidence for different values of frequency factor $(\omega d/\beta_1)$.

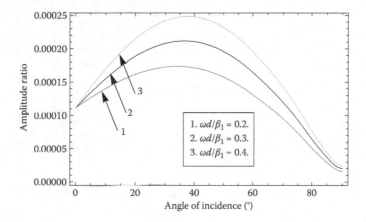

**Figure 16.13** Variation of modulus of amplitude ratio $\left(D_1'/B_0\right)$ with respect to angle of incidence for different values of frequency factor $(\omega d/\beta_1)$.

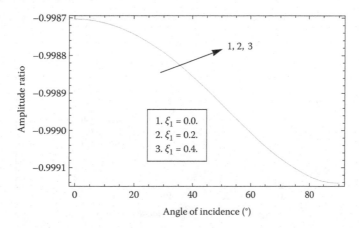

**Figure 16.14** Variation of modulus of amplitude ratio $(B/B_0)$ with respect to angle of incidence for different values of initial stress parameter $(\xi_1)$ associated with the poroelastic half-space.

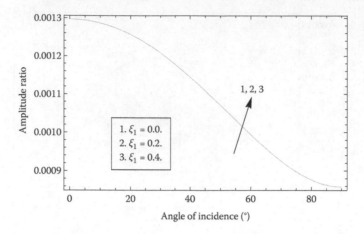

**Figure 16.15** Variation of modulus of amplitude ratio $(D_0/B_0)$ with respect to angle of incidence for different values of initial stress parameter $(\xi_1)$ associated with the poroelastic half-space.

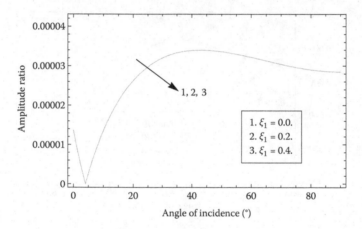

**Figure 16.16** Variation of modulus of amplitude ratio $(B_1/B_0)$ with respect to angle of incidence for different values of initial stress parameter $(\xi_1)$ associated with the poroelastic half-space.

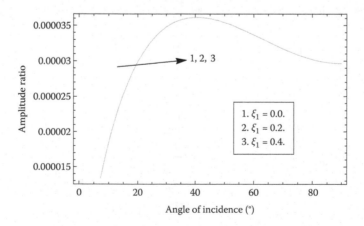

**Figure 16.17** Variation of modulus of amplitude ratio $(D_1/B_0)$ with respect to angle of incidence for different values of initial stress parameter $(\xi_1)$ associated with the poroelastic half-space.

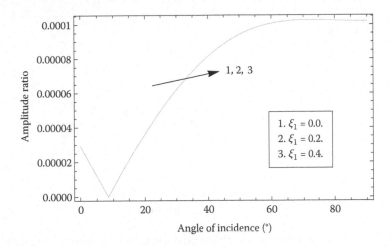

*Figure 16.18* Variation of modulus of amplitude ratio $(B_1'/B_0)$ with respect to angle of incidence for different values of initial stress parameter $(\xi_1)$ associated with the poroelastic half-space.

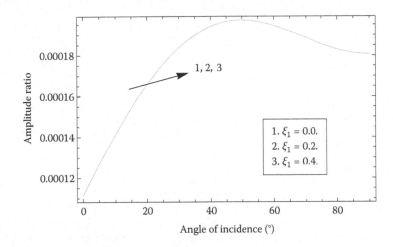

*Figure 16.19* Variation of modulus of amplitude ratio $(D_1'/B_0)$ with respect to angle of incidence for different values of initial stress parameter associated $(\xi_1)$ with the poroelastic half-space.

half-space. The amplitude ratios have varying influence due to the incident angles, but it is very much evident from all the figures, that the initial stress parameters ($\xi_1 = 0.0, 0.2,$ and $0.4$) do not have any influence on the poroelastic half-space as all the curves overlap each other. Thus, it can be concluded that the initial stress parameter related to the poroelastic half-space does not put any prominent effect on the amplitude ratios.

## 16.9.5 Influence of initial stress parameter on highly anisotropic half-space

Figures 16.20–16.25 have been plotted to demonstrate the variation of various amplitude ratios against incidence angle by varying the values of initial stress parameter ($\xi_1 = 0.0,$ $0.2,$ and $0.4$) related to highly anisotropic half-space. Variation of amplitude ratios ($B/B_0$) and ($D_0/B_0$) in Figures 16.20 and 16.21, respectively, appear to be similar, which indicates

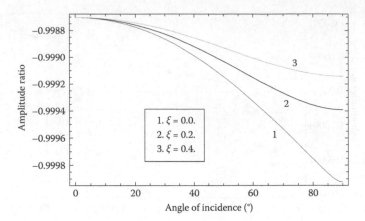

**Figure 16.20** Variation of modulus of amplitude ratio ($B/B_0$) with respect to angle of incidence for different values of initial stress parameter ($\xi$) associated with the highly anisotropic half-space.

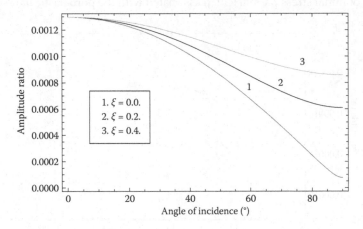

**Figure 16.21** Variation of modulus of amplitude ratio ($D_0/B_0$) with respect to angle of incidence for different values of initial stress parameter ($\xi$) associated with the highly anisotropic half-space.

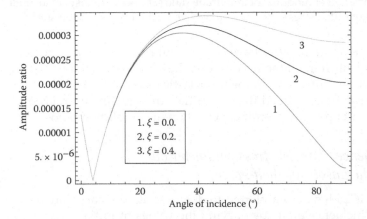

**Figure 16.22** Variation of modulus of amplitude ratio ($B_1/B_0$) with respect to angle of incidence for different values of initial stress parameter ($\xi$) associated with the highly anisotropic half-space.

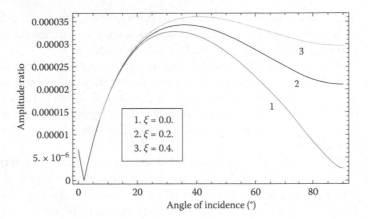

**Figure 16.23** Variation of modulus of amplitude ratio $(D_1/B_0)$ with respect to angle of incidence for different values of initial stress parameter $(\xi)$ associated with the highly anisotropic half-space.

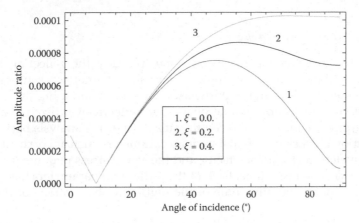

**Figure 16.24** Variation of modulus of amplitude ratio $\left(B_1'/B_0\right)$ with respect to angle of incidence for different values of initial stress parameter $(\xi)$ associated with the highly anisotropic half-space.

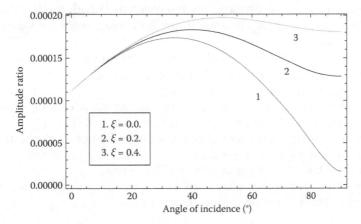

Figure **16.25** Variation of modulus of amplitude ratio $(D_1'/B_0)$ with respect to angle of incidence for different values of initial stress parameter $(\xi)$ associated with the highly anisotropic half-space.

that both ratios experience the same influence. In both graphs, the amplitude ratios have a gradual decrease with increase in incidence angle. But, they are increased due to an initial stress parameter related to the highly anisotropic half-space.

Similarly, $(B_1/B_0)$ in Figure 16.22 and $(B_1/B_0)$ in Figure 16.23 also have identical behaviors. The amplitude ratios start to increase from smaller incidence angle and remain the same and indistinguishable to 15° irrespective of any initial stress parameter value. With a further increase in incidence angle, the effect of initial stress parameter associated with highly anisotropic half-space is clearly recognizable. The ratios have an increasing effect due to the stress parameter.

The amplitude ratio $(B_1'/B_0)$ in Figure 16.24 does not have any considerable effect due to the initial stress parameter until 20°. On further increasing the angle, the effect of the initial stress parameter is clearly visible and $(B_1'/B_0)$ has an increasing effect. Besides, in Figure 16.25, $(D_1'/B_0)$ start to increase from 0° and continue to increase to a certain angle. After that, the amplitude ratios have a gradual decrease. Further, the amplitude ratios have increasing effects due to the increase in the initial stress parameter related to the highly anisotropic half-space.

## 16.10   Concluding remarks

A comprehensive investigation has been done to study the reflection and refraction phenomena of a plane SH-wave through a corrugated interface sandwiched between an initially stressed fluid-saturated poroelastic half-space and a highly anisotropic half-space. Rayleigh's method of approximation has been effectively utilized to derive first- and second-order approximations of the coefficients. A rigorous analysis between the reflection and refraction coefficients against various parameters such as corrugation amplitude, corrugation wavelength, frequency factor and the initial stress parameter associated with both the half-spaces has been done. Each of the individual parameters has been discussed separately and in detail. Finally, some of the highlight observations of the study are as follows:

1. Closed-form expressions were derived for reflection and refraction coefficients by making an effective use of Rayleigh's approximation technique.
2. Some special cases were deduced for the case when both the fluid-saturated poroelastic half-space and highly anisotropic half-space were considered as isotropic elastic half-space.
3. Corrugation amplitude has a positive effect on the amplitude ratios $(B_1/B_0)$ and $(D_1/B_0)$ for most of the incident angles, while it has a negative influence on $(B_1'/B_0)$ and $(D_1'/B_0)$ for the entire range of incident angles.
4. Corrugation wavelength has varied effect on the amplitude ratios $(B_1/B_0)$ and $(D_1/B_0)$. However, it puts a positive influence on $(B_1'/B_0)$ and $(D_1'/B_0)$.
5. The frequency factor parameter puts a favorable influence on all the amplitude ratios.
6. The initial stress parameter associated with the fluid-saturated poroelastic half-space does not have any effect on the amplitude ratios.
7. The initial stress parameter associated with the highly anisotropic half-space has a positive and visible effect on the amplitude ratios.

The critical findings of this descriptive study of the present problem can be worthwhile to the fields of geophysics and geology. This study may furnish some valuable assistance to geoscientists for proper interpretation of geological structures.

# References

Abubakar, I. Scattering of plane elastic waves at rough surfaces – I. *Proc. Camb. Philos. Soc.* 58 (1962a): 136–157.

Abubakar, I. Reflection and refraction of plane SH-waves at irregular interfaces – I. *J. Phys. Earth* 10.1 (1962b): 1–14.

Abubakar, I. Reflection and refraction of plane SH-waves at irregular interfaces – I. *J. Phys. Earth* 10.1 (1962c): 15–20.

Aki, K., Richards, P.G. *Quantitative Seismology*, 2nd ed. University Science Books, Sausalito (2002).

Asano, S. Reflection and refraction of elastic waves at a corrugated boundary surface. Part-I. The case of incidence of SH-wave. *Bull. Earthq. Res. Inst.* 38.2 (1960): 177–197.

Asano, S. Reflection and refraction of elastic waves at a corrugated boundary surface. Part-II. *Bull. Earthq. Res. Inst.* 39.3 (1961): 367–466.

Asano, S. Reflection and refraction of elastic waves at a corrugated interface. *Bull. Seismol. Soc. Am.* 56.1 (1966): 201–221.

Biot, M.A. Theory of elastic waves in a fluid saturated porous solid I. low frequency range. *J. Acoust. Soc. Am.* 28 (1956a): 168–178.

Biot, M.A. Theory of elastic waves in a fluid saturated porous solid II. High frequency range. *J. Acoust. Soc. Am.* 28 (1956b): 179–191.

Biot, M.A. Mechanics of deformations and acoustic propagation in porous media. *J. Appl. Phys.* 33 (1962): 1482–1489.

Biot, M.A. *Mechanics of Incremental Deformation*, Wiley, New York (1965).

Crampin, S. A review of the effects of anisotropic layering on the propagation of seismic waves. *Geophys. J. Int.* 49.1 (1977): 9–27.

Daley, P.F., Hron F. Reflection and transmission coefficients for seismic waves in ellipsoidally anisotropic media. *Geophysics* 44.1 (1979): 27–38.

Deresiewicz, H. The effect of boundaries on wave propagation in a liquid-filled porous solid: II. Love waves in a porous layer. *Bull. Seismol. Soc. Am.* 51.1 (1961): 51–59.

Ewing, W.M., Jardetzky, W.S., Press, F. *Elastic Waves in Layered Media*. Lamont Geological Observatory Contribution. McGraw-Hill, New York (1957).

Fokkema, J.T. Reflection and transmission of elastic waves by the spatially periodic interface between two solids (theory of the integral-equation method). *Wave Motion* 2.4 (1980): 375–393.

Keith, C.M., Crampin, S. Seismic body waves in anisotropic media: Reflection and refraction at a plane interface. *Geophys. J. Int.* 49.1 (1977): 181–208.

Lord Rayleigh, O. M. On the dynamical theory of gratings. *Proc. R. Soc. Lond. A* 79.532 (1907): 399–416.

Pal, A.K., Chattopadhyay, A. The reflection phenomena of plane waves at a free boundary in a pre-stressed elastic half-spaces. *J. Acoust. Soc. Am.* 76.3 (1984): 924–925.

Rokhlin, S.I., Bolland T.K., Adler L. Reflection and refraction of elastic waves on a plane interface between two generally anisotropic media. *J. Acoust. Soc. Am.* 79.4 (1986): 906–918.

Saini, S.L., Singh, S.J. Effect of anisotropy on the reflection of SH-waves at an interface. *Geophys. Res. Bull.* 15.2 (1977): 67–73.

Sato, R. The reflection of elastic waves on corrugated surface. *Zisin* 8.1 (1955): 8–22.

Sharma, M.D., Gogna, M.L. Reflection and refraction of plane harmonic waves at an interface between elastic solid and porous solid saturated by viscous liquid. *Pure and Appl. Geophys.* 138 (1992): 249–266.

Tajuddin, M., Hussaini, S.J. Reflection of plane waves at boundaries of a liquid filled poroelastic half-space. *J. Appl. Geophys.* 58.1 (2005): 59–86.

Thomsen, L. Reflection seismology over azimuthally anisotropic media. *Geophysics* 53.3 (1988): 304–313.

Tomar, S.K., Arora, A. Reflection and transmission of elastic waves at an elastic/porous solid saturated by two immiscible fluids. *Int. J. Solids Struct.* 43 (2006): 1991–2013 [Erratum, ibid 44, 5796–5800 (2007)].

Tomar, S.K., Kaur, J. Reflection and transmission of SH-waves at a corrugated interface between two laterally and vertically heterogeneous anisotropic elastic solid half-spaces. *Earth, Planets and Space* 55.9 (2003): 531–547.

Tomar, S.K., Kaur, J. SH-waves at a corrugated interface between a dry sandy half-space and an anisotropic elastic half-space. *Acta Mech.* 190 (2007): 1–28.

Tomar, S.K., Saini, S.L. Reflection and refraction of SH-waves at a corrugated interface between two dimensional transversely isotropic half spaces. *J. Phys. Earth* 45.5 (1997): 347–362.

Tomar, S.K., Singh, S.S. Quasi-P-waves at a corrugated interface between two laterally dissimilar monoclinic half spaces. *Int. J. Solids Struct.* 44.1 (2007): 197–228.

Tomar, S.K., Kumar, R., Chopra, A. Reflection and refraction of SH-waves at a corrugated interface between transversely isotropic and visco-elastic solid half spaces. *Acta Geophys. Pol.* 50.2 (2002): 231–249.

Tiersten, H.F. *Linear Piezoelectric Plate Vibration*, Plenum Press, New York (1969).

Wu, K.Y., Xue Q., Adler L. Reflection and transmission of elastic waves from a fluid-saturated porous solid boundary. *J. Acoust. Soc. Am.* 87.6 (1990): 2349–2358.

# Index